NanoScience and Technology

NanoScience and Technology

Series Editors:
P. Avouris B. Bhushan D. Bimberg K. von Klitzing H. Sakaki R. Wiesendanger

The series NanoScience and Technology is focused on the fascinating nano-world, mesoscopic physics, analysis with atomic resolution, nano and quantum-effect devices, nanomechanics and atomic-scale processes. All the basic aspects and technologyoriented developments in this emerging discipline are covered by comprehensive and timely books. The series constitutes a survey of the relevant special topics, which are presented by leading experts in the field. These books will appeal to researchers, engineers, and advanced students.

Bharat Bhushan
Harald Fuchs

Applied Scanning Probe Methods XII

Characterization

With 101 Figures and 14 Tables
Including 70 Color Figures

 Springer

Editors
Prof. Dr. Bharat Bhushan
Ohio State University
Nanoprobe Laboratory for Bio- & Nanotechnology
& Biomimetics (NLB^2)
201 W. 19th Ave
Columbus, Ohio 43210-1142
USA
bhushan.2@osu.edu

Prof. Dr. Harald Fuchs
Universität Münster
FB 16
Physikalisches Institut
Wilhelm-Klemm-Str. 10
48149 Münster
Germany
fuchsh@uni-muenster.de

Series Editors
Professor Dr. Phaedon Avouris
IBM Research Division
Nanometer Scale Science & Technology
Thomas J.Watson Research Center, P.O. Box 218
Yorktown Heights, NY 10598, USA

Professor Bharat Bhushan
Nanoprobe Laboratory for Bio- & Nanotechnology
and Biomimetics (NLB^2)201 W. 19th Avenue
The Ohio State University
Columbus, Ohio 43210-1142, USA

Professor Dr. Dieter Bimberg
TU Berlin, Fakutät Mathematik,
Naturwissenschaften,
Institut für Festkörperphysik
Hardenbergstr. 36, 10623 Berlin, Germany

Professor Dr., Dres. h. c. Klaus von Klitzing
Max-Planck-Institut für Festkörperforschung
Heisenbergstrasse 1, 70569 Stuttgart, Germany

Professor Hiroyuki Sakaki
University of Tokyo
Institute of Industrial Science,
4-6-1Komaba, Meguro-ku, Tokyo 153-8505
Japan

Professor Dr. RolandWiesendanger
Institut für Angewandte Physik
Universität Hamburg
Jungiusstrasse 11, 20355 Hamburg, Germany

ISBN: 978-3-540-85038-0   e-ISBN: 978-3-540-85039-7

NanoScience and Technology ISSN 1434-4904

Library of Congress Control Number: 2008933866

Cover design: WMXDesign GmbH, Heidelberg

Printed on acid-free paper

9 8 7 6 5 4 3 2 1

springer.com

Preface for Applied Scanning Probe Methods Vol. XI–XIII

The extremely positive response by the advanced community to the Springer series on Applied Scanning Probe Methods I–X as well as intense engagement of the researchers working in the field of applied scanning probe techniques have led to three more volumes of this series. Following the previous concept, the chapters were focused on development of novel scanning probe microscopy techniques in Vol. XI, characterization, i.e. the application of scanning probes on various surfaces in Vol. XII, and the application of SPM probe to biomimetics and industrial applications in Vol. XIII. The three volumes will complement the previous volumes I–X, and this demonstrates the rapid development of the field since Vol. I was published in 2004. The purpose of the series is to provide scientific background to newcomers in the field as well as provide the expert in the field sound information about recent development on a worldwide basis.

Vol. XI contains contributions about recent developments in scanning probe microscopy techniques. The topics contain new concepts of high frequency dynamic SPM technique, the use of force microscope cantilever systems as sensors, ultrasonic force microscopy, nanomechanical and nanoindentation methods as well as dissipation effects in dynamic AFM, and mechanisms of atomic friction.

Vol. XII contains contributions of SPM applications on a variety of systems including biological systems for the measurement of receptor–ligand interaction, the imaging of chemical groups on living cells, and the imaging of chemical groups on live cells. These biological applications are complemented by nearfield optical microscopy in life science and adhesional friction measurements of polymers at the nanoscale using AFM. The probing of mechanical properties by indentation using AFM, as well as investigating the mechanical properties of nanocontacts, the measurement of viscous damping in confined liquids, and microtension tests using in situ AFM represent important contributions to the probing of mechanical properties of surfaces and materials. The atomic scale STM can be applied on heterogeneous semiconductor surfaces.

Vol. XIII, dealing with biomimetics and industrial applications, deals with a variety of unconventional applications such as the investigations of the epicuticular grease in potato beetle wings, mechanical properties of mollusc shells, electro-oxidative lithography for bottom-up nanofabrication, and the characterization of mechanical properties of biotool materials. The application of nanomechanics as tools for the investigation of blood clotting disease, the study of piezo-electric polymers, quantitative surface characterization, nanotribological characterization of carbonaceous materials, and aging studies of lithium ion batteries are also presented in this volume.

We gratefully acknowledge the support of all authors representing leading scientists in academia and industry for the highly valuable contribution to Vols. XI–XIII. We also cordially thank the series editor Marion Hertel and her staff members Beate Siek and Joern Mohr from Springer for their continued support and the organizational work allowing us to get the contributions published in due time.

We sincerely hope that readers find these volumes to be scientifically stimulating and rewarding.

August 2008   Bharat Bhushan
                                                                    Harald Fuchs

Contents – Volume XII

14   Mechanical Properties of Metallic Nanocontacts

15   Dynamic AFM in Liquids: Viscous Damping and Applications to the Study of Confined Liquids

Contents – Volume XI

Contents – Volume XIII

Contents – Volume I

Contents – Volume II

Contents – Volume III

Contents – Volume IV

Contents – Volume V

Contents – Volume VI

Contents – Volume VII

Contents – Volume VIII

Contents – Volume IX

Contents – Volume X

List of Contributors – Volume XII

N. Agraït
Departamento de Física de la Materia Condensada C-III, Universidad Autonoma de Madrid, Madrid 28049, Spain
e-mail: nicolas.agrait@uam.es

Jean-Pierre Aimé
CPMOH, Université Bordeaux1, 351 Cours de la Libération, 33405 Talence Cedex, France
e-mail: jp.aime@cpmoh.u-bordeaux1.fr

David Alsteens
Unité de Chimie des Interfaces, Université Catholique de Louvain, Croix du Sud 2/18, B-1348 Louvain-la-Neuve, Belgium
e-mail: david.alsteens@uclouvain.be

Guillaume André
Unité de Chimie des Interfaces, Université Catholique de Louvain, Croix du Sud 2/18, B-1348 Louvain-la-Neuve, Belgium
e-mail: guillaume.andre@uclouvain.be

Sophie Bistac
Université de Haute-Alsace, CNRS, 15, rue Jean Starcky, BP 2488, 68057 Mulhouse Cedex, France
e-mail: sophie.bistac-brogly@uha.fr

Massimiliano Bocciarelli
Politecnico di Milano, Dipartimento di Ingegneria Strutturale, piazza Leonardo da Vinci 32, 20133 Milano, Italy
e-mail: bocciarelli@stru.polimi.it

Rodolphe Boisgard
CPMOH, Université Bordeaux1, 351 Cours de la Libération, 33405 Talence Cedex, France
e-mail: r.boisgard@cpmoh.u-bordeaux1.fr

Gabriella Bolzon

Politecnico di Milano, Dipartimento di Ingegneria Strutturale, piazza Leonardo da Vinci 32, 20133 Milano, Italy
e-mail: gabriella.bolzon@polimi.it, bolzon@stru.polimi.it

Enzo J. Chiarullo

Politecnico di Milano, Dipartimento di Ingegneria Strutturale, piazza Leonardo da Vinci 32, 20133 Milano Italy
e-mail: chiarullo@stru.polimi.it

Touria Cohen-Bouhacina

CPMOH, Université Bordeaux1, 351 Cours de la Libération, 33405 Talence Cedex, France
e-mail: t.bouhacina@cpmoh.u-bordeaux1.fr

Etienne Dague

Unité de Chimie des Interfaces, Université Catholique de Louvain, Croix du Sud 2/18, B-1348 Louvain-la-Neuve, Belgium
e-mail: etienne.dague@laas.fr

Jurg Dual

ETH Zentrum, IMES – Institute of Mechanical Systems, CLA J23.2, Department of Mechanical and Process Engineering, 8092 Zürich, Switzerland
e-mail: juerg.dual@imes.mavt.ethz.ch

Yves F. Dufrêne

Unité de Chimie des Interfaces, Université Catholique de Louvain, Croix du Sud 2/18, B-1348 Louvain-la-Neuve, Belgium
e-mail: dufrene@cifa.ucl.ac.be

Vincent Dupres

Unité de Chimie des Interfaces, Université Catholique de Louvain, Croix du Sud 2/18, B-1348 Louvain-la-Neuve, Belgium
e-mail: vincent.dupres@uclouvain.be

Robert H. Eibl

Plainburgstr. 8, 83457 Bayerisch Gmain, Germany
e-mail: robert_eibl@yahoo.com

Grégory Francius

Unité de Chimie des Interfaces, Université Catholique de Louvain, Croix du Sud 2/18, B-1348 Louvain-la-Neuve,
Belgium
e-mail: gregory.francius@uclouvain.be

Hongjun Gao
Nanoscale Physics & Devices Laboratory, Institute of Physics, Chinese Academy of Sciences, P. O. Box 603, Beijing 100080, China
e-mail: hjgao@aphy.iphy.ac.cn

Pietro Giuseppe Gucciardi
CNR-Istituto per i Processi Chimico-Fisici, Salita Sperone c.da Papardo, I-98158 Messina, Italy
e-mail: gucciardi@me.cnr.it

Haiming Guo
Nanoscale Physics & Devices Laboratory, Institute of Physics, Chinese Academy of Sciences, P. O. Box 603, Beijing 100080, China
e-mail: hmguo@aphy.iphy.ac.cn

Cedric Hurth
CPMOH, Université Bordeaux1, 351 Cours de la Libération, 33405 Talence Cedex, France
e-mail: cedric.hurth@asu.edu

Cédric Jai
CPMOH, Université Bordeaux1, 351 Cours de la Libération, 33405 Talence Cedex, France
e-mail: c.jai@free.fr

Udo Lang
ETH Zentrum, IMES – Institute of Mechanical Systems, Department of Mechanical and Process Engineering, 8092 Zürich, Switzerland
e-mail: udo.lang@imes.mavt.ethz.ch

Abdelhamid Maali
CPMOH, Université Bordeaux1, 351 Cours de la Libération, 33405 Talence Cedex, France
e-mail: a.maali@cpmoh.u-bordeaux1.fr

J.J. Riquelme
Departamento de Física de la Materia Condensada C-III, Universidad Autonoma de Madrid, Madrid 28049, Spain
e-mail: juanjo.riquelme@uam.es

Gabino Rubio-Bollinger
Departamento de Física de la Materia Condensada C-III, Universidad Autonoma de Madrid, Madrid 28049, Spain
e-mail: gabino.rubio@uam.es

Marjorie Schmitt
Université de Haute-Alsace, CNRS, 15, rue Jean Starcky, BP 2488, 68057 Mulhouse
Cedex, France
e-mail: Marjorie.Schmitt@uha.fr

Claire Verbelen
Unité de Chimie des Interfaces, Université Catholique de Louvain, Croix du Sud
2/18, B-1348 Louvain-la-Neuve, Belgium
e-mail: claire.verbelen@uclouvain.be

S. Vieira
Departamento de Física de la Materia Condensada C-III, Universidad Autonoma de
Madrid, Madrid 28049, Spain
e-mail: sebastian.vieira@uam.es

Yeliang Wang
Nanoscale Physics & Devices Laboratory, Institute of Physics, Chinese Academy of
Sciences, P. O. Box 603, Beijing 100080, China
e-mail: ylwang@aphy.iphy.ac.cn

List of Contributors – Volume XI

Houssein Awada

Université Catholique de Louvain, Unité de chimie et de physique des hauts polymères (POLY), Croix du Sud 1 – 1348 Louvain-la-Neuve – Belgique (B)
e-mail: houssein.awada@uclouvain.be

Elmar Bonaccurso

Max-Planck-Institute for Polymer Research, Ackermannweg 10, D-55128 Mainz, Germany
e-mail: bonaccur@mpip-mainz.mpg.de

Paolo Bonanno

Department of Biophysical and Electronic Engineering, Unversity of Genova, Via all'Opera Pia 11a, I-16145 Genova, Italy
e-mail: paolo.bonanno@unige.it

Maurice Brogly

Université de Haute Alsace (UHA), Equipe Interfaces Sous Contraintes (ICSI - CNRS UPR 9069), 15 rue Jean Starcky – 68057 Mulhouse Cx – France (F)
e-mail: maurice.brogly@uha.fr

Hans-Jürgen Butt

Max-Planck-Institute for Polymer Research, Ackermannweg 10, D-55128 Mainz Germany
e-mail: butt@mpip-mainz.mpg.de

Lorenzo Calabri

CNR-INFM – National Research Center on nanoStructures and bioSystems at Surfaces (S3), Via Campi 213/a, 41100 Modena, Italy
e-mail: calabri.lorenzo@unimore.it.

M. Teresa Cuberes

Laboratorio de Nanotécnicas, UCLM, Plaza Manuel de Meca 1, 13400 Almadén, Spain
e-mail: teresa.cuberes@uclm.es

Dmytro S. Golovko
Max-Planck-Institute for Polymer Research, Ackermannweg 10, D-55128 Mainz,
Germany
e-mail: golovkod@mpip-mainz.mpg.de

Mykhaylo Evstigneev
Fakultät für Physik, Universität Bielefeld, Universitätsstr. 25, 33615 Bielefeld,
Germany
e-mail: Mykhaylo@Physik.Uni-Bielefeld.De

Harald Fuchs
Physikalisches Institut and Center for Nanotechnology (CeNTech), Universität
Münster, Wilhelm-Klemm-Str. 10, Münster D48149, Germany
e-mail: fuchsh@uni-muenster.de

Thomas Haschke
University of Siegen, Faculty 11, Department of Simulation, Am Eichenhang 50,
D-57076 Siegen, Germany
e-mail: haschke@simtec.mb.uni-siegen.de

Donna C. Hurley
National Institute of Standards & Technology, 325 Broadway, Boulder, Colorado
80305 USA
e-mail: hurley@boulder.nist.gov

Johann Jersch
Physikalisches Institut, Universität Münster, Wilhelm-Klemm-Str. 10, Münster
D48149, Germany
e-mail: jersch@uni-muenster.de

Olivier Noel
Université du Maine, Laboratoire de Phjysique de l'Etat Condensé (CNRS UMR
6087), Avenue Olivier Messiaen – 72085 Le Mans Cx 9 – France (F)
e-mail: olivier.noel@univ-lemans.fr

Stefano Piccarolo
Dipartimento di Ingegneria Chimica dei Processi e dei Materiali, Università di
Palermo, Viale delle Scienze, 90128 Palermo, Italy and INSTM Udr Palermo
e-mail: piccarolo@unipa.it

Nicola Pugno
Department of Structural Engineering, Politecnico di Torino, Corso Duca degli
Abruzzi 24, 10129 Torino, Italy, National Institute of Nuclear Physics, National
Laboratories of Frascati, Via E. Fermi 40, 00044, Frascati, Italy
e-mail: nicola.pugno@polito.it

Roberto Raiteri
Department of Biophysical and Electronic Engineering, Unversity of Genova, Via all'Opera Pia 11a I-16145 Genova, Italy
e-mail: rr@unige.it

Davide Tranchida
Dipartimento di Ingegneria Chimica dei Processi e dei Materiali, Università di Palermo, Viale delle Scienze, 90128 Palermo, Italy and INSTM Udr Palermo

Sergio Valeri
CNR-INFM – National Research Center on nanoStructures and bioSystems at Surfaces (S3), Via Campi 213/a, 41100 Modena, Italy. Department of Physics, University of Modena and Reggio Emilia, via Campi 213/a 41100 Modena, Italy
e-mail: sergio.valeri@unimo.it

Wolfgang Wiechert
University of Siegen, Faculty 11, Department of Simulation, Am Eichenhang 50, D-57076 Siegen, Germany
e-mail: wolfgang.wiechert@uni-siegen.de

List of Contributors – Volume XIII

Francois Barthelat

Department of Mechanical Engineering, McGill University, Macdonald Engineering Building, Rm 351, 817 Sherbrooke Street West, Montreal, Quebec H3A 2K6
e-mail: francois.barthelat@mcgill.ca

Bharat Bhushan

Nanotribology Laboratory for Information Storage and MEMS/NEMS (NLIM), Ohio State University, Columbus, OH 43210, USA

Sophie Bistac

Université de Haute-Alsace, 15, rue Jean Starcky, BP 2488, 68057, Mulhouse Cedex, France
e-mail: Sophie.Bistac-Brogly@uha.fr

Horacio D. Espinosa

Department of Mechanical Engineering, Northwestern University, 2145 Sheridan Rd., Evanston, IL 60208-3111, USA
e-mail: espinosa@northwestern.edu

Stanislav Gorb

Evolutionary Biomaterials Group, Max-Planck-Institut für Metallforschung, Heisenbergstrasse 3, 70569 Stuttgart, Germany
e-mail: s.gorb@mf.mpg.de

Stephanie Hoeppener

Laboratory of Macromolecular Chemistry and Nanoscience, Eindhoven University of Technology, P.O. Box 513, 5600 MB Eindhoven, The Netherlands Center for NanoScience, Lehrstuhl für photonik und Optoelektronik Luduig - Maximilians - Universität München, Geschwister-Scholl platz 1, 80333 München, Germany
e-mail: s.hoeppener@tue.nl

Ingomar L. Jäger

Department of Materials Science, University of Leoben, Jahnstrasse 12, 8700
Leoben, Austria
e-mail: ingomar@unileoben.ac.at

Taekwon Jee

Mechanical Engineering, Texas A&M University, College Station, TX 77843-3123
e-mail: taekwonjee@gmail.com

Hyungoo Lee

Department of Mechanical Engineering, Texas A&M University,
College Station, TX 77843, USA
e-mail: thanku7@gmail.com

Hong Liang

Department of Mechanical Engineering, Texas A&M University,
College Station, TX 77843-3123, USA
e-mail: hliang@tamu.edu

Helga C. Lichtenegger

Institute of Materials Science and Technology E308, Vienna University of
Technology, Favoritenstrasse 9-11, 1040 Wien, Austria
e-mail: helga.lichtenegger@tuwien.ac.at

Shrikant C. Nagpure

Nanotribology Laboratory for Information Storage and MEMS/NEMS (NLIM),
Ohio State University, Columbus, OH 43210, USA
e-mail: nagpure.1@osu.edu

H. Peisker

Evolutionary Biomaterial Group, Max-Plank-Institut für Metallforschung,
Heisenbergstrasse 3, 70569 Stuttgart, Germany

Jee E. Rim

Mechanical Engineering, Northwestern University, 2145 Sheridal Road,
Technological Institute B224, Evanton, IL 60208
e-mail: j-rim@northwestern.edu

Maria Cecília Salvadori

Institute of Physics, University of São Paulo, C.P. 66318, CEP 05315-970, São
Paulo, SP, Brazil
e-mail: mcsalvadori@if.usp.br

Marjorie Schmitt
Université de Haute-Alsace, 15, rue Jean Starcky, BP 2488, 68057, Mulhouse
Cedex, France
e-mail: Marjorie.Schmitt@uha.fr

Matthias Schneider
University of Augsburg, Experimental Physics I, Universitätsstr. 1, 86159 Augsburg,
Germany
e-mail: matthias.schneider@physik.uni-augsburg.de

Thomas Schöberl
Erich Schmid Institute of Materials Science of the Austrian Academy of Sciences,
Jahnstrasse 12, 8700 Leoben, Austria
e-mail: schoeber@unileoben.ac.at

Ulrich S. Schubert
Friedrich-Schiller-Universität Jena
Institute für Organische Chemie and Makromolekulare
Chemie, Humbolattstr. 10, 07743 Jena, Germany

Daniel Steppich
University of Augsburg, Experimental Physics I, Universitätsstr.
1, 86159 Augsburg, Germany
e-mail: daniel.steppich@physik.uni-augsburg.de

Stefan Thalhammer
GSF-Institut für Strahlenschutz, Neuherberg, Germany
e-mail: stefan.thalhammer@gsf.de

D. Voigt
Evolutionary Biomaterials Group, Max-Planck-Institut für
Metallforschung, Heisenbergstrasse 3, 70569 Stuttgart,
Germany
e-mail: voigt@mf.mpg.de

Ke Wang
Department of Mechanical Engineering, Texas A& M University,
College Station, TX 77843, USA
e-mail: ke.phwk@gmail.com

Achim Wixforth
University of Augsburg, Experimental Physics I, Universitätsstr.
1, 86159 Augsburg, Germany
e-mail: achim.wixforth@physik.uni-augsburg.de

9 Direct Force Measurements of Receptor–Ligand Interactions on Living Cells

Robert H. Eibl

Abstract. The characterization of cell adhesion between two living cells at the level of single receptor–ligand bonds is an experimental challenge. This chapter describes how the extremely sensitive method of atomic force microscopy (AFM) based force spectroscopy can be applied to living cells in order to probe for cell-to-cell or cell-to-substrate interactions mediated by single pairs of adhesion receptors. In addition, it is outlined how single-molecule AFM force spectroscopy can be used to detect physiologic changes of an adhesion receptor in a living cell. This force spectroscopy allows us to detect in living cells rapidly changing, chemokine SDF-1 triggered activation states of single VLA-4 receptors. This recently developed AFM application will allow for the detailed investigation of the integrin–chemokine crosstalk of integrin activation mechanisms and on how other adhesion receptors are modulated in health and disease. As adhesion molecules, living cells and even bacteria can be studied by single-molecule AFM force spectroscopy, this method is set to become a powerful tool that can not only be used in biophysics, but in cell biology as well as in immunology and cancer research.

Key words: Single-molecule force measurements, Living cells, Mouse, Human, Cell adhesion receptor, Cell adhesion molecule, Unbinding event, Rupture, AFM, Atomic force microscopy, Force spectroscopy, Lymphocyte homing, Integrins, Integrin activation, VLA-4, VCAM-1, LFA-1, ICAM-1, SDF-1, CD44, Chemokines

Abbreviations

$\alpha_4\beta_1$	Alpha(4)beta(1) (integrin VLA-4)
$\alpha_L\beta_2$	alpha(L)beta(2) (integrin LFA-1)
AFM	Atomic force microscope
Anti-VLA-4	Antibody against VLA-4
BSA	Bovine serum albumin
CD44	Cluster of differentiation 44
ConA	Concanavalin A
CXCL12	CXC chemokine receptor ligand 12 (SDF-1)
CXCR4	CXC chemokine receptor 4 (receptor for SDF-1)
Dynes/cm^2	Unit for shear forces
EDTA	Ethylene diamine tetraacetic acid
EGTA	Ethylene glycol tetraacetic acid
FACS	Fluorescence-activated cell sorter
F_c	Crystallizable fragment (of an antibody)

GRGDSP	Amino-acid sequence, RGD-containing peptide (binding to, for example, integrin $\alpha_5\beta_1$)
ICAM-1	Intercellular cell adhesion antigen 1
IHC	Immunohistochemistry
HEPES	4-(2-hydroxyethyl)-1-piperazineethanesulfonic acid
HSA	Human serum albumin
LFA-1	Lymphocyte function-associated antigen 1 (integrin $\alpha_L\beta_2$)
MAb	Monoclonal antibody
μm	Micrometer
mN/m	Milli Newton per meter
min	Minutes
NaHCO$_3$	Sodium bicarbonate
ng/ml	Nanogram/milliliter
PBS	Phosphate-buffered saline
pN	Pico Newton
RGD	Arginine–glycine–aspartic acid (aminoacid sequence found in several integrin ligands)
RT	Room temperature
SDF-1	Stromal-derived factor 1 (CXCL12)
SFM	Scanning force microscope
SLeX ag	Sialylated Lewis X antigen
STM	Scanning tunnelling microscope
UV	Ultraviolet
VCAM-1	Vascular cell adhesion molecule 1
VLA-4	Very late antigen 4 (integrin $\alpha_4\beta_1$)

9.1
Introduction

In 1986 Binnig, Quate and Gerber introduced the first scanning force microscope (SFM) also known as an atomic force microscope (AFM) [1]. This AFM was derived from its precursor, the scanning tunnelling microscope (STM), developed by Binnig and Rohrer in 1981, for which they won the Nobel Prize in Physics in 1986. At the imaging level the AFM allows a resolution down to single atoms and subatomic features on a surface, i.e. an improvement of several orders of magnitude over the capacity of any conventional light microscopy. The AFM is a mechanical microscope the function of which is not affected by the optical diffraction limit, well known to be the limiting factor in classical light microscopy [2, 3]. Another application of AFM is the measurement of force-distance curves (Fig. 9.1). This force spectroscopic mode of operation has been routinely used to study even very weak forces in the range of pico-Newton (pN) and less. These are forces necessary to break atomic bonds, van der Waals forces, as well as forces needed to stretch and rupture biological molecules.

During the last two decades, the AFM has been applied in specialized fields linked to material sciences, applied physics and industrial processes. During the last

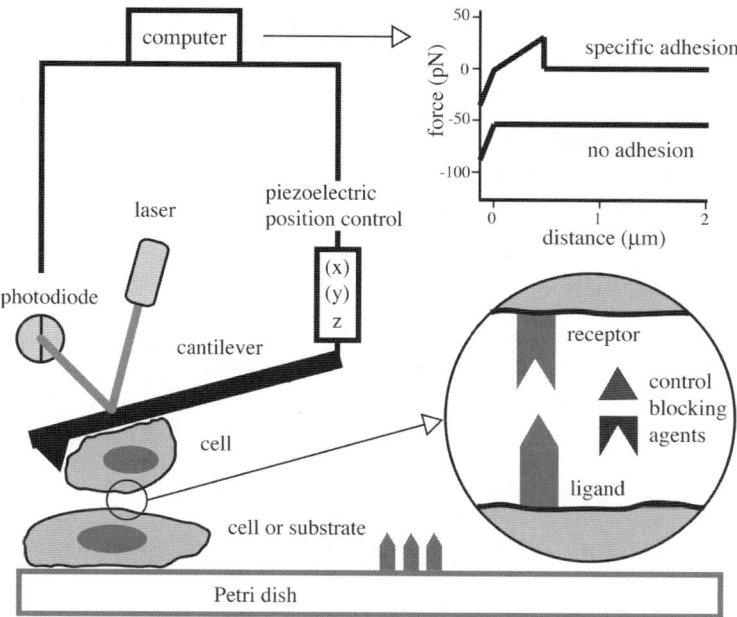

Fig. 9.1. Schematic setup of AFM measurements of a living cell

decade, however, the AFM has become a versatile tool in life sciences, where the measurements have to be adjusted for biological conditions. Applications vary from studies of unbinding forces of double-stranded DNA and stretching characteristics of isolated muscle titin proteins to measurements of viscoelastic properties of a living cell and the characterization of receptor-ligand bonds [4–15]. The specific interaction of biotin with streptavidin or avidin has been studied extensively and can serve as a reference model to guide AFM force measurements of isolated receptor–ligand interactions on a single-molecule level. The results obtained using other techniques estimating on and off-rates and energy landscapes of receptor ligand bonds can also be compared with AFM results as a reference [16–20] (Table 9.1).

Despite its enormous imaging resolution, the power of an AFM to directly measure cell adhesion forces brings the AFM force spectroscope into the focus of cell adhesion researchers, including immunologists and cell biologists, who are in general not very familiar with this technology. Many in the biophysical field still use the AFM on purified and often non-functional molecules which are studied isolated from the complexity of an intact living cell and not as highly motile members of an active cell membrane. For the critical reader it may appear rather challenging to measure the discrete unbinding force of a single pair of two distinct adhesion receptors, regardless whether these receptors were isolated and immobilized to a substrate or within a cell membrane of a large living cell. In 2000, Benoit and coworkers presented AFM measurements of the force needed to unbind the single adhesion bonds of homotypic receptors on a living amoeboid *Dictyostelium* cell with the identical, corresponding receptors on an adjacent amoeboid cell; this at a resolution of 23 pN

Table 9.1. Unbinding forces of individual pairs of cell adhesion receptors. Rupture forces of single adhesion receptor–ligand interactions revealed by AFM force spectroscopy on living cells (cell–cell, or cell–substrate interactions) and between isolated adhesion receptors (substrate–substrate interaction). Although it is not a pair of cell adhesion receptors the (strept)avidin–biotin system is included as a reference for one of the strongest single receptor–ligand interactions in biology, but not directly related to cellular adhesion

Receptor	Ligand	Interacting cell or substrate	Force (pN)	References
		Cell–cell		
VLA-4	VCAM-1	B16 melanoma cell –bEnd.3 endothelial cell	33	[9]
CsA	CsA	*Dictyostelium* amoeboid cell – *Dictyostelium* amoeboid cell	23	[30]
		Cell–substrate		
VLA-4	VCAM-1	RAW lymphoma cell – VCAM-1/F_c fusion protein	21–80	[4, 5, 31]
VLA-4	VCAM-1	B16 melanoma cell – VCAM-1/F_c fusion protein	21–45	Eibl (unpubl.)
VLA-4	VCAM-1	U937 monocytic cell – VCAM-1/F_c fusion protein	25–170	[24]
Integrin $\alpha_5\beta_1$	Fibronectin	K562 myeloid leukemia cell – protein	40–160	[26]
LFA-1	ICAM-1	3A9 lymphoma cell – ICAM-1/F_c fusion protein	100–300	[28]
Carbohydrates	ConA	NIH3T3 fibroblast–lectin	86	[32]
Integrin $\alpha_5\beta_1$	GRGDSP	Osteoclast–RGD peptide	32	[33]
Integrin $\alpha_V\beta_3$	GRGDSP	Osteoclast–RGD peptide	42	[33]
Integrin $\alpha_V\beta_3$	Osteopontin	Osteoclast–protein ligand	50	[33]
Integrin $\alpha_V\beta_3$	Echistatin	Osteoclast–protein ligand	97	[33]
		Substrate–substrate (related to cell adhesion)		
Integrin $\alpha_V\beta_3$	GRGDSP	Protein–RGD peptide	32	[34]
P-selectin	PSGL-1	Protein–protein	115–165	[35]
P-selectin	sLeX	F_c fusion protein – carbohydrate complex	40–200	[25]
VE-cadherin	VE-cadherin	F_c fusion protein – F_c fusion protein	15–150	[15]
Carbohydrate	ConA	Carbohydrate–lectin	35–65	[36]
Proteoglycan	Proteoglycan	Proteoglycan–Proteoglycan (from marine sponge)	400	[10]
		Substrate–substrate (unrelated to cell adhesion)		
HSA	Antibody	Protein–antibody	244	[37]
Avidin	Biotin	Protein–organic molecule	115–170	[17]

Table 9.1. (continued)

Receptor	Ligand	Interacting cell or substrate	Force (pN)	References
Avidin	Biotin	Protein–organic molecule	5–170	[20]
Avidin	Biotin	Protein–organic molecule	200	[38]
Avidin	Biotin	Protein–organic molecule	160	[19]
Avidin	Desthiobiotin	Protein–organic molecule	94	[18]
Avidin	Iminobiotin	Protein–organic molecule	85	[18]
Strepavidin	Biotin	Protein–organic molecule	257	[18]
Strepavidin	Iminobiotin	Protein–organic molecule	135	[18]

per average rupture. In their system, the specificity was obtained through the use of mutant cell lines that either could adhere to all other cells of the same sub line or did not adhere at all [21]. The non-adhering mutants did not show the characteristic rupture events. The authors then concluded that CsA-CsA adhesion receptors are indeed the interacting partners detected at the single-molecule level on the *Dictyostelium* surface. At that time, they could measure similar ruptures between mammalian cells, however, the identification of any specific pairs of mammalian receptors at the single-molecule level was not possible, because the cells studied were not well characterized and expressed more than one type of adhesion receptor on their surface [22]. Using well chosen cell lines and function blocking antibodies, Eibl and Benoit described in 2004 how to identify the unbinding forces of a specific pair of adhesion receptors between two living cells. Experiments described here, succeeded to directly measure—for the first time between two living mammalian cells—the rupture force of single bonds between the integrin very late antigen 4 (VLA-4) and its receptor, the vascular cell adhesion molecule 1 (VCAM-1), either protein expressed on B16 melanoma cells or on bEnd.3 endothelial cells, respectively [9].

It seems technically easier to detect unbinding forces of the interaction between a given pair of adhesion receptors fixed to a solid support, than those necessary to separate the same pair embedded in membranes of living cells. The potential difficulties to confirm the specificity of adhesion events when using two living cells could be reduced by using just one cell together with a corresponding and functional counter-receptor on an opposing surface. Meanwhile, integrins such as LFA-1, VLA-4, integrin $\alpha_5\beta_1$ and other cell adhesion receptors, for example, cadherins, have been studied using just one single cell expressing those receptors and the corresponding ligands either immobilized to the surface of a Petri dish or, alternatively, bound onto the AFM cantilever [15, 23–29] (Table 9.1).

Most of these reports include measurements performed under different conditions, as they are allowing variable off-rates by changing the velocity of the cantilever approaching the probe and retracting from it. As these reports describe, the experimental setup did allow for ambient temperatures and pH changes. Both parameters, temperature and pH, are known to be critical for the integrity of integrins and many other cell adhesion receptors. Such unphysiologic conditions did not prevent the cells adhering and reasonable single bond ruptures could be measured. One could conclude that these conditions may either not be too critical, or may even support the measurement at the single-molecule level, as the cell becomes less sticky with a reduced number of functional receptors on its surface. Furthermore, AFM measurements at room temperature may be much easier to perform than under conditions of higher temperatures, since the cantilever itself is sensitive to changes in temperature and to faster moving water molecules which can result in thermal drifts and higher background noise, respectively. Nevertheless, measuring the physiologic activation of integrin VLA-4 by the chemokine SDF-1 requires a higher temperature close to cell culture conditions of 37 °C as well as a constant and physiologic pH (Eibl, unpubl.) [4, 8]. An intact and viable cell allows for studies of single molecules in their native environment, in which different receptors embedded in the cell membrane can influence each other and be influenced by signalling molecules from outside and from inside the cell, for example, for inside-out and outside-in signalling or integrin–chemokine crosstalk. Purified full-length transmembrane molecules (what integrins are) are often difficult to study in the absence of membrane structures because hydrophobic transmembrane domains can then aggregate and the proteins may become insoluble.

More than often, the proteins loose their functional conformation. It is reasonable to experiment with purified deletion variants of these receptors, or to assemble a system with purified proteins and artificial and supported lipid bilayers. Although it has been shown that integrin receptors [39] as well as chemokine receptors [40] can be integrated in artificial lipid bilayers, it may take many more years to replace a living cell's membrane with an appropriate, integrin-coated artificial cell membrane, which will allow studies of the physiologic process of rapid SDF-1 triggered VLA-4 activation.

This activation step is part of the multi-step mechanism of lymphocyte homing, which has been studied since the 1980s and described independently by Butcher [41, 42] and Springer [43], who both were honoured with the Crafoord Prize of the Royal Swedish Academy of Sciences in 2004. Basically, lymphocytes and other white blood cells can use in health and disease a set of cell adhesion receptors for migrating to distant sites within the body, especially to attach to the blood vessel endothelium and resist shear forces imposed by the blood stream, this, just before they extravasate into the surrounding tissue of, for example, lymph nodes or Peyer's patches. Part of this homing process are discrete steps which comprise the initial tethering and cell rolling on the wall of a blood vessel, then the migrating cells arrest and stick to the endothelial monolayer, before they extravasate into the regional tissue (Fig. 9.2).

The initial tethering and rolling steps are mediated by selectins and a few integrins, such as VLA-4 or LFA-1. These integrins finally can mediate the firm adhesion

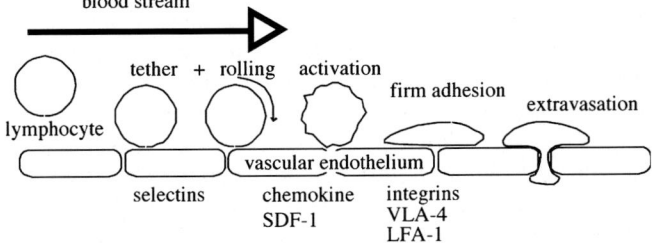

Fig. 9.2. Multi-step adhesion model of circulating and homing lymphocytes

or sticking onto the blood vessel wall. Interestingly, each of these few specialized integrin receptors, in particular VLA-4 and LFA-1 and their cognate ligands, VCAM-1 and ICAM-1, have been implicated in lymphocyte rolling as well as in the subsequent rapid arrest or sticking. It has been speculated that the contribution of integrins to this rolling phenomenon is that the integrin receptor functions as a fast low-affinity/high-affinity switch. A chemokine like SDF-1, usually presented by endothelial cells, can trigger the switch within milliseconds via its surface-bound chemokine receptor CXCR4, what then can lead to the rapid arrest of a rolling lymphocyte [44, 45]. More than a century after rolling cells were discovered in a fish fin, the knowledge about mechanisms and receptors involved in this phenomenon has become immense and has deepened significantly during the past three decades. Today's core methods include parallel-plate or capillary-based flow assays [46] and even intravital microscopy techniques used in mice and rats harbouring fluorescently labelled lymphocytes [47, 48]. The latter assays allow live imaging of rolling and sticking lymphocytes. Of course, many other methods have been developed during the past century to study the expression pattern and to characterize the functions of cell adhesion receptors. Receptor-specific antibodies or labelled ligands are routinely used in immunohistochemistry (IHC) and in cell sorting (e.g. fluorescence-activated cell sorting, FACS) [49, 50]. However, not every receptor specific antibody also blocks an adhesion receptor, therefore, the development of function blocking antibodies for each receptor is important for functional studies, especially on living cells. One functional assay, originally designed by Stamper and Woodruff [51], combines, for example, lymph-node tissue sections and lymphocytes as an in vitro model of lymphocyte homing. Lymphocytes which were layered over fixed sections of lymph nodes adhered selectively to specialized vascular endothelia. This assay can be applied to functionally characterize relevant pairs of homing receptors of circulating lymphocytes by using function-blocking antibodies. Such antibodies inhibit the adhesion of lymphocytes to specialized high endothelial cells on a lymph node section. The unbound or weakly adhering lymphocytes then can't resist applied centrifugation forces and are washed away. In another assay, the parallel-plate flow chamber assay, intact cells have to resist via their adhesion receptors the shear forces which are mainly dependent on the speed of the flow. This force imposed upon a cell is measured in Dynes/cm^2 and is characteristic for a cell and the receptors studied. The migration properties of lymphocytes and other white blood cells can be studied in rolling chamber assays [46]. All red and white blood cells circulating within the blood stream are prevented from adhering to each other and to endothelial cells in order to avoid thrombus formation. However, a prerequisite to evade the blood stream is that actively migrating white blood cells need to use or activate adhesion receptors to initiate homing. An interesting note on the side is that metastatic cancer cells and adult stem cells may also share some of the lymphocyte adhesion receptors for their hematogenous spread to distant sites from a primary tumour or a metastatic lymph node [52, 53]. The sensitivity of all of these methods is far away from single-molecule techniques and usually depends on many functional receptor molecules per cell. To detect rarely expressed molecules like the chemokine receptor CXCR 4 on a cell's surface with immunohistochemistry, an additional signal-amplification process has to be included to increase the signal or signal-to-noise ratio. Although these methods are powerful in the investigation of cell-adhesion receptor function, none of these methods had allowed the direct measurement of cell adhesion forces on a single-molecule or single-receptor level and within the native environment of such a complex system as a living cell. Therefore, single-molecule AFM force spectroscopy is well positioned to become one method of choice when characterizing the regulation of cell adhesion receptors on living cells including migrating lymphocytes, adult stem cells and metastatic cancer cells.

The goal of this chapter is to bring the powerful method of single-molecule AFM force spectroscopy on living cells to a broader audience and to inform interested researchers and students, including cell biologists, cancer researchers and immunologists, in general not familiar with AFM-based force spectroscopy and its recent applications in life sciences. In addition, this chapter is of great interest for advanced AFM and biophysics experts, who have already gained experience in applying AFM to the measurement of single-molecules, but who are looking forward to using a living cell to study and manipulate single molecules within their native environment. Several supplementing step-by-step protocols will also guide and inspire fully advanced AFM and immunology enthusiasts. The protocol for activation measurements of integrin VLA-4 by the chemokine SDF-1 on a living cell and at the single-molecule level, as described in this chapter, can currently be considered as one of the most detailed descriptions of the method published in this field. The protocols are designed to help modify and adapt this method to similar questions for other cell types and adhesion receptors, as well as for many other physiologic or artificial activators or inhibitors. The principles described here need not just be applied to higher eukaryotic cells, but can easily be adapted for use with lower eukaryotic (e.g. yeast) and prokaryotic cells or combinations thereof, like for studies of the binding of bacteria to macrophages.

9.2
Procedure

When designing AFM experiments to measure cell–cell or cell–protein interactions, decisions have to be made about (a) which pair of adhesion receptors or set of molecules are of interest and fit, (b) which cells or cell lines express the relevant receptors on their surface, and (ideally) lack other, possibly interfering, adhesion receptors, (c) whether to use function-blocking antibodies, other inhibitors or activators in function control experiments, or agents in combination, (d) what the optimal experimental conditions (temperature, pH, growth medium) are for the selected cell culture, and (e) what the best AFM setup (including cantilever, its functionalization and retraction distance) is to comply with the cell culture conditions in order to obtain a good signal and signal-to-noise ratio. Note: I recommend refraining from (a) performing experiments with cells and biomolecules at suboptimal ambient temperatures and (b) using growth media with an unphysiologic or changing pH value. The pH of CO_2/HCO_3-buffered media can fluctuate dramatically if the medium is left unchecked for evaporation. Adding different amounts of Hepes buffer does not prevent CO_2 from leaving the media which appears to be critical for some adhesion measurements. Most cell lines can be adapted to more expensive, CO_2-independent medium which I recommend for AFM measurements of living cells.

9.2.1
Principle of AFM Force Spectroscopy

The AFM has the extraordinary capacity to measure single-molecule interaction forces (pico Newton) and displacements (nanometer) at the nanoscale level. The key part of the AFM is a microfabricated spring or cantilever. Usually, sample molecules are coated to the small pyramidal tip at the end of the cantilever and probed against other molecules on an opposing sample stage. Alternatively, a cell can be attached to one or both contacting parts of the AFM, i.e. to the end of the cantilever and/or to the bottom of a cell culture Petri dish (Fig. 9.1). By moving the cantilever (or the stage), both samples can be brought into contact. A piezoelectric element controls the movement of the cantilever in the direction of the z-axis (perpendicular to the sample stage) and is required to detect developing forces. By controlling the scanning in the x and y direction, additional piezoelectric elements provide data for the creation of three-dimensional AFM images. The combined data (position of the tip and measured force) allows pinpointing the location of different adhesion areas on a cell surface. AFMs used in cell adhesion experiments usually need a piezocontrol element that retracts the cantilever by up to 15 µm or more from the sample contact point. Manually controlled micrometer screws can be used for the spatial adjustment of the cantilever tip. A standard light microscope (not shown), to which a video camera is attached, is placed underneath the sample stage of the AFM to monitor the fishing of the cell onto the tip of the cantilever and to focus the laser beam onto the cantilever. The laser is deflected from the flip side of the sensor tip into the centre of a photo detector. The two segments of the photodiode are needed to adjust the laser and to measure the magnitude of deflection to one or the other half during the force scans.

The sample interaction force leads to bending of the cantilever and that alters the deflection angle of the laser beam and the location where the beam hits the photodiode. In order to translate the recorded deflection of the laser beam into a force scale, a proportionality constant—the spring constant of the cantilever—has to be determined. Although the manufacturers provide a nominal value of the spring constant, each cantilever should be calibrated since the spring constant may vary significantly. An estimation method based on the cantilever's thermal fluctuation has been proven useful [6,54]. For experiments with living cells soft and long cantilevers are required to achieve a spring constant between 10 and 15 mN/m. Lower constant values of up to 5 mN/m can be achieved by breaking away one of the two arms of a delta-shaped cantilever using a hand-held forceps under low magnificaton [9,55]. A series of force distance curves is recorded as a loop of forward and backward scans. Two examples of forward and backward scans are shown, one without any rupture event, the other with a single force step of 33 pN (Fig. 9.3). Specific adhesion events can be translated into rupture forces and further analyzed in a force spectrogram.

9.2.2
Cell–Cell Interactions

In order to measure cell–cell adhesion or deadhesion events, cells have to be immobilized on both sides of the probing instrument, for example, onto the tip of the cantilever and on the bottom of the Petri dish (Figs. 9.1 and 9.4). Although it is easy

Fig. 9.3. AFM force scans of cell–cell interactions.
Force-distance diagram of forward and backward scans.
In the case of no adhesion after short contact of a B16
melanoma cell to a bEnd.3 endothelial cell, both, the
backward and forward scans look similar and show the
indentation force (40 pN) superimposed by the thermal
noise, whereas an adhesion event of about 33 pN can be
easily detected by the step in the backward force scan

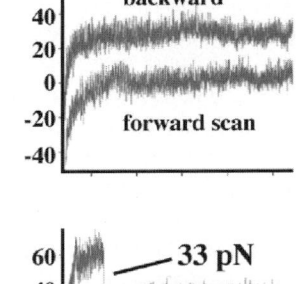

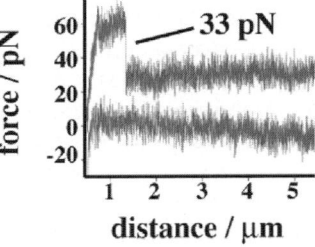

Fig. 9.4. Scanning steps in AFM-based force spectroscopy for
cell–cell adhesion measurements

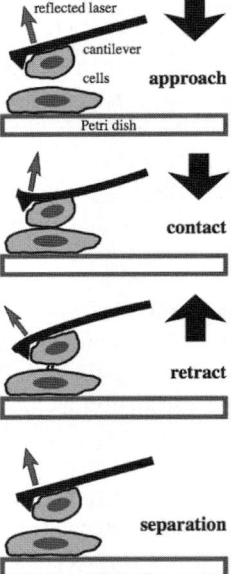

to grow most cell types on a Petri dish to which they adhere, it is more difficult to
attach a living cell onto the tip of a cantilever and to maintain physiologic cell culture
conditions for a cell that, perhaps, remains attached for up to several hours. In many
cases, it has been useful to functionalize the cantilever with poly-D-lysine in order to
catch a single cell onto the tip of the cantilever. A more detailed protocol for functionalizing the cantilever with concanavalin A (conA) is provided at the end of this
chapter. In order to mimic the homing conditions of rolling and rapidly arresting lymphocytes, the ideal cell type for adhesion studies with AFM were several lymphoma

cell lines expressing VLA-4 and an endothelial cell line expressing VCAM-1. The endothelial cells grow and attach strongly onto the plastic surface of a common Petri dish, whereas the lymphoma cell usually grows as a single cell or within a cluster of a few cells in suspension. Some lymphoma cells may grow weakly attached to a cell culture flask but can easily be separated by gentle shaking. Instead of placing a lymphoma cell onto the cantilever it is also possible to use a cell growing firmly attached to a cell culture flask. Then, mild trypsinization, Ca^{2+} ion removal by short exposure to a chelating agent, such as EDTA, or using both methods in combination, can separate even strongly attached cells. After separation of such cells, the attachment of one of these cells onto the tip of the cantilever should be done quickly, since the cell may bind firmly to the bottom of the Petri dish. A few separated cells are placed into the measuring Petri dish and the functionalized cantilever is placed above an unbound cell and smoothly lowered onto the top of the cell, monitored by an optical microscope placed underneath the Petri dish. After a few seconds of contact the cell adheres to the cantilever, which is then retracted and placed above firmly bound endothelial cells or a region of immobilized receptor proteins or a control region, respectively (Fig. 9.5).

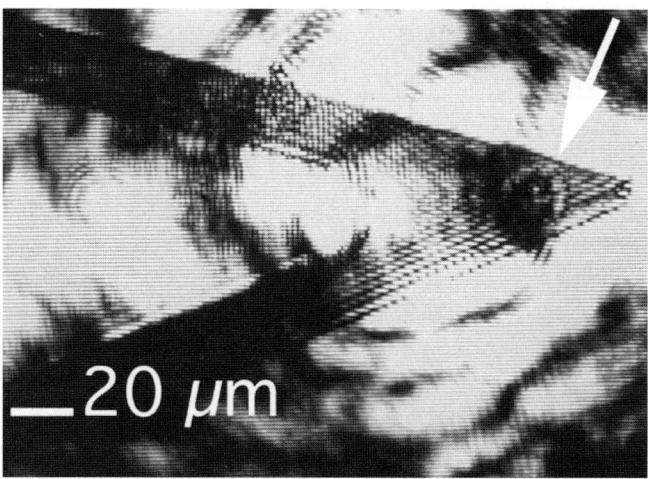

Fig. 9.5. Placing a cell on the cantilever. Photomicrograph of a living cell attached to the tip of an AFM cantilever previously functionalized with conA

9.2.2.1
Multiple Bond Ruptures

Different parameters can influence the quality of cell–cell adhesion contacts. Not only the cell type, the activation state (affinity) or the density (avidity) of the receptors involved, but also the approaching and retracting velocities, the contact time and force, and the detachment force may influence the adhesion strength. Choosing a longer contact time and higher contact forces before separating two cells usually leads to the detection of multiple bond ruptures (Fig. 9.6). This method can already

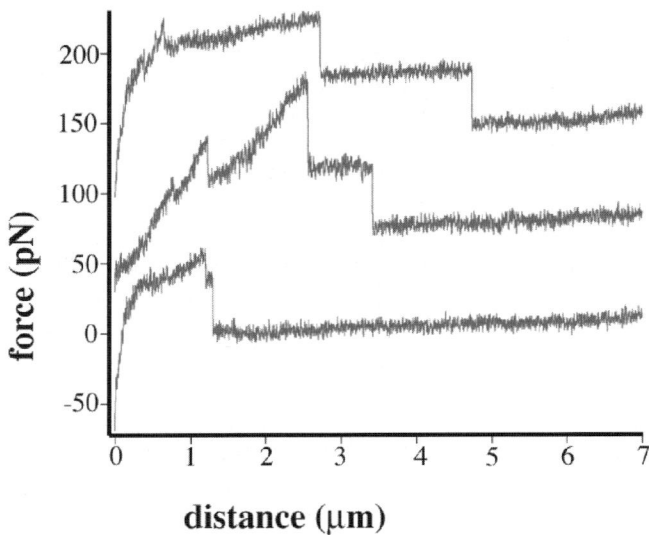

distance (μm)

Fig. 9.6. AFM force scans of cell–cell interactions. Multiple unbinding events are shown in each of the three backward scans, indicating the rupture of more than one pair of VLA-4–VCAM-1 receptors, each of them expressed on the surface of a B16 melanoma cell or a bEnd.3 endothelial cell, respectively

be used to compare differently activated or inhibited cells or any drugs which may influence the adhesive strength of cells [11, 56, 57]. Therefore, the integrated area below the force-distance curve is considered as work force. Lower and higher work forces of differently treated cells allow conclusions about the overall adhesiveness and regulation of adhesion in a given cell.

9.2.2.2
Single Bond Ruptures

To directly measure single-receptor interactions between two living cells, several conditions have to be optimized, but they may vary between cell types and cell activation states, receptor types, receptor activation states and receptor densities on the cell surface. Adapting the contact force and duration of contact (e.g. 30 pN, 200 ms) to conditions which lead to adhesion events of 33% or less are considered to represent a probability of 83% or 91% to detect single-molecule rupture events [18, 21] (Fig. 9.7).

9.2.3
Cell-Substrate Measurements

For most cell adhesion studies, the measurement of receptor–ligand interaction between two living cells appears to be too complicated and not required, especially since a second cell may add a number of possible interaction partners not related

Fig. 9.7. AFM force-distance scans of cell–cell interactions optimized for single-molecule interactions. Ten consecutive adhesion events are shown with only a single rupture (no adhesions left out). The unbinding can occur at different distances reflecting the different length of tethers

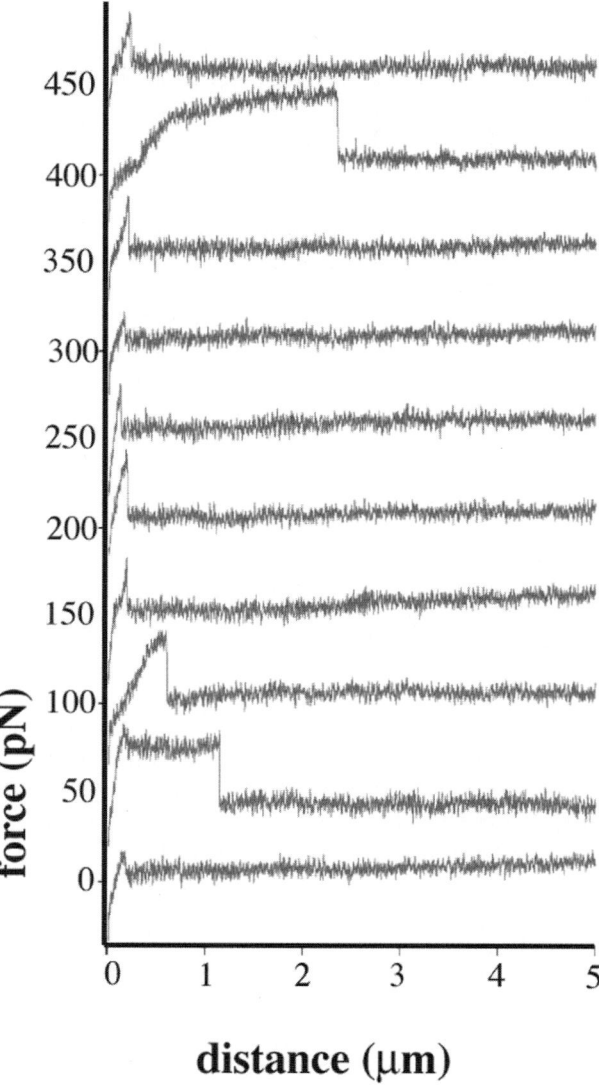

to the receptor of interest on the first cell. Therefore, it is reasonable to study a receptor of interest in its native environment of a living cell and to probe it against a suitable counter-receptor immobilized on a surface (Fig. 9.8). This may help to increase the specificity and sensitivity, since a specific binding of additional ligands to the receptor of interest or binding of additional pairs of receptors is widely excluded, and non-specific binding, for example, of charged proteins to the probing surface is reduced. Nevertheless, it is still possible that a single type of immobilized ligand can recognize more than one or two types of counter-receptors on a cell. There appears to be no need for time-consuming control experiments with blocking antibodies. An experimental approach with just one living cell and immo-

Fig. 9.8. Scanning steps for cell–substrate measurements

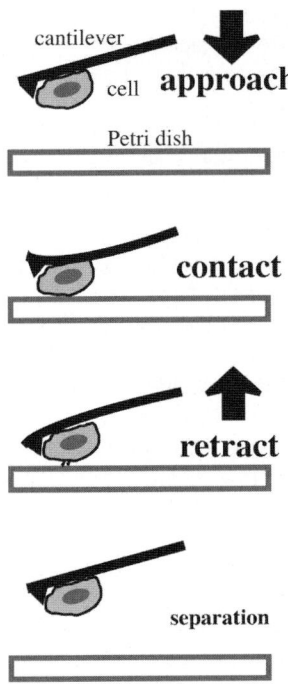

bilized counter-receptor appears to be ideal for most purposes addressing adhesion receptors and signalling cascades, such as the integrin–chemokine crosstalk, which hardly could be addressed in a cell-free system. In principle, the cell could be placed either on the cantilever or on the Petri dish, and vice versa, the isolated counter-receptor. Growing a cell firmly attached on the bottom of a Petri dish and function-alizing the probing cantilever with the suitable counter-receptor has the advantage of reducing cell culture procedures, but may result in otherwise stressed cells, since poking with a pyramidal tip of the cantilever onto the cell can damage the cell. On the other hand, such a procedure is used in studies of cell viscoelasticity, where the cells apparently survive the procedure. It often appears to be much easier and suit-able to functionalize the bottom of a Petri dish with critical proteins and binding the cell to a cantilever rather than funtionalizing the pyramidal tip of a cantilever, since most of the protein receptors rapidly denature or are otherwise hard to bind to the cantilever.

9.2.4
Specificity and Blocking Antibodies

Highly specific function-blocking antibodies towards a number of cell adhesion receptors have been developed and are used to prevent cell adhesion to other cells or to the extracellular matrix. These antibodies, or any other inhibitors, such as small and fully synthetic function-blocking molecules, for example, with a sugar backbone, can bind to receptors, and thereby prevent its binding domain from recognizing and binding to its ligand. They exert their function either by binding directly to the binding domain, by masking the domain, or by inducing conformational changes in the receptor.

As shown with rolling chamber assays, lymphocytes, drawn through the chamber by otherwise cell-free medium, can roll and stick by adhering through their integrin VLA-4 receptors to endothelial VCAM-1 ligands; this can also be observed when analyzing metastasizing B16 melanoma cells (Eibl, unpublished results). In a rolling chamber assay, mimicking physiologic blood stream conditions, not only lymphocytes can roll and stick via their integrin VLA-4 receptors to endothelial VCAM-1 ligands, but also metastasizing B16 melanoma cells (Eibl unpubl.) (Fig. 9.9). Rolling and sticking of B16 melanoma cells was not observed when function-blocking antibodies against either VLA-4 or VCAM-1 were added, but did occur when control antibodies were used. In order to compare the results obtained from the rolling chamber assays, and to highlight the potential of single-molecule AFM force measurements, the complex system of B16 melanoma cells rolling on activated bEnd.3 endothelial cells was reconfigured for the analysis in an AFM.

The B16 melanoma cell can now be fixed onto the tip of an AFM cantilever and can be brought into contact with the endothelial cells on the sample stage for a set period of time. By manipulating cell–cell contact time and cantilever approaching and retracting velocities the touch-and-go rolling characteristic binding and unbinding events can now be re-examined, at least in part, in an AFM.

A typical force spectrum of single interactions between two living cells is shown (Fig. 9.10): a single B16 melanoma cell bound to the cantilever and a single bEnd.3 endothelial cell on the Petri dish. The adhesion frequency was adjusted to 33%, i.e. in 67% of all 960 cell–cell contacts binding and subsequent rupture events were not detected (Fig. 9.3). These 67% of force-distance curves (0–5 pN) showing no apparent rupture events were omitted from the histogram (0–5 pN). In this example the

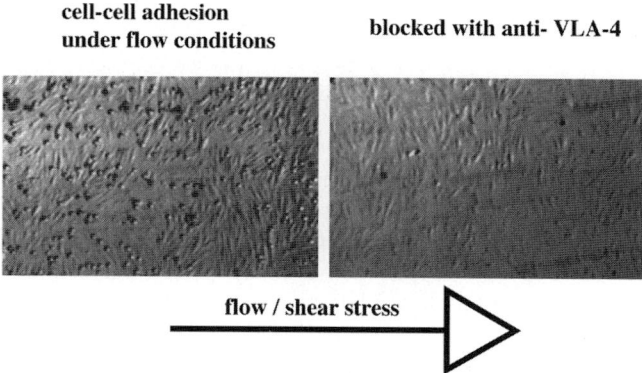

cell-cell adhesion under flow conditions   **blocked with anti- VLA-4**

flow / shear stress

Fig. 9.9. VLA-4–VCAM-1 mediates cell–cell adhesion under flow conditions. Under physiologic shear stress in a parallel-plate flow chamber single B16 melanoma cells roll and stick on a monolayer of activated bEnd.3 endothelial cells (*left* photomicrograph). Function-blocking antibodies against VLA-4 can completely block this receptor–ligand interaction (*right*). Because of the speed of the flow, unbound melanoma cells are only visible as long and weak shadows across each picture frame

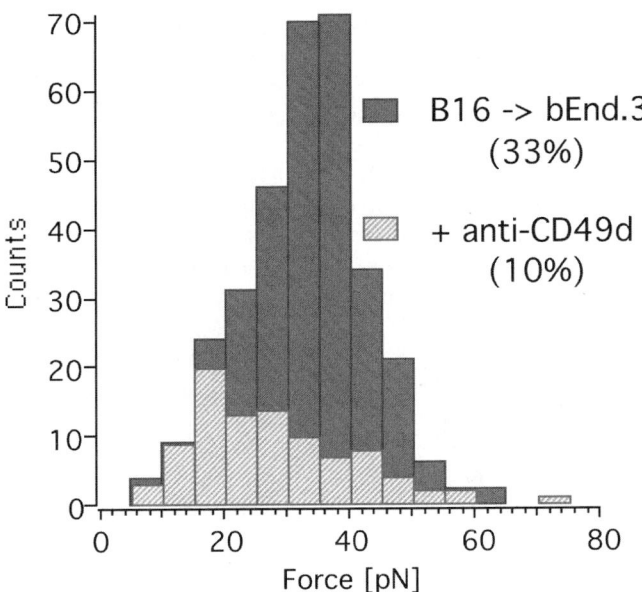

Fig. 9.10. Single-molecule AFM force spectrum of cell–cell interactions identifying VLA-4–VCAM-1 receptor-ligand bonds. Two typical histograms of AFM force measurements are shown, one of them including a function-blocking antibody against the VLA-4 adhesion receptor on the B16 melanoma cell (*lower* colloms) which leads to a significant reduction of binding events to the bEnd.3 endothelial cell

average unbinding force is 33 pN (standard deviation 12 pN). As expected results from blocking experiments performed in rolling assays show that antibodies, blocking either VLA-4 or VCAM-1, respectively, but not control antibodies, reduce the binding frequency and the average rupture force significantly. Blocking anti-VLA-4 antibodies block most of the interactions above 20 pN. In contrast, the average force and frequency below 20 pN appear to be identical. Therefore, most, if not all interactions around 33 pN can be attributed to VLA-4–VCAM-1 interactions. Therefore, as in the rolling chamber experiments, B16 melanoma cells can interact with bEnd.3 endothelial cells via one major pair of receptors, VLA-4 and VCAM-1, expressed on the surfaces of the interacting cells. In both experimental setups, this interaction can be blocked with inhibitors, such as function-blocking antibodies, supporting the ability of AFM force spectroscopy to measure very specific interactions between two living cells. Finally, according to the low frequency of rupture events resulting in a very high probability of detecting ruptures of just one pair of adhesion receptors, these rupture forces can be attributed specifically to VLA-4–VCAM-1 interactions. This view is supported by the lack of lower and higher peaks, which was observed under suboptimal conditions in amoeboid *Dictyostelium* [21].

The identical frequencies of single rupture events around 20 pN and below in both experiments can easily be attributed to non-specific binding of charged proteins on a cell's surface. On the other hand, it may point to another cell adhesion receptor that binds less frequently and at lower forces under the conditions optimized for

VLA-4–VCAM-1 interactions. Interestingly, under certain circumstances, for example, while testing cells with stronger and more rolling receptors such as VLA-4, other rolling mediating receptors known to be weak, such as CD44, may totally be masked in a rolling chamber assay or intravital microscopy experiment [58, 59]. Therefore, some other unknown rolling receptors may be masked in many of the standard experiments, even when blocking of a main receptor leads to a total inhibition of rolling and adhesion under flow conditions. AFM force spectroscopy may help to discover new adhesion receptors or new functions of adhesion receptors that currently escape other methods. Additional AFM force experiments may test some of the paradoxical functions of cell adhesion receptors such as VLA-4 and CD44 in normal development and disease [50, 52, 60–64]. Under some circumstances, VLA-4 mediates homotypic aggregation of cells which may reduce the migration potential of tumour cells to leave the primary tumour, which could result in a lower metastatic potential [62], but tumour cells entering the blood stream and expressing significant amounts of VLA-4 on their surface can use VLA-4 in at least one mechanism of metastasis (Eibl unpubl.) [64]. The AFM force measurements may help to identify a new, but weak receptor or a new and underestimated variation of a CD44 variant as a very common cell adhesion, lymphocyte homing and tumour metastasis receptor. This CD44 receptor is highly expressed on B16 melanoma cells but appears to be totally inactivated on all cells tested in standard rolling experiments (Eibl unpubl.).

9.2.5
Activation by SDF-1

AFM force measurements can be used to directly measure VLA-4–VCAM-1 interactions, either between two living cells [9, 21] (Figs. 9.4, 9.6, 9.7, 9.10), or between one cell and immobilized VCAM-1 fusion proteins [8] (Figs. 9.11, 9.12, 9.13). Rolling experiments identified VLA-4 to mediate the rolling and sticking parts during the multistep mechanism of lymphocyte homing [41, 43, 65]. Usually, the integrin VLA-4 has to be triggered by a chemokine, which binds to its corresponding chemokine receptor, for example, SDF-1 (= CXCL12) and CXCR4, respectively [44, 45]. This chemokine-mediated triggering is expected to change within milliseconds the activation state from a low-affinity to a high-affinity VLA-4 receptor, with both states recognizing their ligand VCAM-1, but with different characteristics. Direct AFM force measurements appear to have a unique potential to study such an immediate activation within the context of a living cell. In the following experiment, a VLA-4 expressing lymphoma cell is bound to the tip of the cantilever (Fig. 9.8) and different areas of the same Petri dish are functionalized with recombinant VCAM-1 fusion protein, as well as with controls, including LFA-1 ligand ICAM-1 and other control substances. Under conditions of higher adhesion frequencies single and multiple rupture bonds appear per backward scan (Fig. 9.11). Adding soluble SDF-1 many more rupture events appear per backward scan, as well as higher rupture forces at the last rupture of each scan (Fig. 9.11). This indicates that soluble SDF-1 can change the adhesive behaviour of a living cell in a setup of AFM force measurements without the need for intravital microscopy in a whole animal or in a parallel-plate chamber

Fig. 9.11. Effects of
chemokine SDF-1 on AFM
force scans of
VLA-4–VCAM-1
cell–substrate interactions.
Before activation (*above*),
and after activation with
the chemokine SDF-1

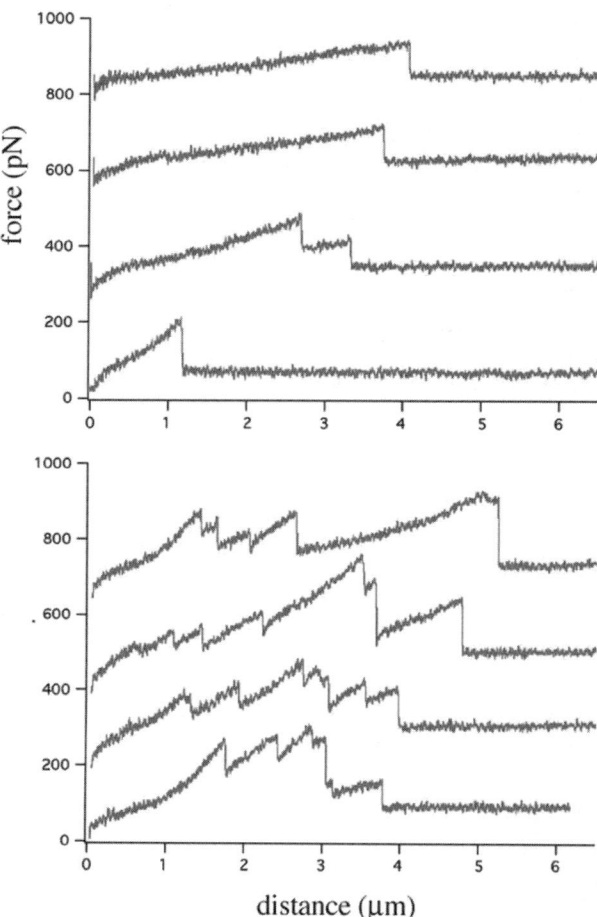

distance (μm)

assay. Since the integrated area below the force-distance curve is larger after the
addition of SDF-1 the mechanical energy required per detachment event is considerably increased. It is beyond the scope of this chapter to discuss more applications
of easier to measure multiple rupture bonds and AFM applications such as biosen-
sors in drug development. Such minimal changes in the workforce detected by AFM
force measurements may be highly indicative for a good candidate in the screening
of substances involved in inflammatory and autoimmune responses.

To preferentially measure the rupture of single VLA-4–VCAM-1 bonds instead
of multiple bond ruptures, the critical parameters have to be adjusted to frequencies less than 33%. For pharmacological dose–response experiments it is reasonable
to choose much lower frequencies of 2–10% as the starting frequency. Otherwise,
SDF-1 may lead to multiple ruptures per bond. From many similar experiments,
one representative experiment with increasing concentrations of SDF-1 is shown
in eight typical force histograms (Fig. 9.12). The average forces were plotted in a
dose–response diagram (Fig. 9.13). This experiment shows a limited number of rupture scans per histogram, since it is optimized for a higher number of increasing

Fig. 9.12. Affinity modulation of single VLA-4 receptors on a living cell by SDF-1. AFM force histograms of a lymphoma cell expressing integrin VLA-4 probed on immobilized VCAM-1 fusion protein. Increasing SDF-1 concentrations show a shift of the rupture forces to higher forces, then a shift back to normal

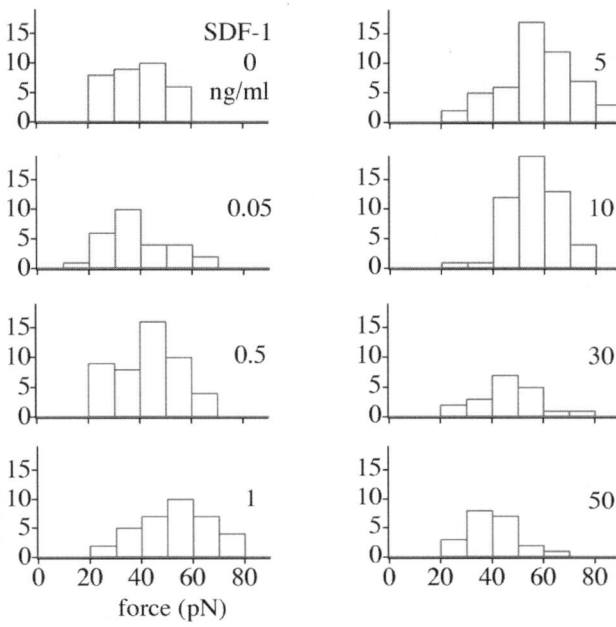

Fig. 9.13. Dose–response effects of chemokine SDF-1 revealed by AFM force spectroscopy. The dose–response curve shows the mean value of forces from the histograms shown in Fig. 9.12. The peak of unbinding single VLA-4–VCAM-1 adhesion bonds at physiologic conditions indicates a physiologic switch to a higher affinity state of the integrin VLA-4

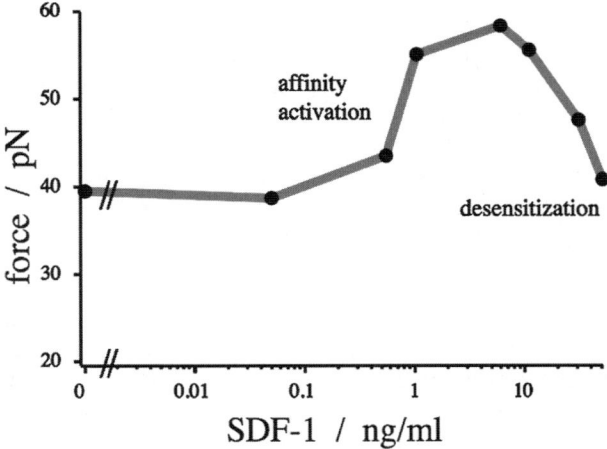

concentrations of SDF-1 to cover several orders of magnitude in the chemokine concentration, from no addition of SDF-1 up to physiologic levels, and above. Resting cells, i.e. cells with no addition of SDF-1, as well as cells at sub-physiologic concentration (0.05 ng/ml) show a force spectrum with rupture forces mainly between 20 and 60 pN. At physiologic concentrations of SDF-1 at 1–5 ng/ml, the rupture forces are shifted to higher levels of 40–80 pN. At higher concentrations of the chemokine, the average rupture forces shift back to 40 pN, i.e. to a level similar to that prior to the activation. Although there is some legitimate scepticism about the interpretation of this observation, which has been reproduced in several laboratories (personal communication), it appears safe to state that an arrest chemokine can induce quick cellular changes under conditions of AFM force spectroscopy. There is robust evidence that the force resolution at a single-molecule or single-receptor level is not only possible between isolated or "dead" receptors, but also between one living cell and an immobilized ligand and even between two living cells. Therefore, it may appear plausible that the force difference of low-affinity versus high-affinity integrin receptors could also be measured by AFM force spectroscopy. Future studies will have to correlate the different aspects of direct force measurements versus energy landscapes and off-rates of receptor-ligand binding, although for the immunologist such comparisons may not be necessary as long as a directly measured force can be translated into the function of an adhesion receptor.

9.3
Protocols

The following five step-by-step protocols should serve the interested student, biophysicist or cell biologist to choose the right experimental setup for their own needs. The protocols are designed to complement each other and should be easy to adapt to similar questions of cell adhesion and candidate drug screening. Starting with the functionalizing of the cantilever to attach a cell of interest onto its tip, the protocols continue with the description of using such an attached cell to probe it on another cell or on counter-receptors immobilized at different densities, and on suitable controls. A further protocol shows how to inhibit a specific cell adhesion binding by antibodies and other inhibitors which might be useful for the identification and prove the existence of specific interacting adhesion receptors. A state-of-the art protocol is included that allows us to measure the effect of potential adhesion activators such as a chemokine, a rare activating antibody or phorbol esters and other substances. Finally, a protocol for cell-free receptor ligand binding is provided.

9.3.1
Cantilever Functionalization

In order to probe a living cell with another cell or against an immobilized substrate the cell should be well attached to the tip of the cantilever (Fig. 9.14). The following protocol is more elaborate than the alternative protocol in which the cantilever is just incubated in poly-D-lysine for 30 min at RT and washed once with PBS. It has been shown that a very careful preparation of the cantilever pays off, and, therefore, this extended protocol is recommended for work on many cell types. However, the protocol includes the use of conA and if the protocol is being used on immune cells, one has to consider that conA has the potential to activate many different pathways in a living cell. Therefore, depending on the cell type and the underlying research question other procedures may need to be established first in the laboratory. This protocol can easily be adapted to functionalization procedures much more specific

Fig. 9.14. Scheme of immobilizing an intact living cell to the AFM cantilever

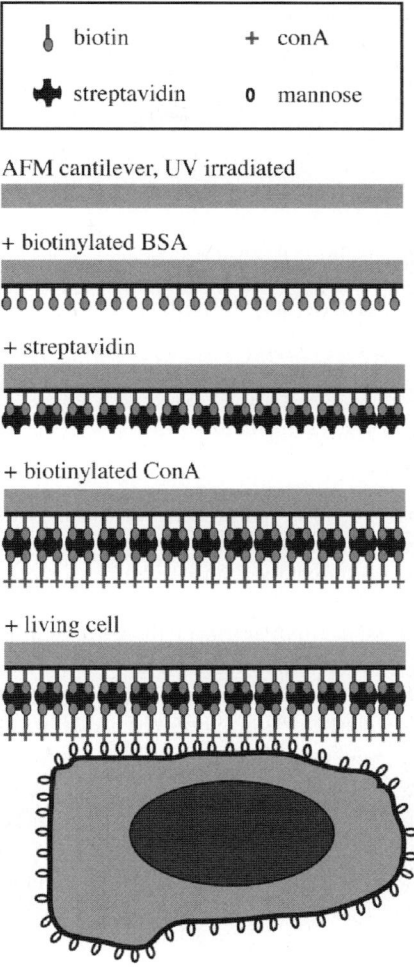

for the cell to be attached. For example, biotinylated conA could be replaced by biotinylated antibodies recognizing and binding to the surface of the investigated cell ubiquitously expressed specific receptors. This may include antibodies against specific adhesion receptors, such as CD43, CD44, and similar receptors on white blood cells, but many of them are also involved in lymphocyte activation processes which may interfere with the original question [66, 67]. Therefore, several strategies should be applied when trying to answer specific questions about lymphocyte activation.

9.3.1.1
Materials

Caution: Refer to your laboratory guidelines for appropriate handling of material marked with <!>, i.e. flammable or hazardous material.

Buffers and Solutions

Acetone <!>
Biotinylated BSA (Sigma)
Streptavidin
Biotinylated concanavalin A (conA; Sigma) <!>
PBS
NaHCO$_3$

Special Equipment

- Unsharpened cantilevers (e.g. model MLCT-AUHW; Thermomicroscopes, Sunnyvale, CA)
- UV-lamp

9.3.1.2
Method

1. Wash the cantilever for 5 min with acetone in a glass (!) Petri dish under a fume hood.
2. Remove the cantilever from acetone bath and irradiate under a UV light for 15 min.
3. Wash the cantilever three times for 5 min in 0.01-M PBS, pH 7.4. Cleaned cantilevers can be stored at 4 °C for several weeks.
4. Wash once for 5 min in 0.1-M NaHCO$_3$, pH 8.6.
5. Incubate the cantilever in 100-μl biotinylated BSA (0.5 mg/ml) for 12–16 h (e.g. overnight) at 4 °C (or for 4 h at RT).
6. Wash three times with 0.01-M PBS, pH 7.4.
7. Incubate the cantilever in 50-μl streptavidin (0.5 mg/ml in 0.01-M PBS, pH 7.4) for 15 min at RT.
8. Wash the cantilever three times in 0.01-M PBS, pH 7.4 before use.
9. A biotinylated ligand recognizing receptors on the cell may be coupled to the streptavidin-functionalized tip (e.g. biotinylated concanavalin A; conA, 0.5 mg/ml, 15 min at RT).

9.3.2
AFM Measurement on Living Cells

This protocol describes in detail the preparations needed to succeed in extremely sensitive cell measurements with living cells. As an example a Petri dish is prepared with different densities of immobilized ligands including VCAM-1 fusion protein and useful controls [6]. Then, the AFM is assembled and the functionalized cantilever from the previous protocol used after its calibration to attach a cell onto the tip of the cantilever. Finally, the AFM can be used to directly detect rupture forces between the cell and the counter-receptors studied. Using physiologic conditions of temperature and pH, this protocol can be used to study chemokine-mediated triggering of

integrin activation on lymphocytes or similar cells or cell lines, such as bone mar-
row stem or leukaemia-derived cells, and eventually metastatic cells. Additionally,
this protocol can easily be adapted from cell–substrate to cell–cell adhesion measurements. In order to identify the interacting receptors, blocking agents such as
function blocking antibodies against the receptors studied may be needed. The right
choice of interacting cell lines is strongly suggested, for example, choosing cell lines
known to bind just via one major pair of functional adhesion receptors and lacking all
others.

9.3.2.1
Materials

Additional Buffers and Solutions

Biotinylated concanavalin A (ConA) (Sigma) $<! >$
Cells expressing cell adhesion receptors of interest, for example, integrin LFA-1 (or
    VLA-4)
Receptor protein(s), for example, recombinant Fc fusion protein with binding
    domains of receptor, ICAM-1 or VCAM-1, or native soluble receptor proteins,
    or other ligands
Cell culture medium
Note: I recommend refraining from (a) performing experiments with cells and
    biomolecules at suboptimal ambient temperatures and (b) using growth media with
    an unphysiologic or changing pH value. The pH of CO_2/HCO_3-buffered media
    can fluctuate dramatically if the medium is left unchecked for evaporation. Adding
    different amounts of HEPES (4-(2-hydroxyethyl)-1-piperazineethanesulfonic
    acid) buffer does not prevent CO_2 from leaving the media which appears to be
    critical for some adhesion measurements. Most cell lines can be adapted to more
    expensive, CO_2-independent medium which I recommend for AFM measure-
    ments of living cells.
Inhibitors of receptor-ligand interaction, for example, function-blocking antibodies
    or peptidomimetic compounds
Activators, for example, Mg^{2+}, Mn^{2+} plus EGTA, activating antibodies, phor-
    bolesters (slow) or chemokines (fast)

Additional Special Equipment

An AFM apparatus with software is needed for the acquisition and analysis of force-
    distant curves along the z-axis. An AFM with an additional x-, y-axis is not really
    necessary for pure force spectroscopic measurements, but may help in combina-
    tion with AFM imaging to locate the binding events on, for example, a cell surface
    (not shown here). Professional AFMs are equipped with software from the manu-
    facturer. Custom AFMs may be used with Image SXM as a Macintosh application
    which is available based on NIH Image. We also use Igor Pro 4.0 for data analysis.
Standard cell culture lab for maintenance of cell lines or primary cells (with incuba-
    tor, laminar flow hood, centrifuge).
Plastic tissue culture dishes (uncoated).

Heating device for physiologic temperature (some experiments may yield suitable results at RT). We developed our own temperature-controlled heating device based on several ceramic resistors and silicone forming a ring-like structure on which the Petri dish was placed.

9.3.2.2
Method

1. Prepare Petri dish with immobilized ligand adsorbed to the surface, for example, adsorb different concentrations of soluble ICAM-1—or VCAM-1— fusion proteins, 25 μl for each droplet in a humid chamber at 4 °C for 12–16 h, wash with PBS, block with 0.5-mg/ml BSA for 30 min, wash with PBS.
2. Functionalize the cantilever tip with conA (see previous Protocol 1).
3. Assemble quickly the AFM with the functionalized cantilever pointing into a cell culture dish with medium (Care: avoid dry-out of proteins on the cantilever).
4. Adjust focused laser path, mirror and photodetector to optimal signal (in our system we optimize the detected laser signal to a value of about 8.0 on a scale from 0 to 10.24 with an equal distribution on both parts of the segmented photodetector).
5. Calibrate cantilever and obtain spring constant (we usually can confirm the manufacturers' value for the spring constant of the largest cantilever in a range of 10–14 mN/m; too low spring constants are likely the result of a broken cantilever, too high could be the result of a wrong setup or wrong cantilever).
6. Position the cantilever over a region without ligand protein and at least 20-μm away from the bottom of the Petri dish. Add 2–10-μl of cells ($\sim 10^5$/ml) close to the cantilever; wait 1–2 min to allow the cells to settle down; locate a single cell with the light microscope, position the tip of the cantilever above the cell; with the positioning micrometer screw carefully lower the cantilever onto the cell; ideally the cell should rest directly behind the pyramid of the cantilever tip. After 1–5 s of smooth contact retract the cantilever slowly, the cell should be firmly attached to the cantilever. (*Note*: avoid pressing the cell too firm against the bottom of the Petri dish, several days/weeks of experience may help to improve this critical step considerably.)
7. Position the cantilever with the attached cell over a region of the Petri dish with the previously adsorbed ligand. Collect "test" force scans from areas with different densities of the ligand; choose a suitable area and density of ligand to adjust the frequency of rupture events to approximately 30% by then adjusting the contact force and contact time. Our typical settings include an approach rate of 1 μm/s, a contact duration of 10 ms and a compression force of 100 pN. (*Note*: for activation experiments the adhesion rate should be adjusted too much less than 30%, for example, 1–10%, for the non-activated cells, in order to remain under optimal conditions of single-rupture events after activation of the cell adhesion receptor.)
8. Collect force scans at the appropriate ligand density and from negative control regions. ICAM-1 – fusion protein serves as an ideal control for VCAM-1 – fusion protein, and vice versa. Additional controls include uncoated areas and areas

which were only blocked with 0.5 mg/ml BSA for 30 min and washed with PBS (see step 1).

9. Collect force scans at different loading rates to determine the force spectrum of the receptor–ligand interactions.

9.3.3
Inhibition with Blocking Antibodies, Peptidomimetic Inhibitors or EDTA

This method is useful to identify the nature of adhesion events. Function blocking antibodies are used in cell biology to inhibit specific interactions of one pair of adhesion receptors preventing a cell to adhere via these receptors. Any remaining strong binding should be due to other pairs of adhesion receptors. Weak additional binding could be due to either unspecific binding of charged protein–protein interactions, or could point to an additional, but otherwise masked specific pair of weak adhesion receptors (Eibl, unpubl.). The protocol could be adapted to combine two antibodies blocking two different pairs of adhesion receptors on one cell. If a combination of two antibodies can block most of the adhesion events, each of the blocking antibodies might be used to study the unblocked pair of adhesion receptors. This appears to be a crude experiment since the blocking antibodies may interfere in an unspecific way with the unblocked pair of adhesion receptors, but for some experiments it may be difficult to find the right set of cell lines and this might be an option, since the antibodies block very specifically one type of receptor without interfering significantly with the other receptors of interest (Eibl, unpubl.).

Of course, smaller peptidomimetic agents mimicking the blocking function [68, 69] can be tried instead of antibodies. The equivalent concentration of such smaller molecules should be translated accordingly. Chelating agents such as EDTA or EGTA can also be used to interfere with the cell adhesive function to be measured with AFM [70], but this appears to be in an unspecific way, since those chelating substances remove ions needed by many different cell adhesion molecules, including integrins, for their function. Usually EDTA is used to remove CA^{2+} ions from integrins to abolish their binding to their receptors.

9.3.3.1
Method

1. Optimize conditions as determined in Protocol 2 (Sect. 9.3.2) for measuring receptor–ligand interactions at an adhesion rate of approximately 30%.
2. Collect at least 100 force scans.
3. Add non-blocking control antibody (or control compound) to cell culture medium to a final concentration of 10–20 μg/ml.
4. After 30 min collect force scans to determine adhesion frequency.
5. Repeat experiment with function-blocking antibodies at a final concentration of 10–20 μg/ml against the expressed receptor (or peptidomimetic compounds at suitable concentrations)—or add EDTA if analyzing a Ca^{2+}-sensitive adhesion receptor.
6. Determine changes in adhesion frequency and force value.

9.3.4
Activation with Mg²⁺, Mn²⁺ Ions, Activating Antibodies, Phorbolester or Chemokines

This protocol should be the first protocol to measure on a single-receptor level the activation of any integrin by any chemokine on any living cell type. It may also be used for many different types of cells, several other integrins and chemokines, or any adhesion receptors which exist in different activation states like CD44 which has at least three different activation states including low- and high-binding affinity as well as a non-binding state to its major ligand hyaluronic acid [53, 71]. Other known activators, for example, of integrins include Mg^{2+} and Mn^{2+} ions, (very rare) activating antibodies and phorbolester.

9.3.4.1
Method

1. Optimize the conditions as described in Protocol 2 (Sect. 9.3.2), but to a much lower frequency of adhesion events, i.e. 1–15%, depending on the type of activators and cells, i.e. very potent activators need a lower adhesion frequency at the start of the experiment.
2. Collect at least 100 force scans, eventually with a suitable negative control (PBS, control antibody, or heat-inactivated chemokine, or inactive recombinant protein).
3. Add the activating substance. In the case of integrins, for example, Mg^{2+}, Mn^{2+} and EGTA, phorbol ester, or a chemokine-like SDF-1 (or a test compound as putative activator). For better dilution of the compounds and avoiding possible desensitization phenomena with high local concentrations of chemokines these biological compounds may be added by replacing part of the cell culture media with the new compound. (*Note*: avoid changes in temperature and pH.)
4. Wait 1–20 min to continue collecting force scans, depending on dilution and time needed for the activating mechanism of the substance.
5. Collect force scans, eventually collect series of force scans every 30 min during longer activation processes.
6. For pharmacological dose–response curves it is possible to add stepwise with 10–100-fold increments biologic active substances like chemokines such as SDF-1 starting from low concentrations below any detectable response up to concentrations covering the physiologic range and resulting in highest activity, and eventually up to doses which lead to desensitization of the chemokine receptor.

9.3.5
AFM Measurement—Cell Free

Although this chapter focuses on cell adhesion measurements of living cells the following short protocol is included for completion of the protocol section as an example that AFM force spectroscopy can also be used for isolated cell adhesion receptors without the need of a living cell [25].

9.3.5.1
Materials

Additional Buffers and Solutions

Receptor and ligand protein(s), one of them should be biotinylated, for example, P-selectin and sLeX-PAA-biotin (high molecular complex of sialyl Lewis X and Biotin)

Hanks buffer with 2-mM Ca^{2+}

Inhibitors of receptor–ligand interaction, for example, function-blocking antibodies (WAPS [72] against human P-selectin) and suitable control antibody.

9.3.5.2
Method

1. Prepare Petri dish with immobilized ligand, for example, adsorb $25\,\mu l$ of P-selectin/Fc fusion protein in a humid chamber at $4\,°C$ for 12–16 h, wash with PBS, block with 0.5-mg/ml BSA for 30 min, wash with PBS.
2. Functionalize cantilever tip with counter-receptor, for example, sLex-biotin (see Protocol 1 (Sect. 9.3.1), instead of biotinylated ConA use a suitable biotinylated ligand for the receptor chosen in the step before).
3. Assemble quickly the AFM with the functionalized cantilever pointing into the Petri dish filled with medium or Hanks' buffer including 2-mM $CaCl_2$. (*Note*: avoid dry-out of proteins on the cantilever.)
4. Adjust focused laser path, mirror and photodetector.
5. Calibrate cantilever and obtain spring constant.
6. Position the cantilever over a region with adsorbed ligand.
7. Collect force scans.
8. As a negative control position the cantilever over an uncoated region and/or over a region with a control protein. Collect force scans.
9. Collect force scans at different loading rates to determine the force spectrum of the receptor–ligand interactions.
10. Add inhibiting antibodies at $10–50\,\mu g/ml$ (e.g. WAPS against human P-selectin) to prove the specificity of the interaction. Control IgG antibodies should not interfere with the adhesion rate and force.
11. Instead of adding specific antibodies, removal of Ca^{2+} ions by adding 5–20-mM EDTA can also interfere with the folding and bond strength of Ca-sensitive receptors.

9.4
Conclusion and Future Developments

Interestingly, function-blocking antibodies have been used in many animal models to prevent a disease or reduce the outcome of a disease. In particular, antibodies against VLA-4 have been shown in mouse models to prevent lung metastasis [73] or multiple sclerosis [74], and therefore provide a new therapeutic approach for many different diseases in humans. In addition to the treatment of autoimmune or inflammatory diseases, the further study of cell adhesion receptors and their regulation and manipulation may lead to new therapies in medicine, including autoimmune diseases such as rheumatoid arthritis, multiple sclerosis, Sjögren's syndrome, type I diabetes [75], and perhaps metastasizing cancer. Combined analysis of cell adhesion functions with the genetic profile of a tumour may help to better understand and prevent the ability of tumour or tumour stem cells to migrate and form distant metastases [49, 76–87]. AFM force experiments on living cells are usually not designed for high-throughput analysis of many samples, but this methodology could still be used do characterize the action of a series of the most promising few candidate drugs and therefore may help to reduce the costs of an otherwise broad pharmaceutical screening of too many candidates.

Using sophisticated AFM force measurements should be useful to screen candidate substances for small changes in cell adhesion properties, perhaps just modulating cell trafficking and with the potential of reducing side effects of otherwise too potent inhibitors. As shown with a highly active substance like SDF-1, even smallest concentrations of 1 ng/ml can have very dramatic effects on a living cell within a short time interval. To measure direct forces in real-time, i.e. in milliseconds may add to the potential value of future experiments. The triggering of rapid arrest appears to be as fast as one can measure. More than a decade ago, it appeared to be very unreasonable to expect such a fast activation process, usually known only from signalling of neurons and cardiomyocytes. Originally, integrin activation was considered to be in a range of dozens of seconds, in the meantime it is very clear that the rapid activation of integrins is triggered by arrest chemokines and that it takes place in a range of a few milliseconds. AFM force measurement may also be helpful to study the kinetics of activation. Interestingly, SDF-1 is usually presented on the surface of endothelial cells, although it can come from other cells than the endothelial monolayer. There are conflicting speculations about the different activity potential of soluble SDF-1 versus SDF-1 molecules bound to and presented on the endothelial cell. It may be hard to fully exclude that even immobilized SDF-1 molecules separate from the endothelial cell and become locally soluble, or vice versa, soluble SDF-1 molecules could attach to the Petri dish or to endothelial cells within an experiment, and therefore could be presented as immobilized molecules. Future studies may address the question whether the postulated difference of soluble versus immobilized SDF-1 can be supported or not. Mouse and human lymphocytes appear to express homologous molecules with identical names and comparable functions: VLA-4, VCAM-1 and SDF-1 and CXCR4. In functional experiments these molecules of the two species are widely interchangeable. The only difference appears to be that human VCAM-1 has a second binding site for VLA-4. It may be interesting to find any differences in the use of these two binding sites in contrast to the single mouse binding site. For many experiments studying human VLA-4, a fusion protein with just one functional domain is used.

Two decades after its invention, AFM applications continue to broaden into very powerful applications in many different disciplines. Although more and more cell biologists appear to find interest in these powerful methods, most cell biologists remain unfamiliar with this methodology. This may be due to the different perceptions of physicists and biologists. Often, the physicist may want to invent a very new derivation or resolution of an AFM or a new combination of an AFM with another type of microscope or apparatus, whereas the biologist wants to use an existing method and can not find the appropriate support from the physicist who is not very interested in the cell biology question. In addition, the cell biologist may not have access to an existing AFM and does not want to invest in such a system without being used to the handling, controls and possible artefacts.

It may help if physicists could provide more access to perhaps older AFMs, or if the manufacturer would offer free trials for biology and cell adhesion labs. This may improve the existing AFMs for future cell adhesion experiments.

Acknowledgments. I gratefully acknowledge Heinz Höfler for his support on initiating the interdisciplinary project between nanotechnology and pathology, Irving Weissman, Eugene Butcher, James Campbell, Sara Michie and Peter Herrlich for helping me to greatly expand my knowledge in the field of lymphocyte rolling and cell adhesion, and Hermann Gaub, Vincent Moy, Xiaohui (Frank) Zhang, Olga Vinogradova and Martin Benoit for expert advice in the field of AFM. I thank Bernhard Holzmann, Horst Kessler and Ronen Alon for making available reagents and cell lines, Guttorm Haraldsen, Anna Müller, Albert Zlotnik, Markus Schneemann and Tim Springer for helpful discussions and Stephan Bärtsch and Matthias Maiwald for critical comments on the manuscript. This work was supported in part by the DFG (Ei378/1 + 2; Deutsche Forschungsgemeinschaft), a Deans fellowship from the Stanford University School of Medicine, a travel stipend of the GSO (German Scholar Organization) and prize money for travelling the author received from the German Cancer Research Centre (DKFZ, Heidelberg). For presenting part of this work the author received the first alumni prize of the German Cancer Research Centre, DKFZ Heidelberg.

References

1. Binnig G, Quate C, Gerber C (1986) Phys Rev Lett 56:930
2. Giessibl F, Hembacher S, Bielefeldt H, Mannhart J (2000) Science 289:422
3. Giessibl FJ (1995) Science 267:68
4. Eibl RH (2004) In: Immunology 2004: cytokine network, regulatory cells, signalling, and apoptosis. (Editor: Skamene E). Medimond, Milano, Italy, 1:115
5. Eibl RH, Benoit M, Weissman IL, Gaub HE, Moy VT (2004) Clinical and Investigative Medicine Suppl. 27:4
6. Eibl RH, Moy VT (2005) Methods Mol Biol 305:439.
7. Eibl RH, Moy VT (2004) In: Recent res. devel. biophys. (Editor: Pandalai, SG), TRN, Trivandrum, India, p 235
8. Eibl RH, Moy VT (2005) In: Protein-ligand interactions (Editor: Nienhaus GU), p 450
9. Eibl RH, Benoit M (2004) IEE Proc Nanobiotechnol 151:128
10. Dammer U, Popescu O, Wagner P, Anselmetti D, Güntherodt HJ, Misevic GN (1995) Science 267:1173
11. Hinterdorfer P, Dufrêne YF (2006) Nat Methods 3:347
12. Rief M, Clausen-Schaumann H, Gaub HE (1999) Nat Struct Biol 6:346
13. Rief M, Gautel M, Oesterhelt F, Fernandez JM, Gaub HE (1997) Science 276:1109
14. Wu HW, Kuhn T, Moy VT (1998) Scanning 20:389
15. Baumgartner W, Hinterdorfer P, Ness W, Raab A, Vestweber D, Schindler H, Drenckhahn D (2000) Proc Natl Acad Sci U S A 97:4005
16. Wong J, Chilkoti A, Moy VT (1999) Biomol Eng 16:45
17. Yuan C, Chen A, Kolb P, Moy VT (2000) Biochemistry 39:10219

18. Moy VT, Florin EL, Gaub HE (1994) Science 266:257
19. Florin EL, Moy VT, Gaub HE (1994) Science 264:415
20. Merkel R, Nassoy P, Leung A, Ritchie K, Evans E (1999) Nature 397:50
21. Benoit M, Gabriel D, Gerisch G, Gaub HE (2000) Nat Cell Biol 2:313
22. Thie M, Röspel R, Dettmann W, Benoit M, Ludwig M, Gaub HE, Denker HW (1998) Hum Reprod 13:3211
23. Zhang X, Wojcikiewicz EP, Moy VT (2006) Exp Biol Med (Maywood) 231:1306
24. Zhang X, Chen A, De Leon D, Li H, Noiri E, Moy VT, Goligorsky MS (2004) Am J Physiol Heart Circ Physiol 286:H359
25. Zhang X, Bogorin DF, Moy VT (2004) Chemphyschem 5:175
26. Li F, Redick SD, Erickson HP, Moy VT (2003) Biophys J 84:1252
27. Chen A, Moy VT (2000) Biophys J 78:2814
28. Zhang X, Wojcikiewicz E, Moy VT (2002) Biophys J 83:2270
29. Baumgartner W, Golenhofen N, Grundhöfer N, Wiegand J, Drenckhahn D (2003) J Neurosci 23:11008
30. Benoit M, Gaub HE (2002) Cells Tissues Organs 172:174
31. Eibl RH (2007) In: Hinterdorfer P, Schuetz G, Pohl P (eds) Advances in single molecule research for biology and nanoscience. Trauner, p 40
32. Chen A, Moy VT (2002) Methods Cell Biol 68:301
33. Lehenkari PP, Horton MA (1999) Biochem Biophys Res Commun 259:645
34. Kokkoli E, Ochsenhirt SE, Tirrell M (2004) Langmuir 20:2397
35. Fritz J, Katopodis AG, Kolbinger F, Anselmetti D (1998) Proc Natl Acad Sci U S A 95:12283
36. Dettmann W, Grandbois M, André S, Benoit M, Wehle AK, Kaltner H, Gabius HJ, Gaub HE (2000) Arch Biochem Biophys 383:157
37. Hinterdorfer P, Baumgartner W, Gruber HJ, Schilcher K, Schindler H (1996) Proc Natl Acad Sci U S A 93:3477
38. Wong SS, Joselevich E, Woolley AT, Cheung CL, Lieber CM (1998) Nature 394:52
39. Stenlund P, Babcock GJ, Sodroski J, Myszka DG (2003) Anal Biochem 316:243
40. Hussain MA, Agnihotri A, Siedlecki CA (2005) Langmuir 21:6979
41. Berlin C, Bargatze RF, Campbell JJ, von Andrian UH, Szabo MC, Hasslen SR, Nelson RD, Berg EL, Erlandsen SL, Butcher EC (1995) Cell 80:413
42. Butcher EC (1991) Cell 67:1033
43. Springer TA (1994) Cell 76:301
44. Campbell JJ, Hedrick J, Zlotnik A, Siani MA, Thompson DA, Butcher EC (1998) Science 279:381
45. Campbell JJ, Haraldsen G, Pan J, Rottman J, Qin S, Ponath P, Andrew DP, Warnke R, Ruffing N, Kassam N et al (1999) Nature 400:776
46. Altevogt P, Hubbe M, Ruppert M, Lohr J, von Hoegen P, Sammar M, Andrew DP, McEvoy L, Humphries MJ, Butcher EC (1995) J Exp Med 182:345
47. Vajkoczy P, Laschinger M, Engelhardt B (2001) J Clin Invest 108:557
48. Warnock RA, Askari S, Butcher EC, von Andrian UH (1998) J Exp Med 187:205
49. Eibl RH, Kleihues P, Jat PS, Wiestler OD (1994) Am J Pathol 144:556
50. Eibl RH, Pietsch T, Moll J, Skroch-Angel P, Heider KH, von Ammon K, Wiestler OD, Ponta H, Kleihues P, Herrlich P (1995) J Neurooncol 26:165
51. Stamper HBJ, Woodruff JJ (1976) J Exp Med 144:828
52. Gunthert U, Hofmann M, Rudy W, Reber S, Zoller M, Haussmann I, Matzku S, Wenzel A, Ponta H, Herrlich P (1991) Cell 65:13
53. Stamenkovic I, Amiot M, Pesando JM, Seed B (1989) Cell 56:1057
54. Hutter J, Bechhoefer J (1993) Rev Sci Instrum 64:1868
55. Benoit M (2002) Methods Cell Biol 68:91
56. Dufrêne YF, Hinterdorfer P (2007) Pflugers Arch 456:237

57. Wojcikiewicz EP, Zhang X, Chen A, Moy VT (2003) J Cell Sci 116:2531
58. DeGrendele HC, Estess P, Picker LJ, Siegelman MH (1996) J Exp Med 183:1119
59. DeGrendele HC, Estess P, Siegelman MH (1997) Science 278:672
60. Gallatin WM, Weissman IL, Butcher EC (1983) Nature 304:30
61. Weissman I (1967) Nature 215:315
62. Qian F, Vaux DL, Weissman IL (1994) Cell 77:335
63. Qian F, Hanahan D, Weissman IL (2001) Proc Natl Acad Sci U S A 98:3976
64. Gosslar U, Jonas P, Luz A, Lifka A, Naor D, Hamann A, Holzmann B (1996) Proc Natl Acad Sci U S A 93:4821
65. Butcher EC, Picker LJ (1996) Science 272:60
66. Mikulowska A, Johnson GG, Berberian JM, Butcher EC, McEvoy LM, Michie SA (1999) Cell Immunol 194:112
67. Johnson GG, Mikulowska A, Butcher EC, McEvoy LM, Michie SA (1999) J Immunol 163:5678
68. Gottschling D, Boer J, Schuster A, Holzmann B, Kessler H (2002) Angew Chem Int Ed 41:3007
69. Locardi E, Boer J, Modlinger A, Schuster A, Holzmann B, Kessler H (2003) J Med Chem 46:5752
70. Wojcikiewicz EP, Zhang X, Moy VT (2004) Biol Proced Online 6:1
71. Lesley J, He Q, Miyake K, Hamann A, Hyman R, Kincade PW (1992) J Exp Med 175:257
72. Picker LJ, Warnock RA, Burns AR, Doerschuk CM, Berg EL, Butcher EC (1991) Cell 66:921
73. Brocke S, Piercy C, Steinman L, Weissman IL, Veromaa T (1999) Proc Natl Acad Sci U S A 96:6896
74. Lauri D, De Giovanni C, Biondelli T, Lalli E, Landuzzi L, Facchini A, Nicoletti G, Nanni P, Dejana E, Lollini PL (1993) Br J Cancer 68:862
75. Michie SA, Sytwu HK, McDevitt JO, Yang XD (1998) Curr Top Microbiol Immunol 231:65
76. Clarke MF, Dick JE, Dirks PB, Eaves CJ, Jamieson CHM, Jones DL, Visvader J, Weissman IL, Wahl GM (2006) Cancer Res 66:9339
77. Reya T, Morrison SJ, Clarke MF, Weissman IL (2001) Nature 414:105
78. Ohgaki H, Eibl RH, Wiestler OD, Yasargil MG, Newcomb EW, Kleihues P (1991) Cancer Res 51:6202
79. Wiestler OD, Aguzzi A, Schneemann M, Eibl R, von Deimling A, Kleihues P (1992) Cancer Res 52:3760
80. von Deimling A, Eibl RH, Ohgaki H, Louis DN, von Ammon K, Petersen I, Kleihues P, Chung RY, Wiestler OD, Seizinger BR (1992) Cancer Res 52:2987
81. Wiestler OD, Brüstle O, Eibl RH, Radner H, Aguzzi A, Kleihues P (1992) Brain Pathol 2:47
82. Wiestler OD, Brüstle O, Eibl RH, Radner H, Von Deimling A, Plate K, Aguzzi A, Kleihues P (1992) Neuropathol Appl Neurobiol 18:443
83. Radner H, el-Shabrawi Y, Eibl RH, Brüstle O, Kenner L, Kleihues P, Wiestler OD (1993) Acta Neuropathol 86:456
84. Ohgaki H, Eibl RH, Schwab M, Reichel MB, Mariani L, Gehring M, Petersen I, Höll T, Wiestler OD, Kleihues P (1993) Mol Carcinog 8:74
85. Louis DN, von Deimling A, Chung RY, Rubio MP, Whaley JM, Eibl RH, Ohgaki H, Wiestler OD, Thor AD, Seizinger BR (1993) J Neuropathol Exp Neurol 52:31
86. Wiestler OD, Brüstle O, Eibl RH, Radner H, Aguzzi A, Kleihues P (1994) Recent Results Cancer Res 135:55
87. Kleihues P, Ohgaki H, Eibl RH, Reichel MB, Mariani L, Gehring M, Petersen I, Höll T, von Deimling A, Wiestler OD, et al. (1994) Recent Results Cancer Res 135:25

10 Imaging Chemical Groups and Molecular Recognition Sites on Live Cells Using AFM

David Alsteens · Vincent Dupres · Etienne Dague · Claire Verbelen · Guillaume André · Grégory Francius · Yves F. Dufrêne

Abstract. Imaging the nanoscale distribution of specific chemical and biological sites on live cells is an important challenge in current life science research. In addition to imaging the surface topography of live cells, atomic force microscopy (AFM) is increasingly used to probe their chemical groups and biological receptors. In chemical force microscopy, AFM tips are modified with specific functional groups, thereby allowing investigators to probe chemical sites and their interactions on a scale of only ~25 functional groups. In molecular recognition imaging, tips are functionalized with specific biomolecules, or samples labeled with immunogold particles, enabling researchers to localize specific receptors. Clearly, these nanoscale investigations provide new avenues in cellular biology and microbiology for elucidating the structure–function relationships of cell surfaces. In this chapter, we discuss the principles of these AFM modalities and their applications in life science research.

Key words: Atomic force microscopy, Cells, Chemical force microscopy, Molecular recognition studies, Single molecules

10.1 Introduction

Specific molecular recognition interactions between receptors and cognate ligands are ubiquitous in life sciences and increasingly used in nanobiotechnology [1–3]. In addition, nonspecific intermolecular interactions, such as hydrophobic and electrostatic forces, also play essential roles in biology since they promote crucial events like protein folding and cell adhesion [4, 5]. Despite the importance of these noncovalent intermolecular forces, their quantification has long been challenging. Force measuring techniques like the surface forces apparatus (SFA) [6] and the optical and magnetic tweezers [7, 8] have been increasingly used in that respect; yet these approaches are limited by their poor lateral resolution which does not provide access to the spatial distribution of (bio)chemical interactions. In this context, atomic force microscopy (AFM) has recently opened remarkable opportunities. On the one hand, chemical force microscopy (CFM), involving modification of AFM tips with specific functional groups, has enabled researchers to map the spatial arrangement of chemical groups and their interactions on organic surfaces and on cell surfaces. This nanoscale, chemically sensitive imaging tool offers two major advantages over classical probing methods: (1) hydrophobic and charged groups, and their interactions, are measured directly and quantitatively on live cells, and (2) nanoscale variations of hydrophobicity and charge can be resolved. On the other hand, single-molecule force spectroscopy with biologically modified tips provides a means to map individual receptors on cellular surfaces and to measure their molecular recognition forces. As a complement, topographic imaging combined with immunogold labeling has also proved useful for localizing molecular recognition sites. In this chapter, we describe the principle and methodology of chemical force microscopy and molecular recognition imaging, and highlight some of their applications.

10.2
Chemical Force Microscopy

10.2.1
Methods

The general idea of CFM is to use AFM tips with well-defined chemistry for measuring adhesion and friction, and/or for imaging surfaces (for recent reviews [6, 9]). AFM cantilevers and tips are usually made of silicon (Si) or silicon nitride (Si_3N_4) using microfabrication techniques. Because the surface chemistry of such commercial tips is poorly controlled and often contaminated with gold and other materials, reliable CFM measurements require functionalizing the tip surface with organic monolayers terminated by specific functional groups (e.g., OH or CH_3). A common method to achieve this is based on the formation of self-assembled monolayers (SAMs) of alkanethiols on gold surfaces. The procedure involves coating, by thermal evaporation, microfabricated cantilevers with a thin adhesive layer (Cr or Ti), followed by a 15–100-nm thick Au layer, immersing the coated cantilevers in dilute (0.1–1 mM) ethanol solutions of the selected alkanethiol, rinsing with ethanol and drying, using a gentle nitrogen flow. Although the protocol is fairly simple, it is important to validate the quality of the surface modification, which can be done by treating model supports (silicon) in the same way as the tips and characterizing them by means of surface analysis techniques (e.g., contact angle measurements or X-ray photoelectron spectroscopy). Another important point is to use the functionalized tips immediately after they are prepared in order to minimize surface contamination and alteration.

The most common application of CFM is the measurement of the adhesion strength between chemical groups via force spectroscopy. Here, the cantilever deflection is recorded as a function of the vertical displacement of the piezoelectric scanner, i.e. as the sample is pushed towards the tip and retracted. This yields a raw "voltage-displacement" curve which can be converted into a "force-distance" curve as follows. Using the slope of the curves in the region where tip and sample are in contact, the voltage can be converted into a cantilever deflection. In order to minimize the possible effects of repulsive surface forces and/or sample deformation, it is recommended to consider the slope of the retraction curve. The cantilever deflection is then converted into a force (F) using Hooke's law: $F = -k \times d$, where k is the cantilever spring constant. The curve can be corrected by plotting F as a function of $(z - d)$. The zero separation distance is then determined as the position of the vertical linear parts of the curve in the contact region. Note that force mapping can be performed by recording spatially resolved force-distance curves in the (x, y) plane. In adhesion measurements, the hysteresis or "pull-off" force observed during retraction is used to estimate the adhesion (unbinding) force between tip and sample. To get an accurate knowledge of the measured forces, it is important to determine actual spring constants experimentally since they may substantially differ from values quoted by the manufacturer [10].

10.2.2
Probing Hydrophobic Forces

A variety of chemical groups (hydrophobic/hydrophilic; charged/uncharged) and their interactions in different solvents have been investigated using CFM [9, 11–16]. Recently, Alsteens et al. [17–19] validated the use of the technique for probing hydrophobic groups and for quantifying short-range hydrophobic forces. Although hydrophobic forces have been known for 70 years and are of prime importance in biology (protein folding and aggregation, membrane fusion and cell adhesion), their detailed mechanisms are not fully understood. Various effects have been proposed to explain these forces, such as entropic effects due to disruption of the arrangement of water near hydrophobic surfaces (solvent organization), separation-induced phase transition (cavitation), hydrodynamic fluctuating correlation, bridging of submicron bubbles, electrostatic effects, correlated charge fluctuations or correlated dipole interactions.

As a proof of concept, SAMs of CH3- and OH-terminated alkanethiols mixed in different proportions were probed using water contact angle measurements and CFM with hydrophobic, CH3-terminated tips (Fig. 10.1). Consistent with the expectation (work of adhesion, Young equation), the contact angle and adhesion force values measured on mixed SAMs increased gradually with the molar fraction of CH3-terminated alkanethiols (Fig. 10.1a, b), yielding a linear relationship between the adhesion force and the cosine of the contact angle (Fig. 10.1c). This excellent agreement demonstrates that the measured adhesion forces reflect surface hydrophobicity, thus hydrophobic forces.

Notably, as we shall see below, interpretation of the data in terms of interfacial thermodynamics reveals that the measured adhesion forces do not originate from true, direct tip–sample interactions, but may rather reflect entropy changes associated with the restructuration of water near hydrophobic surfaces. We start by considering the Johnson–Kendall–Roberts (JKR) model which links the adhesion force (F_{adh}) to the work of adhesion (W_{adh}) [12]:

$$F_{adh} = 1.5\pi R W_{adh} \tag{10.1}$$

where R is the radius of curvature of the AFM tip. The work of adhesion in water is given by [20]:

$$
\begin{aligned}
W_{adh} &= \gamma_{sample,water} + \gamma_{tip,water} - \gamma_{tip,sample} \\
&= W_{tip,sample} - W_{sample,water} - W_{tip,water} + 2\gamma_{water}
\end{aligned} \tag{10.2}
$$

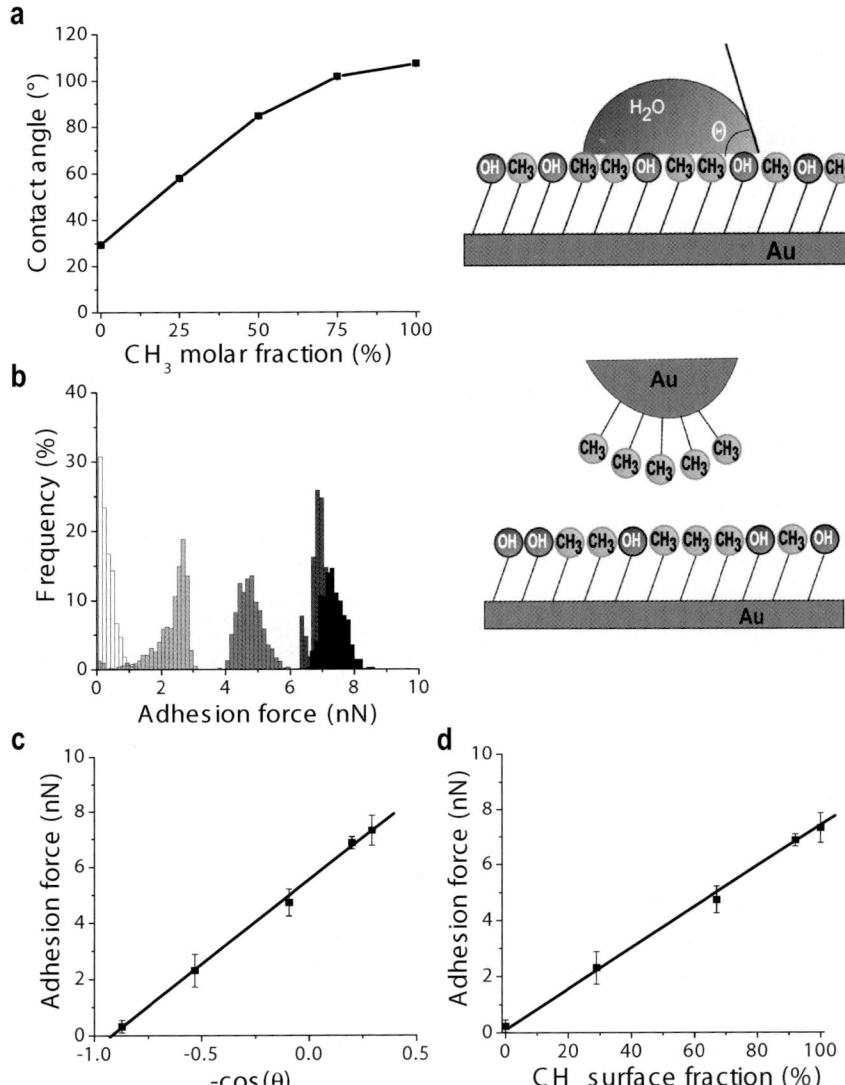

Fig. 10.1. Chemical force microscopy (CFM): principle and application to the probing of hydrophobic forces. (**a**) Water contact angle (θ) values measured for mixed self-assembled mono-layers of CH3- and OH-terminated alkanethiols as a function of the molar fraction of CH3-terminated alkanethiols. (**b**) Histograms of adhesion forces measured on the mixed SAMs using CFM with hydrophobic CH3-tips. (**c**) Variation of adhesion forces as a function of the cosine of the water contact angle. (**d**) Adhesion force as a function of the surface fraction of CH3-terminated alkanethiols computed using Cassie's law. Reprinted with permission from [17]

where $\gamma_{X,Y}$ is the interfacial energy, $W_{X,Y}$ is the work of adhesion in a vacuum, and γ_X is the surface energy. The term $W_{sample,water}$ can be deduced from the water contact angle θ, using the Young equation:

$$W_{sample,water} = \gamma_{sample} + \gamma_{water} - \gamma_{sample,water}$$
$$= \gamma_{water}(1 + \cos\theta) \tag{10.3}$$

Combining Eqs. (10.1), (10.2), and (10.3), yields the following expression:

$$\frac{F_{adh}}{1.5\pi R} = W_{tip,sample} - W_{tip,water} - \gamma_{water}\cos\theta + \gamma_{water} \tag{10.4}$$

Assuming that the $W_{tip,sample}$ and $W_{tip,water}$ values are close, which appears reasonable since the CH_3-modified tip is involved only in dispersion interactions (London forces) (10.4) becomes:

$$\frac{F_{adh}}{1.5\pi R} = \gamma_{water} - \gamma_{water}\cos\theta \tag{10.5}$$

Considering the surface tension of water (72.6 mN/m) and the tip radius (~ 20 nm) then leads to the following linear equation $F_{adh} = 6.84 - 6.84\cos\theta$, which agrees remarkably well with the regression equation of Fig. 10.1c, i.e.: $F_{adh} = 5.60 - 6.08\cos\theta$, where the uncertainties on slope and intercept are 0.14 and 0.10 nN, respectively. The agreement between experimental and theoretical adhesion forces supports the hypothesis that $W_{tip,sample}$ and $W_{tip,water}$ are similar, thus that only dispersion forces are responsible for the measured forces.

Accordingly, this reasoning indicates that the measured adhesion forces do not originate from true, direct tip–sample interactions, but may rather reflect entropy changes associated with the restructuration of water near hydrophobic surfaces. This work is also of practical interest since it shows that CFM can be used for quality control of chemically modified tips, which is not feasible by common analytical techniques. As we shall discuss below, further interpretation of cellular data implied expressing the measured adhesion force as a function of the surface fraction of CH_3-terminated alkanethiols, determined using Cassie's law and contact angle values (Fig. 10.1d).

In earlier work, AFM was also used to probe charged groups and their electrostatic forces [12, 13, 21]. For instance, force-distance curves were recorded at different pH values between tips and surfaces modified with carboxylic groups [22]. The curves showed differences according to pH (Fig. 10.2): while no adhesion was observed at pH 10, significant adhesion forces, of 1.4-nN magnitude, were measured at pH 3. Curves recorded in the 3–10 pH range showed adhesion forces for pH values smaller than 5. In addition, at pH 10 a long range repulsion force was seen starting at 40 nm, while a jump to contact was noted at pH 3. The repulsion force at high pH was attributed to electrostatic double layer forces between negatively charged COO^-/COO^- surfaces. The jump to contact at low pH was due to attractive van der Waals forces and indicated a lack of repulsion between uncharged COOH/COOH surfaces. In the same way, the strong adhesion observed at low pH values could be attributed to hydrogen bonding between COOH/COOH groups while the lack of adhesion at high pH was suggested to reflect the electrostatic repulsion

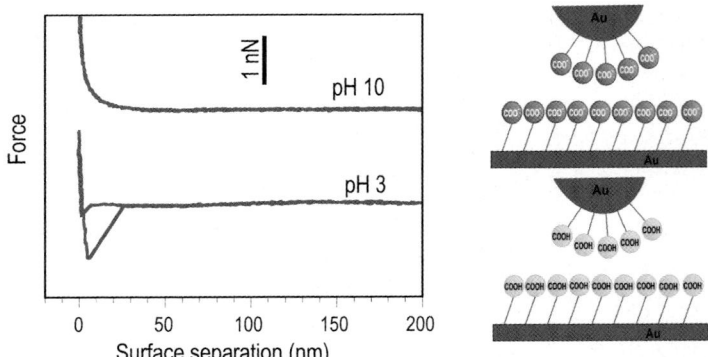

Fig. 10.2. AFM force-distance curves recorded at pH 10 and pH 3 between modified tips and model surfaces terminated with COOH groups. As shown in the drawing the differences observed according to pH can be related to a change of ionization state of the surfaces. Adapted with permission from [22]

between the negatively charged COO^-/COO^- surfaces. Hence, the above studies show that functionalized tips are very sensitive to hydrophobic and charged groups.

10.2.3
Chemical Force Microscopy of Live Cells

For the first time, we showed that the CFM technique can probe chemical groups and their interactions on live cells on a scale of only ∼25 functional groups [17–19, 23]. Dague et al. [19] used CFM with methyl-terminated tips to measure the hydrophobicity of the human opportunistic pathogen *Aspergillus fumigatus*. Topographic images revealed the presence of regularly arranged rodlets on *A. fumigatus* conidia (Fig. 10.3a). These structures are composed of hydrophobins, a family of small, moderately hydrophobic proteins that favor spore dispersion by air currents and mediate adherence to host cells. Force curves recorded across these surfaces with a hydrophobic tip showed large adhesion forces, of ∼3,000 pN magnitude (Fig. 10.3b, c). Comparison with the data obtained on reference surfaces (Fig. 10.1d) indicated that the conidial surface has a marked hydrophobic character, corresponding to a surface composed of ∼10 CH_3 and ∼15 OH groups, which is fully consistent with the presence of an outermost surface layer of hydrophobins and provides direct indications as to their putative functions as dispersion and adherence structures. In agreement with the uniform surface structure, adhesion maps were rather homogeneous (Fig. 10.3b), supporting the idea that the *A. fumigatus* conidial surface is homogeneously hydrophobic.

Two hydrophilic controls confirmed that the measured hydrophobic properties are associated with hydrophobins [19]. First, topographic images recorded on conidia treated with NaOH, a procedure that removes all cell wall proteins including rodlet proteins, revealed a surface lacking any rodlet, but exposing underlying cell wall polysaccharides (Fig. 10.3d, e, f). These structural changes were correlated with profound decrease of cell surface hydrophobicity, the adhesion force towards the hydrophobic tip being only ∼300 pN. Second, similar results were obtained on the

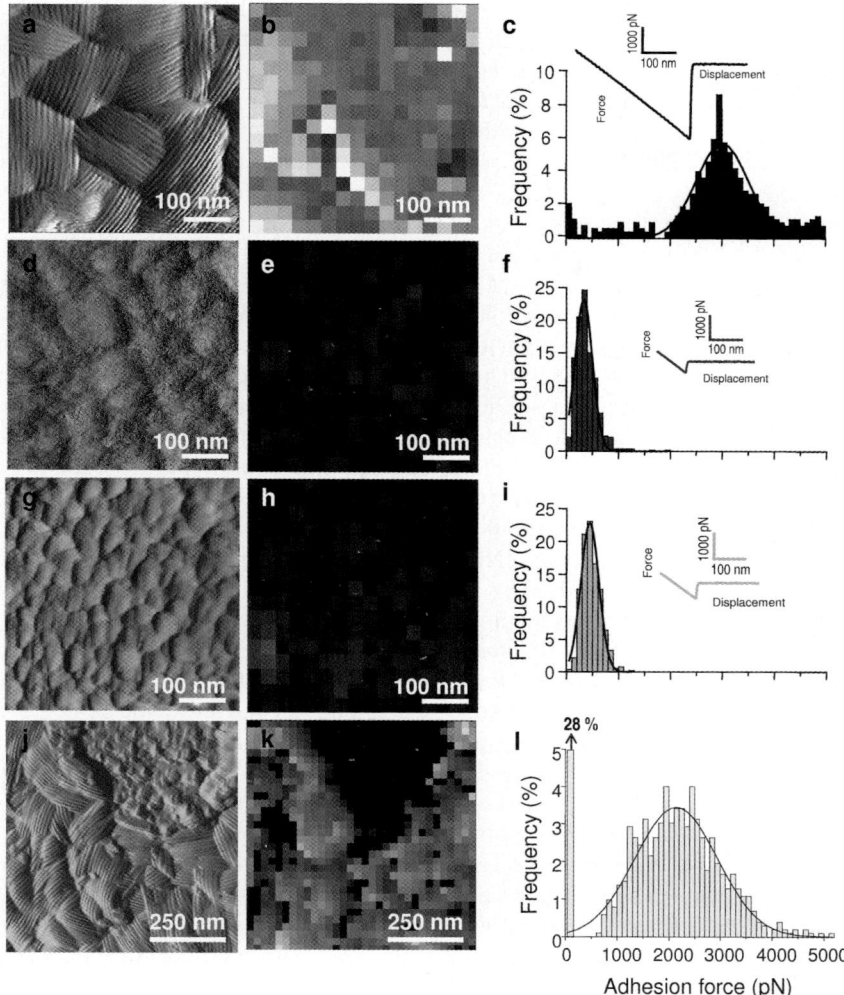

Fig. 10.3. CFM of *Aspergillus fumigatus*. (**a**) High-resolution image of a wild-type conidial surface in aqueous solution revealing rodlets. (**b**) Adhesion force map (z-range: 6 nN) and (**c**) adhesion force histogram (n = 512) recorded with a hydrophobic tip, indicating that the rodlet surface is uniformly hydrophobic. (**d, e, f, g, h, i**) High-resolution images, adhesion force maps and histograms (n = 512) obtained on NaOH-treated conidia (**d, e, f**) and on the ΔrodA ΔrodB double mutant (**g, h, i**). The purely hydrophilic surface is attributed to cell wall polysaccharides. (**j, k, l**) High-resolution image, adhesion force map and histogram obtained on SDS-treated conidia, revealing highly correlated structural and hydrophobic heterogeneities. Reprinted with permission from [19]

ΔrodA ΔrodB double mutant which does not produce rodlets (Fig. 10.3g, h, i). Considering the above reference SAMs surfaces (Fig. 10.1d), it was concluded that the two conidial surfaces are purely hydrophilic, corresponding to purely OH-terminated surfaces, which agrees well with the exposure of cell wall carbohydrates. Notably, nanoscale variations of hydrophobic properties were also resolved on SDS-treated conidia, for which rodlet patches are missing in very localized regions (Fig. 10.3j, k, l). These nanoscale structural heterogeneities were directly correlated with differences in hydrophobicity, the rodlet and polysaccharide regions displaying contrasted hydrophobic and hydrophilic characters.

Remarkably, real-time CFM with a temperature-controlled stage enables researchers to probe not only structural, but also chemical dynamics on cells. In one such study, the changes of A. *fumigatus* conidia occurring during germination at 37 °C were tracked in real time (Fig. 10.4, [23]). Images of the same spore after 20 min, 60 min, and 120 min clearly revealed significant structural alterations, the rodlet layer changing into a layer of amorphous material, presumably reflecting the underlying polysaccharides. Consistent with this, adhesion maps with hydrophobic tips revealed a dramatic loss of hydrophobicity with time. After 2-h of germination, heterogeneous hydrophobic contrast was observed, reflecting the coexistence of hydrophobic rodlets and hydrophilic polysaccharides. These data nicely demonstrate that CFM is capable of resolving submicron chemical heterogeneities on live cells as they grow.

Of particular interest in the medical context, is the possibility to probe differences of chemical properties and interactions on cells following treatments with drugs. For instance, Alsteens et al. [18] showed that the surface of mycobacteria has a remarkably strong hydrophobic character due to the presence of an outermost layer of hydrophobic mycolic acids. This finding is of biomedical relevance since these hydrophobic constituents are thought to represent an important permeation barrier to common antibacterial agents. By contrast, treatment of the cells with two antibiotics, isoniazid and ethambutol, led to a dramatic decrease of cell surface hydrophobicity, attributed to the removal of the mycolic acid layer [18, 19]. Hence, the combination of topographic imaging with CFM provides unique opportunities to gain insight into the action modes of antibiotics as well as into the nanoscale organization of bacterial cell walls.

CFM has also been used to probe charged groups and their interactions on live cells, which is particularly relevant since most microorganisms possess a negative surface charge under physiological conditions due to the presence of anionic surface groups such as carboxyl and phosphate. The resulting cell surface charge plays an important role in controlling cell adhesion and aggregation phenomena, as well as antigen–antibody, cell–virus, cell–drug and cell–ions interactions. An example of this is provided by the work of Ahimou et al. [22] who used AFM tips functionalized with ionizable carboxyl groups to probe the electrostatic properties of S. *cerevisiae*. In Fig. 10.5, one can see that force-distance curves recorded in these conditions were strongly influenced by pH: while no adhesion was measured at neutral/alkaline pH,

▶

Fig. 10.4. Tracking the structural and chemical dynamics of germinating A. *fumigatus* cells. Series of high-resolution deflection images (*left*) and adhesion force maps (*right*) recorded on a single spore during germination. Within less than 3 h, the crystalline rodlet layer changed into a layer of amorphous material, presumably reflecting inner cell wall polysaccharides. After 2 h, both rodlet and amorphous regions were found to coexist (separated by *dashed line*). Consistent with this structural dynamics, substantial reduction of adhesion contrast was noted with time (*right* images), reflecting a dramatic decrease of hydrophobicity. After 2 h, heterogeneous contrast was observed in the form of hydrophobic patches (*dashed line*), surrounded by a hydrophilic sea. Reprinted with permission from [23]

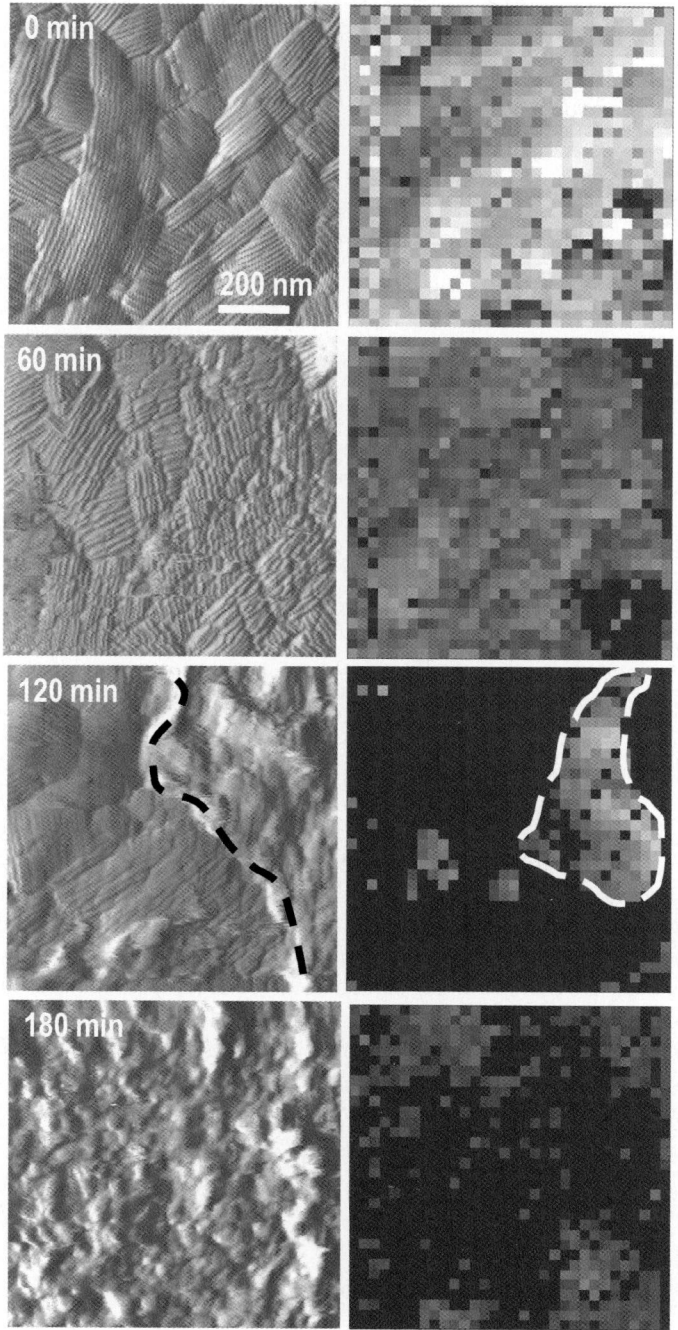

Fig. 10.4.

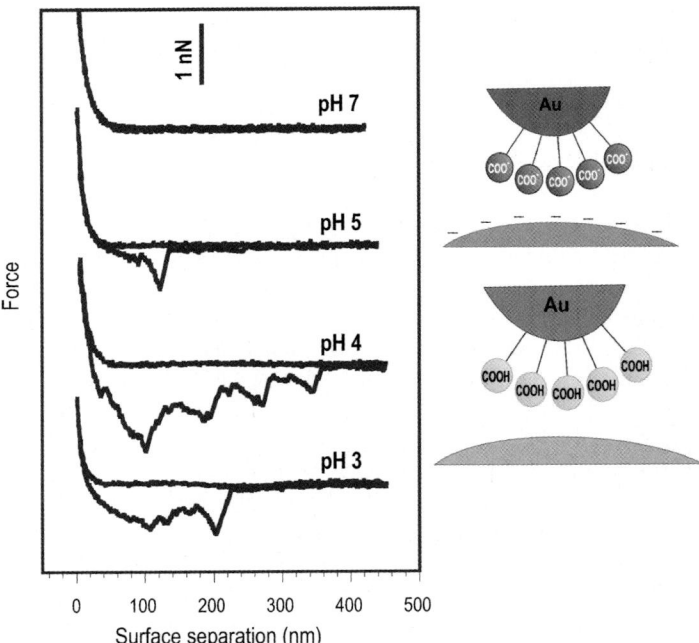

Fig. 10.5. Force-distance curves recorded in solutions of varying pH between the surface of *Saccharomyces cerevisiae* and an AFM tip functionalized with carboxyl groups. The differences in adhesion forces observed with pH were related to a change of ionization state of the cell surface. Adapted with permission from [22]

reflecting electrostatic repulsion between the negatively charged surfaces, multiple adhesion forces were recorded at pH $\leq$ 5 which we attribute to hydrogen bonding between the protonated tip surface and cell surface macromolecules. These changes were shown to be related to differences in the ionization state of the cell surface functional groups: the adhesion force *vs.* pH curve was correlated with microelectrophoresis data, the pH of the largest adhesion force corresponding to the cell isoelectric point (pH 4).

To summarize, the above data demonstrate that the CFM method allows investigators to detect specific chemical groups and measure their interaction forces on live cells, thereby complementing methods currently available for assessing surface properties. Also, the technique can resolve nanoscale chemical heterogeneities while the cells grow or interact with drugs.

10.3
Molecular Recognition Imaging

Single-molecule force spectroscopy (SMFS) with biologically modified tips enables us to measure molecular recognition interactions at the level of single molecules, providing valuable information on the molecular dynamics within the complexes, and to map individual receptors on surfaces. In parallel, molecular recognition sites may also be detected using topographic imaging combined with immunogold labeling.

10.3.1
Spatially Resolved Force Spectroscopy

So-called "affinity imaging" using spatially resolved SMFS provides unique possibilities for the localization of specific receptors on cell surfaces. The method implies recording multiple force curves between the modified tip and sample, assessing the unbinding force between complementary receptor and ligand molecules from the adhesion "pull-off" force observed upon retraction, and displaying the values as gray pixels [24]. The measured unbinding forces are typically in the 50–400-pN range, depending on the experimental conditions, i.e. number of interacting molecules and the loading rate. The method has been applied to different cell types, including red blood cells [25], osteoclasts [26], and endothelial cells [27].

The power of the approach in microbiology is illustrated in Fig. 10.6 with three key examples. Flocculation (i.e. aggregation) of yeast cells in fermentation technology, such as brewing and wine-making, is mediated by specific lectin–carbohydrate interactions. To measure these interactions on *S. carlsbergensis*, AFM tips were functionalized with the plant lectin concanavalin A (Con A). As can be seen in Fig. 10.6b, the force curves obtained on top of a single yeast cell showed single or multiple unbinding forces of respectively $53 \pm 6\,pN$, $94 \pm 12\,pN$ and $145 \pm 17\,pN$ magnitude reflecting the specific interaction between one, two, and three lectins and cell surface mannose or glucose residues. In addition, the homogeneous adhesion maps suggested that the distribution of carbohydrate residues was homogeneous.

In the medical context, molecular recognition events mediate the interaction between adhesins on bacterial pathogens and host receptors. A prominent example is *M. tuberculosis* which adheres to heparan sulfates on epithelial cells via the heparin-binding hemagglutinin (HBHA). Spatially resolved SMFS was used to map the distribution of single HBHA on live mycobacteria (Fig. 10.6c, d; [28]). While high-resolution topographic images of the cells revealed a smooth and homogeneous surface, affinity maps recorded with a heparin-modified tip were highly contrasted (Fig. 10.6d), adhesion events (bright pixels) being observed in about half

---

▶

Fig. 10.6. Molecular recognition imaging of live cells using spatially resolved SMFS. (a) AFM image of a *Saccharomyces carlsbergensis* cell showing a homogeneous, smooth morphology. (b) Adhesion force map (*gray scale*: 200 pN) recorded with an AFM tip functionalized with the plant lectin concanavalin A, revealing that mannose/glucose residues are homogeneously distributed on the surface. (c) AFM topographic image recorded in PBS showing two *M. bovis* BCG cells on a polymer substrate. (d) Representative adhesion force map (*gray scale*: 100 pN) recorded with a heparin-modified AFM tip. Adhesion events (*bright pixels*) reflect the detection of single cell adhesion proteins (HBHA), which apparently are concentrated into nanodomains. (e) AFM image of *Lactococcus lactis* cells during the course of the division process, showing a well-defined division septum rich in nascent peptidoglycan. (f) Adhesion force map (gray scale: 100 pN) recorded with a vancomycin tip on the septum region. Adhesion events were essentially located in the septum region (*dashed line*), suggesting newly formed peptidoglycan is inserted there. Reprinted with permission from [28] and from [29]

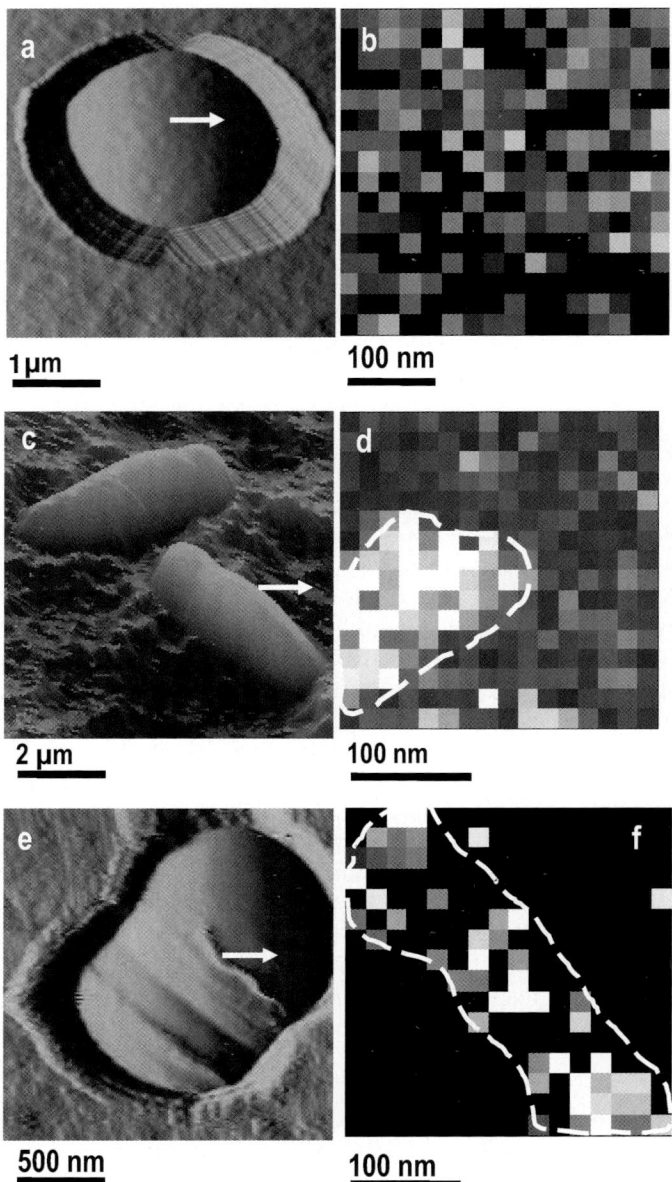

Fig. 10.6.

of the locations. The adhesion force magnitude was very close to the value of single HBHA–heparin interactions, supporting the notion that single adhesins were detected. This was further confirmed by showing that a mutant strain lacking HBHA did not bind the heparin tip. Interestingly, the HBHA distribution was not homogeneous, but apparently concentrated into nanodomains which may promote adhesion to target cells by inducing the recruitment of receptors within membrane rafts. In the future, these molecular recognition studies may help in the development of new drugs capable to block bacterial adhesion.

More recently, spatially resolved SMFS with antibiotic-modified tips was used to map individual binding sites on live bacteria (Fig. 10.6e, f; [29]). Fluorescence microscopy with a fluorescent vancomycin probe was used to visualize D-Ala-D-Ala sites of nascent peptidoglycan in the cell wall of dividing *Lactococcus lactis* cells. Fluorescence staining of the wild-type strain was found around the septum, while no fluorescent labeling was detected for a mutant strain producing peptidoglycan precursors ending by D-Ala-D-Lac instead of D-Ala-D-Ala. AFM topographic images of *L. lactis* cells revealed a smooth and elongated cell morphology as well as a well-defined division septum (Fig. 10.6e). Ring-like structures were seen at a certain distance from the septum, presumably formed by an outgrowth of the cell wall. Notably, adhesion force maps demonstrated that binding sites were essentially located in the septum region, and more specifically on the equatorial rings (Fig. 10.6f), suggesting that newly formed peptidoglycan was inserted in these regions. This study shows that AFM with vancomycin tips is a complementary approach to fluorescent vancomycin to explore the architecture and assembly process of peptidoglycan during the cell cycle of Gram-positive bacteria. While fluorescence microscopy generates microscale images allowing the localization of peptidoglycan in the entire cell wall, AFM adhesion force mapping reveals the distribution of single peptidoglycan molecules on the outermost cell surface.

While spatially resolved SMFS provides a quantitative analysis of unbinding forces, it is limited by its time resolution. The time currently required to record a map is on the order of 2–15 min depending on the acquisition parameters, which is much greater than the time scale at which dynamic processes usually occur in biology. An exciting alternative in this context is dynamic recognition force mapping (TREC) [30] which records topography and recognition images at the same speed as that used for conventional topographic imaging, typically 1–2 images per minute. Recently, Chtcheglova et al. [31] applied TREC imaging to microvascular endothelial cells to demonstrate that cadherins, involved in homophilic cell-to-cell adhesion, are organized into nanodomains ranging from 10 to 100 nm in diameter. Clearly, a challenging issue for future research would be to apply TREC imaging in the microbiological context.

10.3.2
Immunogold Imaging

A different way of exploiting molecular recognition for AFM imaging is to use immunogold labels as cell-surface markers, as is traditionally used in electron microscopy. Here, cells are first incubated with monoclonal antibodies directed against specific cell wall constituents, then further incubated with the corresponding gold-conjugated secondary antibodies, and finally imaged using topographic imaging. In pioneering work [32], application of the approach to dried immunogold-labeled human lymphocytes enabled researchers to resolve the location of antigens on the cell surface. Similarly, types I and II collagen fibers were revealed on dried rat fibroblasts and human chondrosarcoma cells [33]. For the first time, immunogold

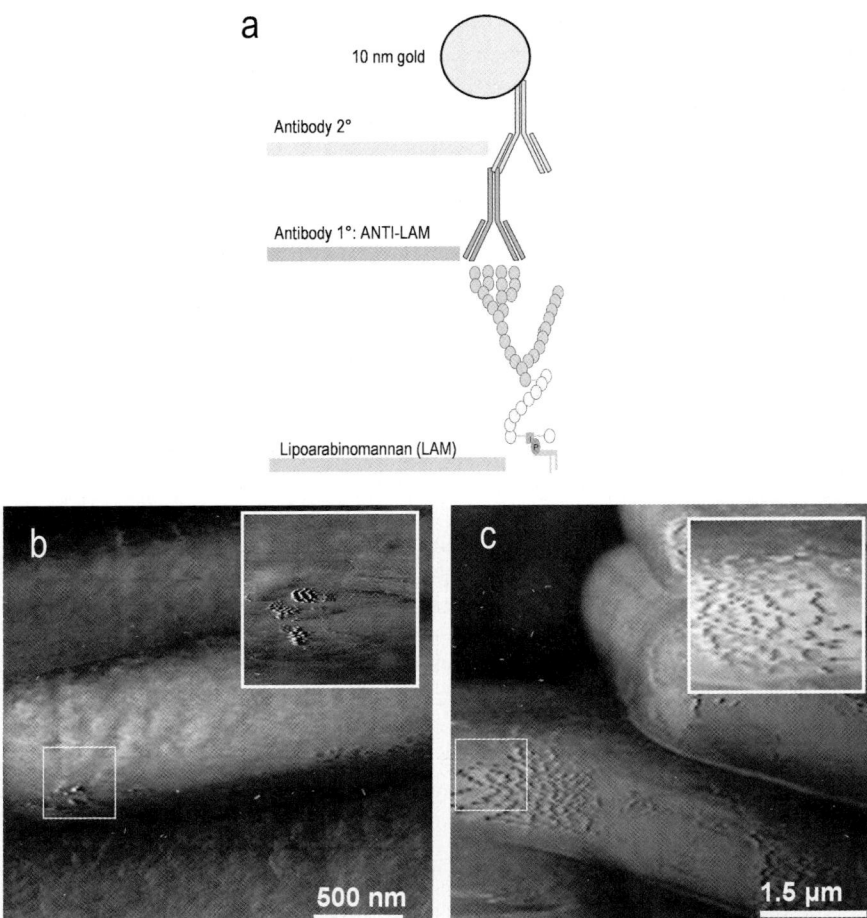

Fig. 10.7. Molecular recognition imaging of live cells using immunogold labeling. (**a**) *M. bovis* BCG cells were incubated with monoclonal anti-LAM antibodies, followed by another incubation with the corresponding gold-conjugated secondary antibodies. (**b, c**) AFM tapping mode (phase) images of immunogold-labeled cells prior (**b**) and after (**c**) treatment for 24 h with isoniazid, revealing that the drug induces the massive exposure of LAM at the surface. Reprinted with permission from [18]

AFM imaging was used to detect and localize lipoarabinomannan (LAM) on the surface of live, hydrated mycobacteria (Fig. 10.7) [18]. Contact mode and tapping mode images were obtained prior and after treatment with two antibiotics, INH and EMB. Gold particles were never seen in contact mode, emphasizing the need to use tapping (phase) mode for such in situ immunogold studies. Using tapping mode, the surface of native cells showed essentially no labeling, suggesting that LAM is not exposed at the surface. This finding was consistent with the uniform distribution of cell surface hydrophobicity measured on native cells. By contrast, INH and EMB-treated cells revealed a large coverage of gold particles, indicating that LAM was exposed. This observation, which correlates with topographic and CFM data, provides direct evidence that the two drugs lead to the massive exposure of LAM at the cell surface. In summary, this study showed that combining immunogold detection with topographic imaging and CFM allows us to shed new light into the 3-D organization of bacterial cell walls.

10.4
Conclusions

Owing to its ability to observe and manipulate biosystems under physiological conditions, AFM is revolutionizing the way in which today researchers explore live cells. Here, we have shown that two AFM modalities, CFM and molecular recognition imaging, offer fascinating prospects for mapping the chemical and biochemical properties of live cells. These nanoscale analyses should have an important impact on future life science and biomedical research, particularly to understand the molecular bases of cell–drug and host–pathogen interactions. Yet, it must be realized that procedures for attaching chemical groups, biomolecules and cells to AFM cantilevers remain labor intensive and require specific expertise that is usually not found in biology laboratories. In the future, defining simple standard protocols for tip functionalization and making them readily available to the biological community should contribute to spread the use of force spectroscopy in the various life science disciplines.

Acknowledgments. This work was supported by the National Foundation for Scientific Research (FNRS), the Région wallonne, the Foundation for Training in Industrial and Agricultural Research (FRIA), the Université Catholique de Louvain (Fonds Spéciaux de Recherche), the Federal Office for Scientific, Technical and Cultural Affairs (Interuniversity Poles of Attraction Programme), and the Research Department of Communauté Française de Belgique (Concerted Research Action). Y.F.D. and D.A. are Research Associate and Research Fellow of the FRS-FNRS, respectively.

References

1. Fritz J, Baller MK, Lang HP, Rothuizen H, Vettiger P, Meyer E, Guntherodt HJ, Gerber C, Gimzewski JK (2000) Science 288:316
2. Turner APF (2000) Science 290:1315
3. Niemeyer CM, Mirkin CA (2004) Nanobiotechnology: Concepts, applications and persepectives. VCH (ed). Wiley, Weinheim, p 469.
4. Jahn TR, Radford SE (2005) FEBS J 272:5962
5. Doyle RJ (2000) Microb Infect 2:391
6. Noy A (2006) Surf. Interface Anal 38:1429
7. Smith SB, Finzi L, Bustamante C (1992) Science 258:1122
8. Ashkin A, Schutze K, Dziedzic JM, Euteneuer U, Schliwa M (1990) Nature 348:346
9. Vezenov DV, Noy A, Ashby P (2005) J Adhes Sci Technol 19:313

10. Burnham NA, Chen X, Hodges CS, Matei GA, Thoreson EJ, Roberts CJ, Davies MC, Tendler SJB (2003) Nanotechnology 14:1
11. Noy A, Frisbie CD, Rozsnyai LF, Wrighton MS, Lieber CM (1995) J Am Chem Soc 117:7943
12. vanderVegte EW, Hadziioannou G (1997) Langmuir 13:4357
13. Vezenov DV, Noy A, Rozsnyai LF, Lieber CM (1997) J Am Chem Soc 119:2006
14. Frisbie CD, Rozsnyai LF, Noy A, Wrighton MS, Lieber CM (1994) Science 265:2071
15. Sinniah SK, Steel AB, Miller CJ, ReuttRobey JE (1996) J Am Chem Soc 118:8925
16. Cappella B, Dietler G (1999) Surf Sci Rep 34:1
17. Alsteens D, Dague E, Rouxhet PG, Baulard AR, Dufrêne YF (2007) Langmuir 23:11977
18. Alsteens D, Verbelen C, Dague E, Raze D, Baulard AR, Dufrêne YF (2008) Eur J Physiol 456:117
19. Dague E, Alsteens D, Latge JP, Verbelen C, Raze D, Baulard AR, Dufrêne YF (2007) Nano Lett 7:3026
20. Dupont-Gillain CC, Nysten B, Hlady V, Rouxhet PG (1999) J Colloid Interface Sci 220:163
21. Ducker WA, Senden TJ, Pashley RM (1991) Nature 353:239
22. Ahimou F, Denis FA, Touhami A, Dufrene YF (2002) Langmuir 18:9937
23. Dague E, Alsteens D, Latgé JP, Dufrêne YF (2008) Biophys J 94:656
24. Ludwig M, Dettmann W, Gaub HE (1997) Biophys J 72:445
25. Grandbois M, Dettmann W, Benoit M, Gaub HE (2000) J Histochem Cytochem 48:719
26. Lehenkari PP, Charras GT, Nykänen A, Horton MA (2000) Ultramicroscopy 82:289
27. Almqvist N, Bhatia R, Primbs G, Desai N, Banerjee S, Lal R (2004) Biophys J 86:1753
28. Dupres V, Menozzi FD, Locht C, Clare BH, Abbott NL, Cuenot S, Bompard C, Raze D, Dufrêne YF (2005) Nat Methods 2:515
29. Gilbert Y, Deghorain M, Wang L, Xu B, Pollheimer PD, Gruber HJ, Errington J, Hallet B, Haulot X, Verbelen C, Hols P, Dufrene YF (2007) Nano Lett 7:796
30. Raab A, Han WH, Badt D, Smith-Gill SJ, Lindsay SM, Schindler H, Hinterdorfer P (1999) Nat Biotechnol 17:902
31. Chtcheglova LA, Waschke J, Wildling L, Drenckhahn D, Hinterdorfer P (2007) Biophys J 93:L11
32. Putman CAJ, Degrooth BG, Hansma PK, Vanhulst NF, Greve J (1993) Ultramicroscopy 48:177
33. Arntz Y, Jourdainne L, Greiner-Wacker G, Rinckenbach S, Ogier J, Voegel JC, Lavalle P, Vautier D (2006) Microsc Res Tech 69:283

11 Applications of Scanning Near-Field Optical Microscopy in Life Science

Pietro Giuseppe Gucciardi

Abstract. Scanning Near-Field Optical Microscopy (SNOM) is capable of attaining sub-diffraction resolution by exploiting nanoscopic light sources to illuminate the samples, such as nanoapertures or sharp metallic tips. In this chapter, after describing the basic principles of SNOM, we review the state-of-the-art in the experimental implementations and applications of fluorescence, infrared and Raman SNOM for the detection and imaging of DNA oligonucleotides, viruses and cellular membrane proteins.

Key words: Scanning near-field optical microscopy, Tip-enhanced Raman spectroscopy, Near-field fluorescence microscopy, Protein localization, Protein co-localization, DNA, Cells, Dendritic cells, Human skeletal cells, Malaria, Actin, Integrin, Sarcoglycan

Abbreviations

AFM	Atomic force microscopy
CCD	Charge coupled device
CLM	Confocal laser microscopy
DC	Dendritic cells
FITC	Fluorescein
GFP	Green fluorescent protein
HIV	Human immunodeficiency virus
imDC	Immature dendritic cells
IR	Infrared
MESA	Mature parasite-infected erythrocyte surface Antigen
NA	Numerical aperture
NIR	Near infrared
PIA	Polysaccharide intercellular adhesin
PBS	Phosphate buffered saline
PMT	Photomultiplier tube
PfHRP1	*Plasmodium falciparum* histidine rich protein
SEM	Scanning electron microscopy
SERS	Surface-enhanced Raman scattering
SM	Single molecule
SNOM	Scanning near-field optical microscopy
STM	Scanning tunneling microscopy

TEM   Transmission electron microscopy
TERS   Tip-enhanced Raman scattering
TMV   Tobacco mosaic virus
TR   Tetramethylrhodamine
WD   Working distance

11.1
Introduction

Spectroscopic imaging techniques, among which are included absorption, fluorescence and Raman imaging, are non-invasive investigation means with a broad field of application in the life sciences. Fluorescence and Raman microscopy, in particular, have gained a firm position among the most important research tools in cell biology. They offer chemically specific contrast and allow for the study of living cells. To date fluorescence microscopy remains as the most widely used technique for cell imaging. Membrane components can be directly visualized via specific dye-conjugated antibodies or direct fusion with green fluorescent proteins (GFP) [1]. The employment of ultra-sensitive optical detectors, such as cooled charge coupled devices (CCD), photomultipliers (PMT) or avalanche photodiodes makes fluorescence imaging possible even at the single molecule (SM) level. SM fluorescence imaging [2, 3] offers a set of advantages compared to ensemble-averaged experiments, among which are (1) the study of the static and dynamic heterogeneity of a population of molecules and its relation to the specific environment; (2) the possibility to obtain information over the function of molecules on cell membranes; (3) the tracking of the conformational motion of biological macromolecules [4]; (4) the study of membrane transport mechanisms [5] as well as the structural change of proteins [6] and DNA [7].

Fluorescence microscopy at room temperature, however, suffers a principal drawback due to photobleaching. Photobleaching includes all those photochemical reactions occurring to the molecules while in the excited state which irreversibly alter the chemical structure of the molecules, causing the extinction of the photoluminescence after the emission of $\sim 10^6$ photons. The label-free, direct characterization of biomolecules is, therefore, of real importance since it avoids the use of an intermediary molecular species, the dye. Such a direct characterization is made possible by the use of vibrational spectroscopies, which are IR absorption and Raman scattering. Vibrational spectroscopies are powerful tools for label-free characterization of biological species since the vibrational modes are actual fingerprints of the whole molecular species or of the local binding between biomolecules. Moreover, they can be combined with standard microscopy techniques to obtain information on the local scale. Unfortunately, IR absorption and Raman scattering cross-sections are far too low ($\sim 10^{-24}$ and $\sim 10^{-30}$ cm^2, respectively) for fast imaging applications. Furthermore, the spatial resolution of these techniques, as well as for fluorescence microscopy, is limited by diffraction to about $\lambda/2NA$ [8], where λ is the wavelength of the radiation and NA the numerical aperture of the illumination/collection optics employed. Using far-field optics, such as microscope objectives in the visible and a Cassegrain objective or a parabolic mirror in the IR, the spatial resolution is always $> 300\,\mathrm{nm}$. Although new techniques, such as 4Pi microscopy and stimulated-emission depletion microscopy, have recently been introduced [9, 10] demonstrating an increased spatial and axial resolution, their applicability is limited to fluorescence microscopy and to the use of specific classes of dyes.

Scanning Near-Field Optical Microscopy (SNOM) [11–14] exploits a nanometric probe to confine the tip–sample optical interaction to scales between 10 and 100 nm, thus circumventing the diffraction limit (see [15, 16] for reviews). Probes can be either nanoapertures or sharp metallic tips [17]. Together with an unprecedented spatial resolution, Near-Field Microscopy offers several unique advantages: (1) similarly to other scanning probe techniques, it allows recovery of the surface morphology together with its optical properties, with a spatial resolution always in the 10-nm range; (2) being an optical microscopy technique, it allows implementation of all those contrast mechanisms available in confocal microscopy, such as elastic scattering, dichroism/birefringence, fluorescence and Raman; (3) owing to the exponential decay of the illumination field, SNOM is a surface-sensitive technique; (4) finally, due to the intriguing field enhancement effects provided by metallic tips, the Tip-Enhanced SNOM configuration is capable of locally amplifying the optical interaction by several orders of magnitude (2–6) making real absorption and Raman imaging with 10-nm spatial resolution possible (for reviews see [18, 19]).

Application possibilities of SNOM in life science are numerous. Fluorescence SNOM finds natural application for the identification of membrane lipid and proteins, for protein localization and co-localization on the cellular membranes and for the study of ion channel clusters. Tip-Enhanced Raman Spectroscopy (TERS) has recently demonstrated enormous potential applications for label-free detection of membrane proteins and DNA nucleobases.

The review of some of the most relevant experiments is the objective of this chapter. The chapter is organized as follows: in Sect. 11.2 we provide a general introduction to Near-Field experimental techniques; in Sect. 11.3 we give a review of the applications of SNOM in life-sciences; in Sect. 11.4 we draw conclusions on the state-of-art in bio-SNOM and propose some perspectives for the future.

11.2
Experimental Techniques in Near-Field Optical Microscopy

11.2.1
Principles of Near-Field Optical Microscopy

Optical microscopy has a prominent role in modern cell biology [20]. Confocal Laser Microscopy (CLM, Fig. 11.1a) is the most widely used technique, allowing reduction of the detection volume and reconstruction of three-dimensional images of the sample. In CLM a laser beam is focused by means of a high NA objective and scanned all over the sample surface. The back-reflected fluorescence is collected by the same objective and sent to the detector which reconstructs point-by-point the sample's image. Optical sectioning of the sample is achieved by using a spatial pinhole to

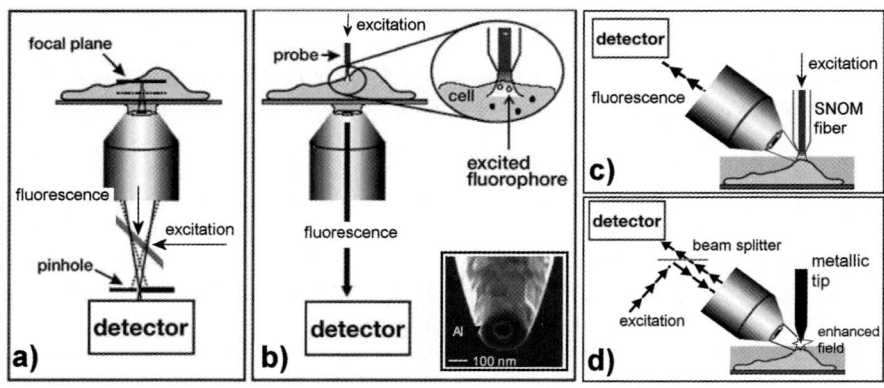

Fig. 11.1. Schematic diagrams of confocal (**a**) and of aperture SNOM microscopy working in transmission (**b**), and reflection (**c**) mode. The *inset* of (**b**) shows a SEM micrograph of the apical aperture of a SNOM fiber probe (reprinted from [25] with permission of the author). (**d**) Schematic of a TE-SNOM working in reflection mode

eliminate out-of-focus light in specimens that are thicker than the focal plane. CLM is, however, a far-field technique, therefore the lateral resolution is diffraction-limited to length scales larger than 250 nm. The microscope objective is located several wavelengths distant from the sample, in the so-called far-field. As a consequence, only propagative electromagnetic waves are detected in the light–matter interaction. In fact, due to the finite *NA* of the objective only waves with a transverse wavevector $k_\perp$ whose components span in the interval $k_\perp = -NA\,(\omega/c)\ldots + NA\,(\omega/c)$ are collected. The spatial resolution, given by the Heisenberg principle as $|\Delta r_\perp|\cdot|\Delta k_\perp| \sim 2\pi$, therefore will be $|\Delta r_\perp| \sim \lambda/2NA$. The enhanced spatial resolution of SNOM relies in the exploitation of evanescent, non-propagative waves in the tip–sample optical interaction [21]. Evanescent waves are characterized by having a transverse wavevector whose modulus $|k_\perp| > \omega/c$. The angular spectrum of the radiation by an aperture of diameter $a \ll \lambda$, will contain components with a transverse wavevector roughly spanning between $k_\perp = -\pi/a \ldots + \pi/a$. The expected spatial resolution of aperture-SNOM will therefore be $|\Delta r_\perp| \sim a \ll \lambda/2NA$. We note that evanescent waves, for which $|k_\perp| > \omega/c$, have components of the wavevector parallel to the propagation direction $k_{//} = \sqrt{(\omega/c)^2 - k_\perp^2}$ that are purely imaginary. The corresponding electric field is exponentially attenuated during propagation. This makes SNOM a purely surface analysis technique. Moreover, due to the rapid decay length of the near-field (100 nm typically) it is imperative to keep the aperture in proximity of the sample surface (< 10 nm) during the scan. This is accomplished usually by means of an independent shear-force detection system [22, 23] controlling and stabilizing the tip–sample distance. Such a system, forcing the tip to follow the surface profile during the scan, allows recovery of the topography map of the sample simultaneously with its optical map.

Figure 11.1b shows the schematics of a so-called "aperture-SNOM" employing optical fiber probes. The sample excitation is accomplished through the nanometer-sized aperture (see inset of Fig. 11.1b) located at the apex of a tapered, metal-coated optical fiber [24, 25]. The light scattered by the sample is subsequently collected in the far-field. Depending on whether the sample is transparent or not, the transmitted (Fig. 11.1b) or the reflected (Fig. 11.1c) components of the scattered fields are collected. The transmission mode configuration takes advantage of the high light-gathering capabilities of oil-immersion objectives, while reflection mode SNOM is a more versatile configuration being independent of the opaqueness of the sample.

Different types of probes have been introduced in aperture-SNOM (see [17] for a review). Tapered, metal-coated optical fibers are still the most widely used sensors due to their versatility, especially in liquids, and good spatial resolution (50–100 nm). Nevertheless, they suffer several limitations among which are fragility, a very low light throughput (only a fraction between 10^{-3} and 10^{-6} of the light injected into the fiber is transmitted by the nanoaperture), and heating of the metallization due to absorption of the injected radiation [26–28] that limits to a few mW the maximum amount of light that can be coupled into the optical fiber before thermal damage of both the fiber and the sample under investigation can occur [29, 30]. The effectiveness of tapered uncoated fibers is still debated. Although it is recognized that elastic scattering images acquired with such probes are mainly dominated by artifacts, applications in photoluminescence detection of semiconductor heterostructures have demonstrated spatial resolution capabilities of the order of 150 nm [31,32]. More recently, cantilevered hollow tips [33] have demonstrated a higher robustness, a higher resistance to light injection, and a good polarization preservation [34]. Moreover, cantilevered hollow tips can be easily fitted to an Atomic Force Microscope (AFM) [35].

SNOM configurations using sharp metallic tips are referred to as "apertureless-SNOMs," "scattering-SNOM," (s-SNOM) or "Tip-Enhanced SNOM" (TE-SNOM) [16]. The Tip-Enhanced SNOM has attracted much attention due to the capability to push the resolution limit of aperture-SNOM even further, demonstrating elastic scattering, fluorescence and Raman imaging with 10-nm resolution [36]. In TE-SNOM the tip plays a dual role, confining the illumination field on length scales comparable to its radius of curvature and acting, at the same time, as a nanoantenna capable of collecting the near-fields scattered by the surface, radiating them in the far-field. As shown in Fig. 11.1d light is focused on a metallic tip which can be the metallic tip of a Scanning Tunneling Microscope (STM) [37] or the metalized cantilevered tip of an AFM. The backscattered light is collected, spectrally filtered and detected to reconstruct the object's image.

Among the TE-SNOM techniques, Tip-Enhanced Fluorescence [38] and Raman Spectroscopy [39, 40] promise the most interesting applications in terms of single molecule sensitivity.

11.2.2
Fluorescence Near-Field Optical Microscopy

Near-Field fluorescence imaging on cells is mainly carried out by using aperture-SNOM configurations. Here the tapered optical fiber acts as a light guide funneling the excitation to a spot whose width is given by the aperture diameter, well below the diffraction limit. SNOM configurations working in transmission-mode benefit from

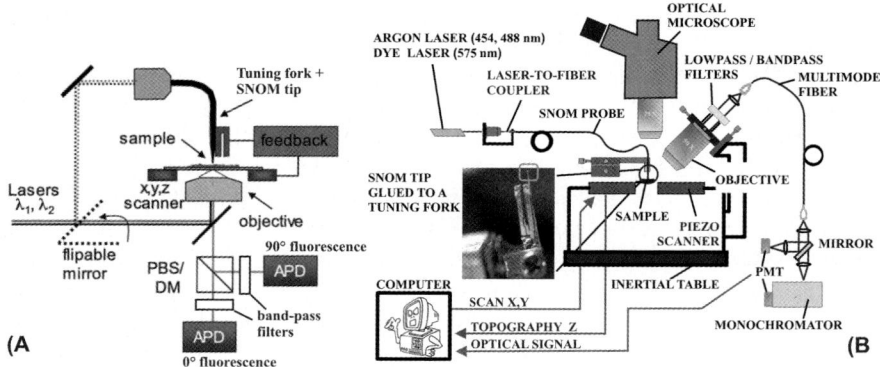

Fig. 11.2. (a) Layout of a confocal/SNOM apparatus with single-molecule fluorescence sensitivity working in transmission mode (reprinted from [85] with permission of the author). (b) Sketch of a SNOM apparatus for fluorescence imaging working in reflection mode

the large numerical aperture of immersion-objectives that allow the light radiated in the far-field to be gathered, thus featuring unique single-molecule sensitivity. A typical apparatus using optical fiber probes is depicted in Fig. 11.2a [41]. The system is integrated into an inverted microscope, and a flappable mirror enables switching between SNOM and confocal operation. One or more excitation laser beams (Ar^+/Kr^+, 457–647 nm) are used for localization or co-localization experiments. In the latter case both beams are controlled in terms of power level and polarization state, and then recombined by a beam splitter prior to injection into the Al-coated SNOM fiber probe. The tapered extremity of the probe is glued to a quartz tuning-fork [23], providing the shear-force signal that drives the tip–sample distance control feedback circuit. The shear-force detection works by dithering the tip parallel to the sample surface through a piezoelectric-slab and detecting the piezo-current generated by the tuning fork which is proportional to the tip oscillation amplitude. This oscillation amplitude decreases almost linearly when decreasing the tip–sample distance on sub-100-nm scales. Therefore, it can be used as the set parameter in a feedback loop to stabilize the tip–sample distance during the scan. The sample, typically fixed to 170-μm-thick glass cover slides, is scanned by a piezotable in all the three spatial dimensions. The emitted fluorescence is collected by an oil-immersion objective (1.4 NA) and spectrally selected by suitable band-pass filters. The fluorescence is then separated into two channels by means of a polarizing beam-splitter cube and detected by two single photon-counting avalanche photodetectors. Polarization-sensitive detection allows for simultaneous monitoring of the relative intensities of the polarized fluorescence signals, and is used to reconstruct the in-plane orientation of the emission dipole moment of the single molecules.

The reflection-mode SNOM configuration is essential when scattering or re-absorption of the radiation inside the sample are critical issues, preventing an efficient light collection "from below." Light is gathered on the same half-plane where excitation occurs through long working-distance objectives (WD = 10–15 mm, NA = 0.3–0.55). We have developed a reflection-mode aperture-SNOM for biological purposes, whose design is inspired by some well assessed commercial models [42,43], implementing the electronic control and the software of a previous SNOM model employed for Raman spectroscopy and imaging [44, 45]. The SNOM head is placed underneath an optical microscope (see Fig. 11.2b) working with long working-distance objectives (Olympus APOPLAN 20X, 50X). This allows for a first inspection of the sample and for the exact localization of the point we want to investigate by SNOM. We use commercial CrAl metal-coated fiber probes (Nanonics) with nominal apical apertures of 100 nm. The probe is glued to a quartz tuning-fork for non-optical shear-force detection, aimed at implementing the tip–sample distance control. The tuning fork is anchored, through a properly designed holder, to an XYZ micrometric translation stage used for coarse tip movement. The excitation radiation is provided by blue laser diodes (403 and 417 nm), Argon ion (454–515 nm) or HeNe (633 nm) lasers and is coupled to the SNOM fiber probe, used to illuminate the sample. The scattered light is collected by a long working-distance microscope objective (Olympus 50X, WD 10.6 mm, NA 0.52) rested at an angle of about 45°. The light is thus coupled to a multimode optical fiber and driven to a monochromator (Jobin-Yvon, Triax 190) for the spectral analysis of the fluorescence light. For imaging purposes, instead, light is spectrally filtered by means of appropriate band- or low-pass filters placed right after the collection objective. The fluorescence light is thus coupled to the multimode fiber and directly driven to the photomultiplier. The sample is rested on a 2.54-cm diameter piezo-tube capable of a $32 \times 32 \times 8$-μm^3 scan. A further piezo-scanner can be used for scans up to $80 \times 80 \times 10 \, \mu$m^3. An aspheric lens ($d = 10$ mm, $f = 8$ mm, NA $= 0.5$) is mounted inside the 2.54-cm diameter piezo-tube (not shown in the picture), and can be used to collect the transmitted light. Maps of 128×128 and 256×256 pixels are typically carried out, with 10 or 20 ms integration time per point.

11.2.3
Near-Field Optical Microscopy in Liquid

Live cell imaging with SNOM requires the scanning tip to be immersed in liquid. Nevertheless, the shear-force feedback must still be able to detect pN interaction forces with soft cell membranes in order to control and stabilize the tip–sample distance during the scan, without damaging the sample. Several approaches have been proposed with optical shear-force detection [46], piezo-feedback mechanisms [47,48], or non-optical tuning-fork systems [49,50]. The main problem when working in liquid concerns the lowering of the oscillator Q-factor due to liquid viscosity, which decreases the sensitivity of the force feedback mechanism. An estimate of the interaction force is, in fact, given by $F = (kx_{res}/Q)\Delta\varphi$ where $\Delta\varphi$ is the phase shift caused to the force interaction, k is the spring constant of the tuning fork and x_{res} is the oscillation amplitude at resonance. With Q-factors of 700 and oscillation amplitudes of 0.1 nm, typical interaction forces in the 100-pN range can be detected, suitable for sensing soft samples such as cells membranes without damage. Figure 11.3a shows the "diving bell" concept introduced by Koopman et al. [50] in which the tuning fork and its holder are embedded into a glass tube. Through the air sealing of the holder, the glass tube acts as a diving bell for the tuning-fork, keeping the

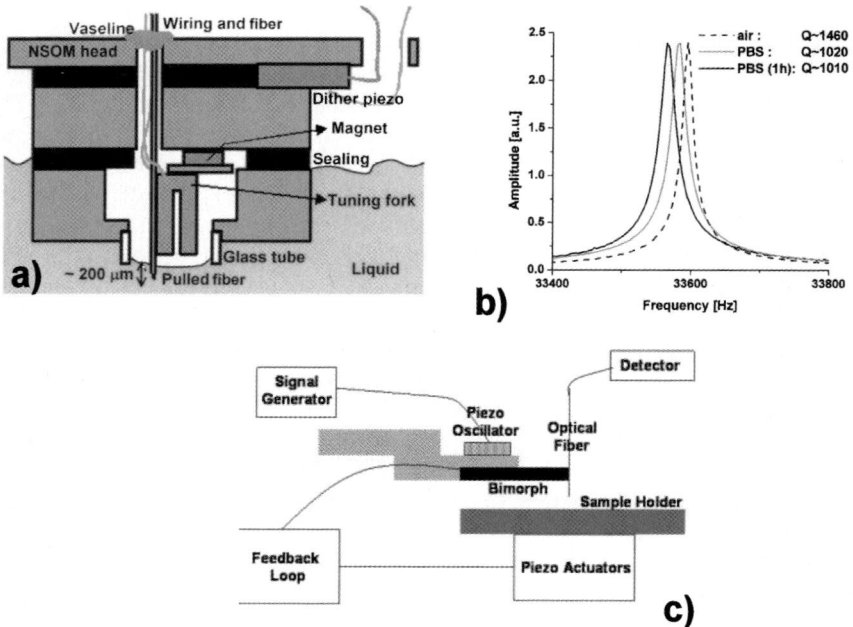

Fig. 11.3. SNOM configurations working in liquid. (**a**) Sketch of the "diving bell" concept. (**b**) Frequency resonance spectra of the tip in air (*dotted line*), in PBS (*solid gray line*), and in liquid after 1 h (*solid black line*), (reprinted from [50] with permission of the author). (**c**) Sketch of a tapping-mode SNOM configuration operating in liquid (reprinted from [48] with permission of the author)

air–liquid interface at a fixed level independently of the amount of liquid used. The optical fiber probe is glued to the tuning fork with the tip end protruding ∼500 μm from the prong's end. The length of the glass tube is chosen such that a fiber length of ∼200 μm protrudes outside the bottom plane of the glass tube. In such a way the tuning fork and its wiring are always dry, while only the tip apex penetrates the liquid surface. Figure 11.3b shows the tuning-fork resonance spectrum in air (dotted line). The resonance frequency is slightly lowered with respect to the free fork value (33.595 kHz against 32.768 kHz) due to the stiffening effect of the glass fiber on the prong. The Q-factor in air is 1,460, while immersion in phosphate-buffered saline solution (PBS) reduces the Q to 1,020 (solid gray line). Immersion for longer times (solid black line) only changes the Q by 1%.

Tapping-mode SNOM, in which the tip is vertically dithered, allows reduction of the tip–sample interaction forces to the pN range, and is therefore well suited for imaging soft samples. The schematics shown in Fig. 11.3c illustrate an apparatus operating in intermittent contact using straight optical fiber probes [48]. The tip is vertically vibrated by a piezoelectric tablet that produces the sinusoidal oscillation transmitted to the fiber probe through a long bimorph. The oscillation amplitude is detected with an external electronic circuit and is maintained constant by the tapping-AFM feedback loop.

11.2.4
Tip-Enhanced Near-Field Optical Microscopy

TE-SNOM is of particular interest due to possible implementation of IR and Raman contrast mechanisms which enables label-free imaging at the nanoscale. In TE-SNOM the use of metallic tips [51–53] allows one to take advantage of several physical processes such as resonant excitation of localized plasmons or antenna effects. A strong electric field is in fact created at the tip apex [54] and high field-enhancement factors are locally achieved [55]. The key to the success of TE-SNOM has been the discovery of different methodologies to extract the tiny near-field scattering signal from the huge far-field background source of artifacts. For IR absorption and scattering microscopy this is done through interferometric techniques [56–59]. Figure 11.4a shows a typical TE-SNOM apparatus for absorption and scattering measurements. The apparatus implements homodyne interferometric detection [60], and is mounted on a tapping-mode AFM framework. The microscope uses commercial Pt-covered, cantilevered Si tips with tapping frequency $\Omega = 33$ kHz and amplitude $= 25$ nm. The tip is illuminated by focused infrared radiation, typically from continuous wave CO laser beams spanning the spectral range around 9.6 and 10.6μm. The backscattered light is superimposed on a reference beam in a Michelson interferometer using a ZnSe partial reflector. The signal of a HgCdTe detector is demodulated at higher harmonics $n \Omega$ by a lock-in amplifier. Signal demodulation at a harmonic $n \Omega$ of the tapping frequency is required, along with interferometric detection, to eliminate the problem of background scattering artifacts [18, 61, 62]. The reference beam's phase is regularly shifted to separate and acquire simultaneously the amplitude and phase information.

TERS is an analytical technique for high-sensitivity Raman spectroscopy and nanometer-scale resolution imaging capabilities. Apparatus for TERS have been developed in both the transmission [39] and reflection configuration [40, 55, 63, 64]. The main difficulty in TERS is the extraction of the localized near-field tip–sample optical interaction from the huge far-field background coming from the sample's scattering from the diffraction-limited illumination spot. In TERS the Raman

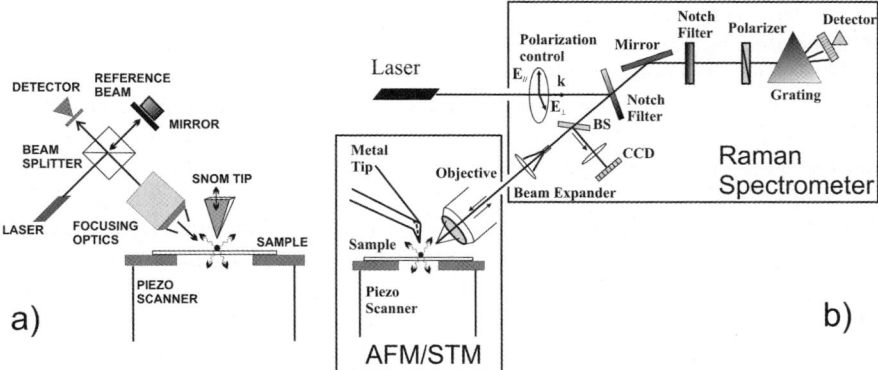

Fig. 11.4. Layout of TE-SNOM apparatus for (**a**) IR scattering and (**b**) Tip-Enhanced Raman scattering experiments

scattering intensity is approximately proportional to $\sim\xi^4\ a^2$ where ξ is the field enhancement factor provided by the tip, and a the radius of curvature of the tip. This value has to be compared to the far-field background proportional to $\sim(\lambda/NA)^2$. To allow for the near-field extraction, one can reduce the far field background by increasing the NA of the illumination objective and then limiting the size of the illumination spot area as much as possible. In the transmission configuration such a task is accomplished by using oil-immersion microscope objectives. In TERS setups with side illumination (reflection-mode configuration) oil immersion objectives cannot be used and the NA is limited to values close to 0.6. In such a case, first experiments suggest that a background reduction can be achieved by exploiting the local light depolarization induced by small particles [65] or metallic tips [66, 67]. Figure 11.4b shows a typical apparatus for TERS in reflection. It basically consists of a Raman spectrometer coupled to an AFM mounting a gold-coated cantilevered tip. Alternatively a shear-force microscope using metallic tips can replace the AFM. The system works in backscattering. Ar-ion, HeNe or diode lasers are used to cover the UV-NIS-NIR spectral range. Polarization control of the excitation is achieved through a set of quarter- or half waveplates. The beam is reflected by a notch filter centered at the laser wavelength and, after expansion, focused to the tip apex through long working-distance objectives. A proper imaging system is needed for precise focusing. The backscattered light is filtered by a double notch system and sent to the detector. An analyzer is typically used to single out the Raman scattering of a definite polarization state. Photomultipliers or CCD cameras are used to detect the radiation.

11.3
Applications of Near-Field Optical Microscopy in Life Science

11.3.1
Infrared Imaging of Tobacco Mosaic Virus with Nanoscale Resolution

Label-free compositional identification of cellular micro-organisms and proteins is a great challenge in biology. The highly confined field present at the apex of a metallic probe in s-SNOM, in particular, permits the mapping of the optical scattering of nanostructures, and to reconstruct the complex dielectric function of the material. This is of utmost interest in the infrared part of the electromagnetic spectrum, where spectroscopic s-SNOM can assess the optical dispersion and therefore one can recognize the infrared vibrational fingerprint of molecules, assigning its chemical composition. Brehm and co-workers [68] have applied IR s-SNOM to image individual tobacco mosaic viruses (TMV) demonstrating, at the same time, that its chemical fingerprint can be retrieved by this technique. For this experiment the s-SNOM (see Sect. 11.2.4 for details) has been interfaced to a CO laser tunable in the region around $\lambda = 6\,\mu m$ where TMV has distinct vibrational resonances. Figure 11.5 shows (a) the IR scattering amplitude and (b) the phase maps at different energies of an individual TMV cast on Si. It is observed that the contrast between virus and substrate varies with the energy. The scatter-

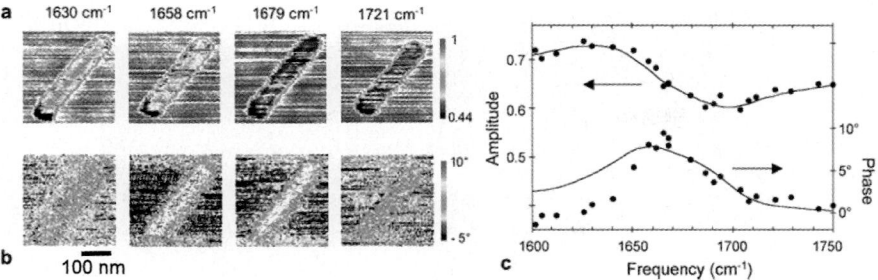

Fig. 11.5. IR scattering amplitude (**a**) and phase (**b**) maps of single TMV nanoparticles imaged at different energies. (**c**) Amplitude (*top plot*) and phase (*bottom plot*) spectra of a TMV particle showing typical dispersion behavior (reprinted from [68] with permission of the author)

ing amplitude shows a more pronounced absorption contrast at $1,679\,\mathrm{cm}^{-1}$, while the phase map shows appreciable contrast only between $1,658$ and $1,679\,\mathrm{cm}^{-1}$. To confirm that the contrast observed in the images is indeed due to the amide I vibrational resonance of the peptide bonds of the virus shell proteins resonance spectra have been acquired on a single TMV. Figure 11.5c (black circles) shows the experimental data of the scattering amplitude and phase acquired as a function of the excitation energy in the $1,600$–$1,750$-cm^{-1} range. The near-field spectra display a typical dispersion-like trend around the amide I vibrational resonance energy, and are well fitted (Fig. 11.5c, black lines) by a simple point-dipole model [16,69].

11.3.2
Co-Localization of Malarial and Host Skeletal Proteins in Infected Erythrocytes by Dual-Color Near-Field Fluorescence Microscopy

One of the most interesting aspects in biology relates to the organization of cells and how interactions between proteins determine important cellular processes such as signal transduction and receptor–ligand binding. Experiments to identify interactions in cellular systems are mostly performed through the co-localization imaging of two or more proteins, carried out by Fluorescence CLM using fluorescence antibody probes. Enderle and co-workers [70] have applied Fluorescence SNOM to study the co-localization of malaria (*Plasmodium falciparum* histidine rich protein, PfHRP1) and host membrane (protein 4.1, and mature parasite-infected erythrocyte surface antigen, MESA) proteins in infected red blood cells. Purified knob structures contain PfHRP1, erythrocyte skeletal proteins, spectrin, actin, dematin, and protein 4.1. The MESA has been localized on the cytoplasmic face of knobs by immune-electron microscopy, but analysis on purified knobs failed to identify MESA, suggesting the need of high spatial resolution immune-fluorescence investigations for a better understanding of these associations. Figure 11.6a shows the fluorescence image of an infected erythrocyte labeled with antibodies against PfHRP1 and immunolabeled with tetramethylrhodamine (TR) conjugated secondary antibody. Experiments

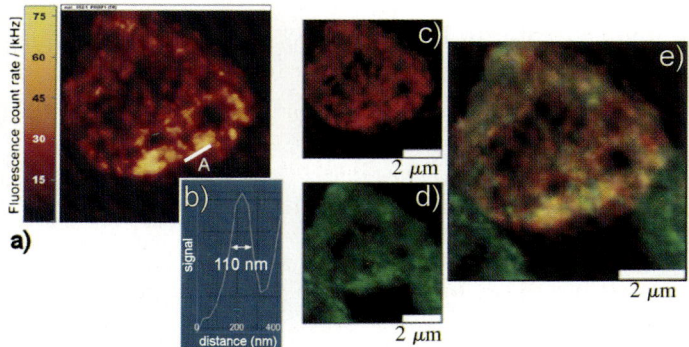

Fig. 11.6. (**a**) SNOM fluorescence map of an infected erythrocyte labeled with antibodies against PfHRP1. The cross-section in (**b**) demonstrates a resolution of the order of 100 nm. Co-localization of PfHRP1and protein 4.1 on a single knob: (**c**) SNOM fluorescence map of PfHRP1, (**d**) analogous for protein 4.1, (**e**) co-localization map of the PfHRP1/protein 4.1 pair (reproduced from [70] with permission of the authors)

were carried out with an aperture-SNOM working in transmission. The sample was excited with a SNOM fiber probe coupled to a 568-nm Ar/Kr laser, using a 100X 1.4 NA oil-immersion objective to gather the fluorescence which was finally detected by an avalanche photodiode. Fluorescence spots of 100–150 nm are typically resolved, as shown by the line profile in Fig. 11.6b. Their average density corresponds to the surface density of knobs deduced by electron microscopy, suggesting that the fluorescent spots originate from PfHRP1 in individual knobs on the infected erythrocyte membrane. Dual-color fluorescence SNOM was then used to study the co-localization of two host–parasite protein pairs. A thin blood smear was infected, fixed and reacted with either mouse monoclonal or rabbit polyclonal antibodies against the proteins under study. Secondary antibodies labeled with Fluorescein (FITC) and TR were used to stain the proteins. Two beams, the 488-nm line of an Ar laser and the 568-nm line of an Ar/Kr laser were collinearly combined and simultaneously injected into the SNOM fiber probe. Suitable band-pass filters were used to reject the excitation and to single out the two fluorescence signals, simultaneously acquired point-by-point. Figures 11.6c and d show the two pseudo-colored images belonging respectively to the TR and to the FITC channels for the host–parasite PfHRP1/protein 4.1 pair. Secondary antibodies conjugated with TRITC and TR were used to localize respectively protein 4.1 and PfHRP1. Figure 11.6e is the overlay of the two images obtained by mixing the two color channels. The map displays sparse green and red spots, arising from the separate emission of protein 4.1 and PfHRP1, and only few yellow zones indicating the simultaneous emission from the PfHRP1/protein 4.1 pair. The map provides clear evidence that the two proteins are only poorly co-localized and demonstrates that there is basically no association between PfHRP1 and protein 4.1 on the membrane of infected red blood cells. In particular this result suggests that, while both proteins are present on the knob structures, they do not directly interact as does MESA with protein 4.1.

11.3.3
Co-Localization of α-Sarcoglycan and β1D-Integrin in Human Muscle Cells by Near-Field Fluorescence Microscopy

The interaction between the extracellular matrix and the sarcolemma-associated cytoskeleton, leading to the dynamic of human muscle fibers, is mediated by the two protein complexes: the dystrophin-glycoprotein complex (DGC) and the vinculin-talin-integrin system [71–73]. The DGC contains, among other proteins, also dystrophin and the sarcoglycans sub-complex. These proteins play a key role in the pathogenesis of many muscular dystrophies and link the cytoplasmic myofibrillar contractile elements to the signal transducing molecules of the extracellular matrix, also providing structural support to the sarcolemma [74, 75]. The vinculin-talin-integrin system connects some components of the extracellular matrix with intermediate filaments of desmin, forming transverse bridges between Z and M lines [76]. Integrins are a family of transmembrane heterodimeric receptors which include at least 14 distinct α subunits and eight β subunits [77]. β1D-integrin, in particular, associates with at least ten α subunits to form distinct integrin dimers, capable of interaction with various extracellular matrix molecules as well as some cell adhesion molecules. The different spatial distribution of the DGC and vinculin-talin-integrin systems has been studied by immunofluorescence investigations using CLM, verifying the exact arrangement of the costameric bands of each protein, the co-localization of all proteins with each other, and, finally, the exact localization of the costameres [78]. The assessment of protein co-localization is, however, related to the spatial resolution capabilities of the microscope employed. We have studied the staining patterns of α-sarcoglycan and β1D-integrin in normal human skeletal muscle using Fluorescence-SNOM, demonstrating that SNOM can be fruitfully used to perform single and double localization of different protein complexes on human muscle cells with sub-diffraction resolution [79]. We have investigated samples of human skeletal muscle, obtained from vastus lateralis muscle biopsies and prepared following the protocol used to carry out indirect immune-fluorescence. In all reactions, TRITC-conjugated IgG anti-mouse in goat was used as the first fluorochrome. Sections were incubated with a second antibody, conjugated with FITC fluorochrome. Samples were then mounted on a reflection-mode SNOM (see. Sect. 11.2.2) and observed. Figure 11.7a shows the spectra of samples incubated either with TRITC (black circles) or with FITC (solid line) showing that the two red and green fluorescence emissions are well spectrally separated. The natural fluorescence of the tissues (gray line) is typically 10-times smaller than the FITC or the TRITC signals. Double localization of β1D-integrin with α-sarcoglycan was carried out by performing two subsequent scans of the same zone, looking each time for a single protein. Figures 11.7b and c display respectively the β1D-integrin and the α-sarcoglycan fluorescence maps, evidencing an alternation of green/red (β1D-integrin/α-sarcoglycan) and black bands (due to the absence of any reaction) at regular intervals. The co-localization of the two proteins was assessed by superimposing the two fluorescence maps, using green color tones to highlight the FITC fluorescence signal and red color tones for the TRITC one. The two topographies were exploited to compensate for lateral drifts between the two scans. The co-localization map (Fig. 11.7d) shows the alternation of yellow and black bands, indicating the spatial overlapping (co-localization) of

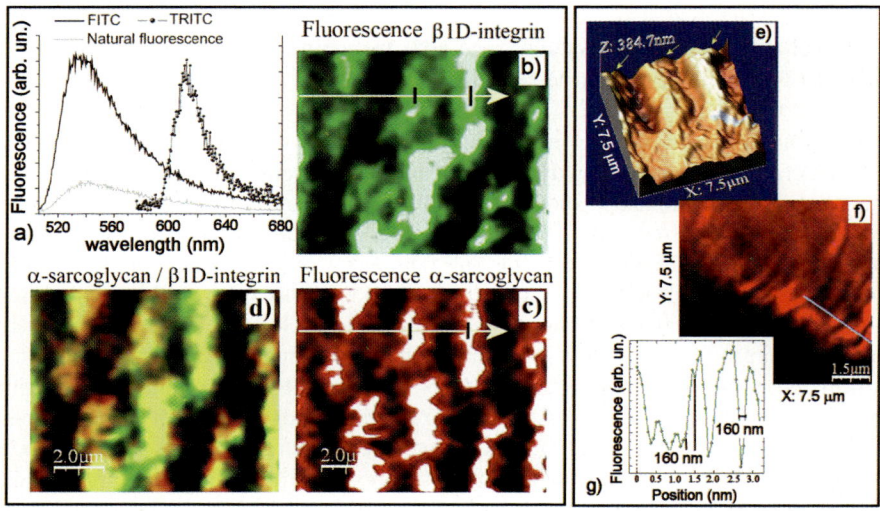

Fig. 11.7. (**a**) Fluorescence spectra of specifically bound FITC (*black line*) and TRITC (*black circles*) compared to natural tissues fluorescence (*gray line*) of human muscle cells immuno-labeled against β1D-Integrin and α-sarcoglycan. SNOM fluorescence maps of (**b**) β1D-Integrin and (**c**) α-sarcoglycan. (**d**) Co-localization map obtained by superposing (**b**) and (**c**) using yellow color tones to code co-localization. (**e**) Topography of the costameric structure of a cell, acquired simultaneously to the fluorescence (**f**) after incubation with anti- α-sarcoglycan antibodies. The dye is concentrated into small channels separated by typically 150–200 nm as shown by the *line* profile in (**g**) (reprinted from [79] with permission of the author)

the two proteins. Figure 11.7e, f and g show the topography (e) and the fluorescence map (f), together with the fluorescence line profile (g) of a sample incubated with anti-α-sarcoglycan/TRITC. In the upper part of the 3D topography image we see costameric reliefs (yellow arrows) with an almost flat fluorescence counterpart, suggesting that the optical signal in this case is either due to natural tissue fluorescence, or to unspecific binding of the anti-α-sarcoglycan. We find, instead, a strongly modulated fluorescence in the central part of the map, showing strict channel-like structures, separated by narrower dark areas whose width is ∼160 nm. This value indicates the spatial resolution attainable in our measurements.

11.3.4
Single Molecule Near-Field Fluorescence Microscopy of Dendritic Cells

Dendritic cells (DCs) play a central role in both innate and adaptive immune response, since they are equipped with a variety of dynamically regulated pathogen-recognition receptors [80]. The surface receptor DC-SIGN is a C-type lectin exclusively expressed on DCs, playing an important role during the immune response and featuring extraordinary pathogen-recognition capabilities. In addition to a high-affinity binding to the human immunodeficiency virus (HIV-1) envelope glycoprotein gp120, DC-SIGN has been shown to mediate binding to several other viruses and microorganisms [81–83]. Still, it remains largely obscure how this receptor is capable to efficiently recognize such a large variety of pathogen agents. Immuno Electron Microscopy evidenced that DC-SIGN organizes in microdomains on the cell membrane of immature DCs (imDCs), and that thus this type of arrangement is crucial for binding HIV-1 particles in situ [84]. Garcia-Parajo and co-workers applied fluorescence-SNOM with single molecule sensitivity to investigate the organization of the pathogen receptor DC-SIGN on the membrane of intact DCs, with single molecule sensitivity [85, 86]. An apparatus working in liquid (see Sects. 11.2.2 and 11.2.3) has been developed in order to operate in physiological conditions and avoid potential drying artifacts. Standard immuno-fluorescence protocols were followed to prepare the specimens. In particular the cells were first incubated with primary monoclonal antibodies AZN-D1, and then a secondary incubation was performed with goat anti-mouse Cy5 to allow fluorescence detection of the anti-DC-SIGN antibodies. Cy5 was excited at 647 nm. Individual cells were first selected by bright field illumination and then analyzed by confocal and SNOM fluorescence microscopy. Figure 11.8a is the confocal fluorescence image of a 32-μm^2 area of a DC expressing DC-SIGN on the surface as round-shaped domains. Figures 11.8b and c are the simultaneous topography and SNOM images on the 16.5-μm^2 area highlighted in Fig. 11.8a. The topography image permits us to precisely identify the cell boundaries and is particularly useful to exclude fluorescence mis-assignments due to unspecific antibody bindings. The SNOM image and, in particular, the zoom shown in Fig. 11.8d (7μm^2) evidence well-separated DC-SIGN domains with FWHM of 90 nm and 120 nm as measured from the line profiles (Fig. 11.8e and f)

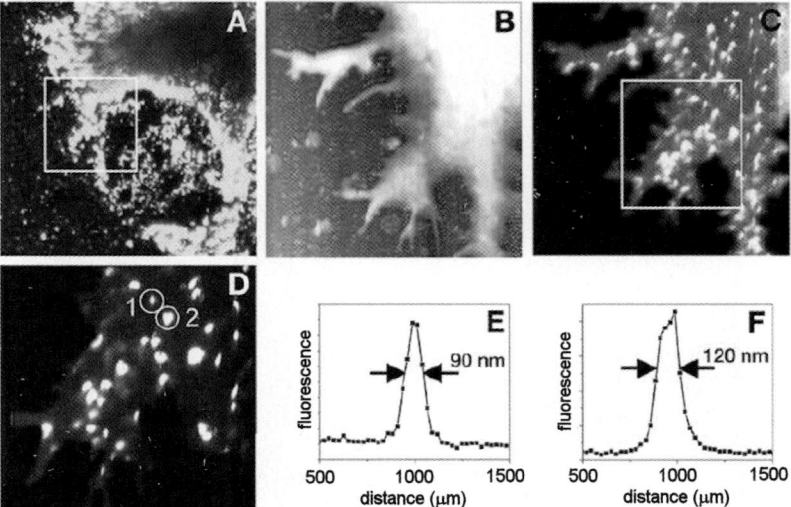

Fig. 11.8. (a) Confocal fluorescence image of an immature Dendritic cell in buffer solution. Simultaneous topography (b) and SNOM fluorescence map (c) of the zone *highlighted* in (a). (d) Close-up of the area *highlighted* in (c) clearly showing clusters of DC-SIGN with typical dimensions of 90 and 120 nm, as retrieved from the *line* profiles in (e) and (f) (reprinted from [86] with permission of the author)

relative to the domains highlighted in the Fig. 11.8d. The total photon counting rate on each spot is indeed proportional to the number of Cy5 molecules. Single-molecule sensitivity combined with quantitative analysis of the fluorescence intensity permits therefore to estimate the different number of DC-SIGN molecules within each domain. This gave an average of 5–10 molecules of DC-SIGN per cluster.

11.3.5
Chemical Information of Bacterial Surfaces and Detection of DNA Nucleobases by Tip-Enhanced Raman Spectroscopy

Label-free detection of membrane proteins by enhanced Raman spectroscopy is gaining considerable attention in biochemistry and biophysical sciences. Surface-Enhanced Raman Scattering (SERS) takes advantage of the optically resonant nanocavities formed in rough metallic substrates or colloids that are effective in providing Raman enhancement factors up to 10^{11}–10^{14} [87, 88]. Tip-Enhanced Raman Spectroscopy combines the extreme sensitivity of SERS with the possibility to confine the effective enhancement on length scales smaller than 50 nm. So far experiments have been carried out on specific probe molecules such as rhodamine 6G [89], benzenethiol [90], malachite green isothiocyanate [40], or on single-wall carbon nanotubes [39], demonstrating enhancement factors between 10 and 10^3. Deckert and co-workers reported the first TERS on bacterial cells [91]. Gram-positive *Staphylococcus epidermidis* bacteria were investigated, relevant as the cause of major infections associated with implanted medical devices. The cytoplasm membrane of Gram-positive bacteria is surrounded by a peptidoglycan layer, pervaded by other polysaccharides and a variety of surface proteins. Depending on the growth conditions, a biofilm consisting mainly of the two polysaccharides teichoic acid and PIA (polysaccharide intercellular adhesin) surrounds *Staphylococcus epidermidis* to a certain extent. TERS experiments were carried out on a system coupling an AFM with an inverted Raman microscope taking advantage of high numerical aperture objectives (NA 1.45) to maximize light collection efficiency. Figure 11.9a shows a

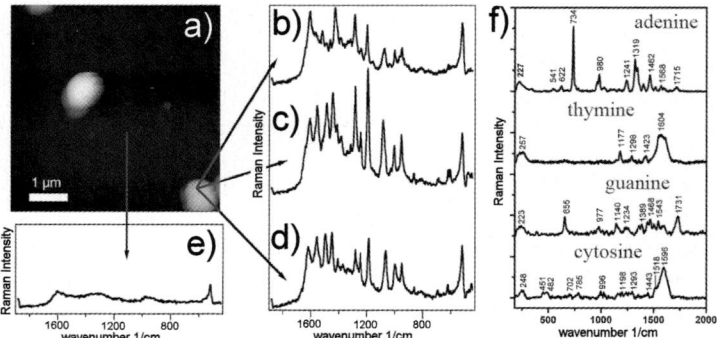

Fig. 11.9. (**a**) Topography of *Staphylococcus epidermidis* cells and corresponding TERS spectra of the membrane proteins (**b, c, d**). (**e**) Reference spectrum on glass (reprinted from [91] with permission of the author). (**f**) TERS spectra of the four DNA nucleobases deposited on Au(111) single crystals (reprinted from [92] with permission of the author)

topography map in which the *Staphylococcus epidermidis* cells show up as round-shaped features about 1 μm in diameter. Figures 11.9b, c, d and e show point TERS spectra recorded on the cell surface (b, c, d) and one (e) on the glass surface showing no signal. Most of the Raman bands are due to contributions from peptides and polysaccharides present on the cells surface. Sugar derivatives dominate with a peptidoglycan layer which is pervaded by teichoic acid and covered with PIA. Fluctuations observed in intensity and peak positions have been attributed to the persistent dynamics of the cell surface.

TERS allows as well for a highly sensitive vibrational analysis of molecules adsorbed in (sub)-monolayer quantities at atomically smooth surfaces. Pettinger and co-workers [92] have succeeded to obtain TERS spectra of (sub)-monolayers of the four DNA bases adenine, guanine, thymine, and cytosine adsorbed at Au(111) single crystals in picomole quantities. Experiments were carried out with a Raman/STM apparatus similar to the one described in Sect. 11.2.4, using gold tips as probes and exciting with the 633 nm of a HeNe laser (2 mW). The nucleobases were adsorbed at a freshly annealed Au(111) single crystal from 1-mM ethanolic solutions, rinsing ethanol to remove multilayers. According to the molecular dimensions the largest molecular density of adsorbed nucleobases can be estimated to around $2–2.5$ molecules/nm^2 for closest packing in a monolayer, which corresponds to a density of only 420 pmol/cm^2 of target species present in the enhanced field region under investigation. Figure 11.9f shows the Raman vibrational fingerprints of monolayers of the four DNA bases after spectral background correction. These measurements, according to the authors, show that label-free sensing of picomole quantities of small, non-resonant biomolecules is reproducible and straightforward with TERS.

11.4
Conclusions

In the last decade Near-Field Optical Microscopy has provided us with eyes to look into the NanoWorld. Different experimental concepts have been proposed and some specific configurations are now well assessed tools for analytical investigations in materials science, nanophotonics, and nanobiotechnology. In the biological field, aperture-SNOM turns out to be perfectly suited for fluorescence mapping of membrane proteins due to its high sensitivity (single molecule), high spatial resolution (100 nm), and small penetration depth (< 100 nm). On the other hand the emerging Tip-Enhanced SNOM started providing convincing examples of the possibility to achieve label-free vibrational imaging of nano-organisms and important functional biomolecules. These experiments, in particular, show the great potential of tip-enhanced Raman spectroscopy for biological and biomedical applications. The technique permits us to improve the sensitivity of non-resonant Raman spectroscopy by several orders of magnitudes. As a non-invasive technique, TERS can characterize the chemistry at specific sites on biological samples with a spatial resolution down to a few tens of nm. This method might even be used to follow the dynamics on cell surfaces. For the future fluorescence-SNOM, IR-SNOM, and TERS, could represent key techniques for a detailed understanding of biological processes and mechanisms occurring on cell surfaces, and might give insights into the interaction of bacteria with their surroundings, including hosts and antibiotics with unprecedented sensitivity and spatial resolution.

Acknowledgments. P. Jones and M. Lamy de La Chapelle are acknowledged for critically reading the manuscript.

References

1. Stephens DJ, Allan VJ (2003) Science 300:82
2. Sako Y, Yanagida T (2003) Nature Cell Biol Suppl S, SS1.
3. Garcia-Parajo MF, Veerman JA, Bouwhuis R, Vallee R, van Hulst NF (2001) Chem Phys Chem 2:347.
4. Edman L, Mets U, Rigler R (1996) Proc Natl Acad Sci USA, 93:6710
5. Schmidt T, Schutz GJ, Baumgartner W, Gruber HJ, Schindler H (1995) J Phys Chem 99:17662
6. Lu HP, Xun L, Xie XS (1998) 282:1877.
7. Harada Y, Funatsu T, Murakami K, Nonoyama Y, Ishiama A, Yanagida T (1999) Biophys J 76:709
8. Abbe E (1873) Arch Mikroskop Anat 9:413
9. Hell SW, Stelzer EH (1992) J Opt Soc Am 9:2159
10. Klar TA, Kakobs S, Dyba M, Egner A, Hell SW (2000) Proc Natl Acad Sci USA 97:8206
11. Pohl DW, Denk W, Lanz M (1984) Appl Phys Lett 44:651.
12. Lewis A, Isaacson M, Harootunian A, Murray A (1984) Ultramicroscopy 13:227
13. Zenhausern F, O'Boyle MP, Wickramasinghe HK (1994) Appl Phys Lett 65:1623
14. Bachelot R, Gleyzes P, Boccara AC (1995) Opt Lett 20:1924
15. Labardi M, Gucciardi PG, Allegrini M (2001) La Rivista del Nuovo Cimento 23:1.
16. Patanè S, Gucciardi PG, Labardi M, Allegrini M (2004) La Rivista del Nuovo Cimento 27:1
17. Cefalì E, Patanè S, Spadaro S, Gardelli R, Albani M, Allegrini M (2007) In: Bushan B, Fuchs H, Tomitori M (eds) Applied scanning probe methods, vol VIII. Springer, Berlin, p78
18. Gucciardi PG, Bachelier G, Stranick S, Allegrini M (2008) In: Bushan B, Fuchs H, Tomitori M (eds) Applied scanning probe methods, vol VIII. Springer, Berlin, p 1
19. Gucciardi PG, Trusso S, Vasi C, Patanè S, Allegrini M (2007) In: Bhushan B, Fuchs H, Kawata (eds) Applied scanning probe methods, vol V. Springer, Berlin, p 287
20. Pawley JB (2006) Handbook of biological confocal microscopy, 3rd edn. Springer, Berlin
21. Novotny L, Hecht B (2006) Principles of nano-optics. Cambridge University Press, Cambridge
22. Betzig E, Finn P, Weiner JS (1992) Appl Phys Lett 60:2484
23. Karrai K, Grober RD (1995) Appl. Phys. Lett. 66:1842
24. Betzig E, Trautman JK, Harris TD, Weiner JS, Kostelak RL (1991) Science 251:1468
25. Veerman JA, Otter AM, Kuipers L, van Hulst NF (1998) Appl Phys Lett 72:3115
26. Gucciardi PG, Colocci M, Labardi M, Allegrini M (1999) Appl Phys Lett 75:3408
27. Ambrosio A, Allegrini M, Latini G, Cacialli F (2005) Appl Phys Lett 87:033109
28. Dickenson NE, Erickson ES, Mooren OL, Dunn RC (2007) Rev Sci Instrum 78:053712
29. Erickson ES, Dunn RC (2005) Appl Phys Lett 87:201102
30. Gucciardi PG, Patanè S, Ambrosio A, Allegrini M, Downes AD, Latini G, Fenwick O, Cacialli F (2005) Appl Phys Lett 86:203109
31. Intonti F, Emiliani V, Lienau Ch, Elsaesser Th, Noetzel R, Ploog KH (2001) Phys Rev B 63:075313

32. Intonti F, Emiliani V, Lienau Ch, Elsaesser Th, Noetzel R, Ploog KH (2001) Phys Rev B 64:155316
33. Mihalcea C, Scholz W, Werner S, Muenster S, Oesterschulze E, Kassing R (1996) Appl Phys Lett 68:3531
34. Biagioni P, Polli D, Pucci A, Ruggeri G, Labardi M, Cerullo G, Finazzi M, Duò L (2005) Appl Phys Lett 87:223112
35. Binnig G, Quate CF, Gerber C (1986) Phys Rev Lett 56:930
36. Novotny L, Stranick SJ (2006) Annu Rev Phys Chem 57:303
37. Binnig G, Rohrer H, Gerber C, Weibel E (1982) Phys Rev Lett 49:57
38. Ma Z, Gerton JM, Wade LA, Quake SR (2006) Phys Rev Lett 97:260801
39. Hartschuh A, Sanchez EJ, Xie S, Novotny L (2003) Phys Rev Lett 90:095503
40. Pettinger B, Ren B, Picardi G, Schuster R, Ertl G (2004) Phys Rev Lett 92:096101
41. De Bakker BI, de Lange F, Cambi A, Korterik JP, van Dijk EM, van Dijk EM, van Hulst NF, Figdor CG, Garcia-Parajo MF (2007) Chem Phys Chem 8:1473
42. Gucciardi PG, Vinattieri A, Colocci M, Damilano B, Grandjean N, Semond F, Massies J (2001) J Microscopy-Oxford 202:212
43. Gucciardi PG, Princi P, Pisani A, Favaloro A, Cutroneo G (2005) J Kor Phys Soc 47:S86
44. Gucciardi PG, Labardi M, Gennai S, Lazzeri F, Allegrini M (1997) Rev Sci Instrum 68:3088
45. Gucciardi PG, Trusso S, Vasi C, Patanè S, Allegrini M (2002) Phys Chem Chem Phys 4:2747
46. Lambelet P, Pfeffer A, Sayah A, Marquis-Weible F (1998) Ultramicroscopy 71:117
47. Brunner R, Hering O, Marti O, Hollricher O (1997) Appl Phys Lett 71:3628
48. Longo G, Girasole M, Cricenti A (2005) Phys Stat Sol B 15:3070
49. Höppener C, Molenda D, Fuchs H, Naber A (2003) J Microsc-Oxford 210:288
50. Koopman M, de Bakker BI, Garcia-Parajo MF, van Hulst NF (2003) Appl Phys Lett 83:5083
51. Ren B, Picardi G, Pettinger B (2004) Rev Sci Instrum 75:837
52. Billot L, Berguiga L, de la Chapelle ML, Gilbert Y, Bachelot R (2005) Eur Phys J-Appl Phys 31:139
53. Bonaccorso F, Calogero G, Di Marco G, Maragò OM, Gucciardi PG, Giorgianni U, Channon K, Sabatino G (2007) Rev Sci Instrum 78:103702
54. Novotny L, Bian RX, Xie XS (1997) Phys Rev Lett 79:645
55. Mehtani D, Lee N, Hartschuh RD, Kisliuk A, Foster MD, Sokolov AP, Caiko F, Tsukerman I (2006) J Opt A Pure Appl Opt 8:S183
56. Hillenbrand R, Keilmann F (2000) Phys Rev Lett 85:3029
57. Hillenbrand R, Keilmann F (2002) Appl Phys Lett 80:25
58. Labardi M, Patanè S, Allegrini M (2000) Appl Phys Lett 77:621
59. Taubner T, Keilmann F, Hillenbrand R (2004) Nanolett 4:1669
60. Taubner T, Hillenbrand R, Keilmann F (2004) Appl Phys Lett 85:5064
61. Gucciardi PG, Bachelier G, Allegrini M (2006) J Appl Phys 99:124309
62. Gucciardi PG, Bachelier G, Allegrini M, Ahn J, Hong M, Chang S, Jhe W, Hong SC, Baek SH (2007) J Appl Phys 101:064303
63. Ossikowski R, Nguyen Q, Picardi G (2007) Phys Rev B 75:045412
64. Gucciardi PG, Lopes M, Lamy de La Chapelle M (2008) Thin Sol Films 516:8064
65. Poborschii V, Tada T, Kanayama T (2005) Jpn J Appl Phys 44:L202
66. Mehtani D, Lee N, Hartschuh RD, Kisliuk A, Foster MD, Sokolov AP, Maguire JF (2005) J Raman Spectrosc 36:1068
67. Gucciardi PG, Lopes M, Deturche R, Julien C, Barchiesi D, Lamy de La Chapelle M (2008) Nanotechnol 19:215702
68. Brehm M, Taubner Th, Hillenbrand R, Keilmann F (2006) Nanolett 6:1307
69. Knoll B, Keilmann F (1999) Nature 399:134
70. Enderle Th, Ha T, Ogletree DF, Chemla DS, Magowan C, Weiss S (1997) Proc Natl Acad Sci USA 94:520

71. Ervasti JM, Ohlendieck K, Kahl SD, Gaver MG, Campbell KP (1990) Nature 345:315
72. Yoshida M, Ozawa E (1990) J Biochem 108:748
73. Mondello MR, Bramanti P, Cutroneo G, Santoro G, Di Mauro D, Anastasi G (1996) Anat Rec 245:481
74. Ohlendieck K (1996) Eur J Cell Biol 69:1
75. Yoshida M, Suzuki A, Yamamoto H, Noguchi S, Mizuno Y, Ozawa E (1994) Eur J Biochem 222:1055
76. Lazarides E (1980) Nature 283:249
77. Hynes RO (1992) Cell 69:11
78. Anastasi G, Amato A, Tarone G, Vita G, Monici MC, Magaudda L, Brancaccio M, Sidoti A, Trimarchi F, Favaloro A, Cutroneo G (2003) Cells Tissues Organs 175:151
79. Anastasi A, Cutroneo G, Pisani A, Bruschetta D, Milardi D, Princi P, Gucciardi PG, Bramanti P, Soscia L, Favaloro A (2007) J Microsc-Oxford 228:322
80. Akira S (2003) Curr Opin Immunol 15:5
81. Geijtenbeek TB, Torensma R, van Vliet SJ, van Duijnhoven GCF, Middel J, Cornelissen I, Nottet H, KewelRamani VN, Littman DR, Figdor CG, van Kooyk (2000) Cell 100:587
82. Alvarez CP, Lasal F, Carrillo J, Muniz O, Corbi AL, Delago R (2002) J Virol 76:6841
83. Pöhlmann S, Zhang J, Baribaud F, Chen Z, Leslie GJ, Lin G, Granelli-Piperno A, Doms RW, Rice CM, McKeating JA (2003) J Virol 77:4070
84. Cambi A, de Lange F, van Maarseveen NM, Nijhuis M, Joosten B, van Dijk EM, de Bakker BI, Fransen JAM, Bovee-Geurts PHM, van Leeuwen NF, van Hulst NF (2004) J Cell Biol 164:145
85. De Bekker BI, de Lange F, Cambi A, Korterik JP, van Dijk EM, van Hulst NF, Figdor CG, Garcia-Parajo MF (2007) Chem Phys Chem 8:1473
86. Garcia-Parajo MF, De Bekker BI, Koopman M, Cambi A, de Lange F, Figdor CG, van Hulst NF (2005) NanoBiotechnology 1:113
87. Kneipp K, Wang Y, Kneipp H, Perelman LT, Itzkan I, Dasari R, Feld MS (1998) Phys Rev Lett 78:1667
88. Nie SM, Emory SR (1997) Science 275:1102
89. Watanabe H, Hayazawa N, Inouye Y, Kawata S (2005) J Phys Chem B 109:5012
90. Ren B, Picardi G, Pettinger B, Schuster R, Ertl G (2005) Angew Chem 117:141
91. Neugebauer U, Rösch P, Schmitt M, Popp J, Julien C, Rasmussen A, Budich Ch, Deckert V (2006) Chem Phys Chem 7:1428
92. Domke KF, Zhang D, Pettinger B (2007) J Am Chem Soc 129:6708

12 Adhesion and Friction Properties of Polymers at Nanoscale: Investigation by AFM

Sophie Bistac · Marjorie Schmitt

Abstract. Friction and adhesion of model elastomers are quantified at two different scales: at nanoscopic scale, using atomic force microscopy (AFM) and at macroscopic scale, using a tack test and a translation tribometer. The objective is firstly to find a correlation between nanoscale adhesion and friction, by comparing the influence of structural parameters such as crosslinking degree and presence of free chains. The scope is also to verify if friction and adhesion behaviors are comparable at both scales. Experimental results underline the major role of molecular parameters such as degree of crosslinking and length und content of free and pendant chains. However, their effect on nano and macroscale properties is different. Conceptual schemes are proposed to describe the complex interfacial molecular mechanisms, especially chain adsorption onto the tip which seems to govern nanoscale friction and adhesion. Further friction studies, performed on polymers in contact with hydrophilic and hydrophobic tip and substrate allow us to underline the complex competition between interfacial interactions and polymer surface rheological behavior.

Key words: Adhesion, Friction, AFM, Interfacial interactions, Polymers, Elastomers

12.1 Introduction

Atomic force microscopy is a fruitful tool to solve adhesion and friction problems for polymer materials. It can help to develop fundamental understanding of interfacial molecular phenomena. AFM can be firstly used to image the surface in order to obtain information about surface properties such as topography, structure, and organization. These surface characteristics will then be correlated to macroscale adhesion and friction behaviors. Such imaging experiments can also be performed after adhesion or friction tests, in order to verify the surface changes induced by interfacial stresses (change in roughness, anisotropy, and crystallization induced by friction, analysis of the transfer layer, wear tracks and scratch, etc.). Also, friction and adhesion are very sensitive to chain mobility, which will affect viscoelastic dissipation. Atomic force microscopy is able to probe the mechanical response of the surface [1,2]. AFM is also used for investigations of thermo-mechanical properties of polymer surfaces, by performing indentation experiments. Local properties such as surface modulus or glass transition temperature (Tg) are able to differ from the bulk, and this change can modify adhesion and friction behavior. AFM can detect a change in chain mobility, due, among others, to a confinement in thin films [3–5].

Moreover, a change in surface modulus or Tg can have a chemical origin: the chemical composition of the polymer surface can differ from the bulk, due to a preferential migration of a given component (e.g., small size molecules) or a segregation. Process conditions used to obtain the polymer material (temperature, pressure, time, chemical nature of the mould, etc.) are able to affect the surface chemical composition, by increasing or decreasing the diffusion of species, or by modifying the orientation of polar or non-polar groups on the top surface. Elasticity maps can also be used to identify relative distribution of soft/rigid domains in semi-crystalline polymers, copolymers or polymer blends and composites [6]. By using phase contrast mode, it is also possible to obtain phase-angle imaging maps able to underline differences in viscoelastic response [7].

Lateral force microscopy (LFM), sometimes referred to as frictional force microscopy (FFM), measures the lateral force between a surface and a sliding probe tip on the nanometer scale [8]. LFM can be used to obtain information about molecular motions of polymers [9, 10]. Lee has studied the relationship between glass transition temperature and lateral force of polymer blends on the basis of lateral force microscopy [11]. The results suggested that the lateral force of a miscible polymer blend reflects the sum of each component contribution, depending on the surface structure of the blend. Therefore, information about surface structure can be obtained from the measured lateral force behavior. However, LFM is more frequently used to quantify nano-friction properties. Scanning probe methods have been indeed applied to the investigation of tribology properties on the nanometer and nanonewton scale [12–16]. The result of nano-friction measurements will be a better understanding of macrotribology properties, through the parallel investigation of identical tribosystems on the macro and nanoscales. However, friction values can be scale-dependant, and the correlation between macro and nanoscale results could be delicate. The contact area and velocities are indeed in somewhat different ranges.

In addition, the interest of LFM is also to provide a single asperity (tip) in contact with the polymer surface, able to simulate what is happening during friction in nanostructures used in magnetic storage or microelectromechanical systems (MEMS) [17, 18].

Nanoscale adhesion level can also be quantified by AFM, through force-distance curve measurements, which provide detailed information on local elastic and adhesion properties of the polymer surface [19–22].

At this scale, humidity and presence of adsorbed water (especially on hydrophilic surfaces, tip or material) can have a major influence on nano-adhesion due to capillary effects [23].

The correlation between nano-friction and nano-adhesion is interesting for polymers, which usually exhibit significant adhesive properties compared to other solid materials [6, 24, 25]. Tribology of polymers is indeed complex, due to their specific properties induced by chain mobility, at small or larger scales. Chain motions allow relaxation mechanisms and energy dissipation, leading to a viscoelastic behavior (time and temperature dependences, usually linked by the time-temperature equivalence). Moreover, adhesion and friction sciences are both characterized by their multidisciplinarity, involving various scientific domains such as chemistry and physics, rheology, fracture mechanics, etc.

Adhesion reflects the total energy of the substrate/adhesive interfacial bonds and depends directly on the nature and density of these interactions. The adherence value is usually greater than the presumed adhesion value, because during separation, a part of the energy is dissipated by internal molecular motions (chain extension, disentanglement, etc.). Other experimental parameters will unfortunately complicate the solving of adhesion and friction problems. Changing the type of the adherence or friction test or the sample geometry can for example influence the measured values.

The relation between adhesion and friction has been investigated at macroscale [26, 27]. Even if the correlation between adhesion and friction is usually delicate, it could help to better predict tribological behavior, in order to control and design smart surfaces able to present specific properties. Some authors have defined some interesting relations between interfacial interactions and friction [28–33]. Surface force apparatus is often used to investigate micro/nanoscale adhesion and friction of polymers or surfactants [29].

The role of chain mobility in macroscale adhesion and friction has also been studied in the case of elastomers in contact with a silicon wafer covered by a grafted layer [34–38]. Results are interpreted in terms of the orientation and relaxation of polymer chains [35].

The effect of molecular weight [39, 40], friction speed, and normal load [41, 42] on friction and wear of polymers has been investigated in the literature. It has been shown that the terminal chain group chemistry of rubbed surfaces and polymer chain mobility can have a significant impact on friction [43].

Moreover, rheology of the polymer surface can be different from the bulk [44, 45]. Surface force apparatus allows us to investigate rheology and tribology of thin liquid films between shearing surfaces, with the measurement of previously inaccessible parameters during frictional sliding, such as the real area of contact, the local asperity load and pressure, and the sheared film thickness [46, 47].

Elastomers present interesting tack and friction properties. They are generally crosslinked, in order to assure elastic behavior and a sufficient cohesion. The crosslinking reaction also allows us to modulate properties, especially mechanical properties, by varying the crosslinking density, i.e. the molecular weight between chemical nodes. Polydimethylsiloxane (PDMS) elastomers are often used as model elastomers, due to the possibility to obtained crosslinked networks in ambient conditions, with a good control of the crosslinking density (molecular weight between crosslinks) by varying the initial molecular weight of the chains. However, the final networks are imperfect, with the presence of pendant chains (which are linked to the network only by one extremity) and free chains (which are not at all chemically bonded to the network). When the initial molecular weight of PDMS is increased, a lower crosslinking density (lower modulus), but also a greater quantity of free and pendant chains are obtained. Free chains, which are soluble, can be extracted by immersion in a good solvent (toluene). Crosslinked elastomers can then also be studied after extraction of free chains.

Common phenomena (interfacial interaction and dissipation) therefore control polymer adherence and friction behaviors. However, the relation between both phenomena is still unclear. The first objective of this work is then to compare nanoscale adherence and friction of model elastomers (PDMS networks) in order to try to establish relations between both properties. AFM will be used to quantify adherence and friction between an AFM tip and the PDMS surface. The second objective of this study is then to compare adherence and friction behaviors of PDMS at two scales (macro and nano), in order to verify if the involved mechanisms are similar and if the nanoscale approach is able to improve the understanding of what is happening at a macroscopic scale. Influence of structural parameters (crosslinking degree and presence of free chains) and experimental factors (speed, normal force) will be analyzed. Further results obtained for other systems will confirm the complex role of interfacial interactions on polymer friction.

12.2
Experimental Part

Two PDMS elastomers with different initial molecular weight ($Mw = 6,000\,g/mol$, called E1, and $Mw = 17,200\,g/mol$, called E2) are used. They are vinyl-terminated and crosslinked with tetrakis(dimethylsiloxy)silane. Free chains can be extracted by immersion in a toluene. Elastomers were also studied after extraction of free chains (corresponding to samples called E1' and E2'). The glass temperature of all PDMS networks is equal to $-123\,°C$.

Nanoscale adhesion and friction of PDMS films are quantified with an AFM (D3000 from Digital Instruments), in contact mode with a commercial silicon tip on a 100-μm triangular cantilever (spring constant $= 0.58\,N/m$). Force distance curves, measured during a loading–unloading experiment, allow the determination of nano-adhesion, which is directly proportional to the maximum cantilever deflection reached during the separation of the tip from the polymer surface, as illustrated in Fig. 12.1.

Nano-friction is quantified in torsion mode. The friction force is proportional to the TMR value (Trace Minus Retrace in volt), which corresponds to the difference between lateral forces scanning left-to-right and right-to-left. Friction forces in Newton can be calculated through calibration methods but such techniques have not been applied for this comparative study. Adhesion force and friction (TMR values) were determined for elastomers before and after free chain extraction.

Nanoscale results have been compared to macroscopic adhesive and tribological properties. Macro-adhesion is quantified by using a tack test (Fig. 12.2).

The substrate used for tack experiments is a cleaned glass plate (roughness $= 2\,nm$). Glass presents several advantages: high stiffness, low roughness, a chemical composition close to the AFM tip and a transparency (contrary to silicon wafers)

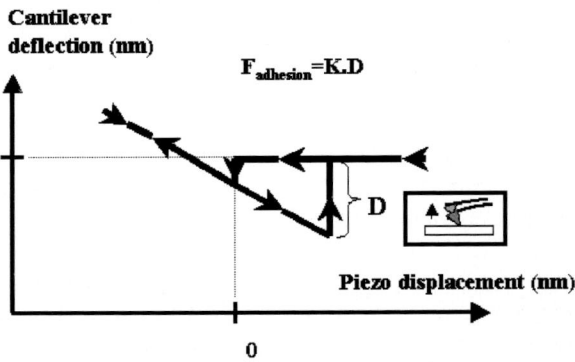

Fig. 12.1. Force-distance curve obtained by AFM (nano-adhesion). The adhesion force F is equal to the maximum deflection D multiplied by a constant K

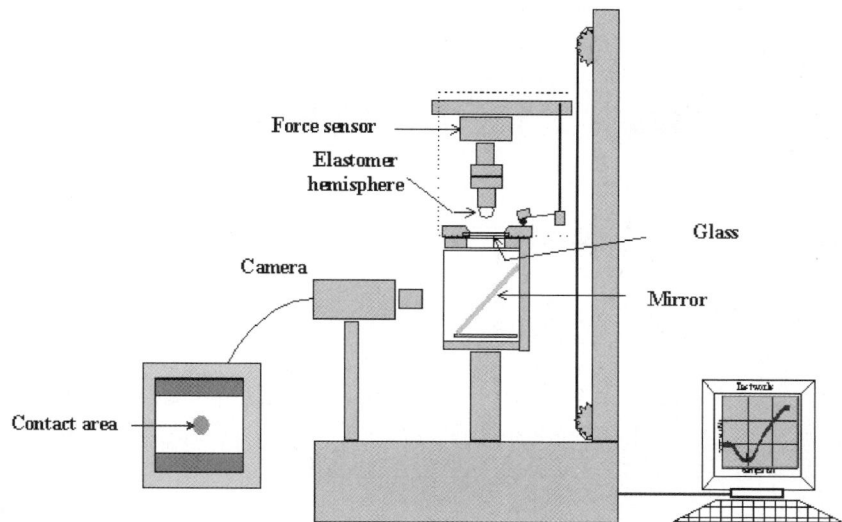

Fig. 12.2. Scheme of the tack apparatus

that allows us to measure the contact area during the experiment. During the tack experiment, a PDMS hemisphere (diameter $= 16$ mm) is put into contact (with an approach speed of 10 mm/min) with the glass plate, during a given contact time and under a controlled normal load. Elastomer and glass are then separated at a given speed, and the separation force is measured. The apparatus also measures the apparent contact area between the substrate and the elastomer, with a video camera placed under the transparent glass plate. The adherence energy, in J/m^2, is calculated by dividing the integral of the separation force versus distance by the contact area corresponding to the maximum force measured during separation.

Macro-friction properties of PDMS hemispheres in contact with a glass plate are measured by using a translation tribometer (Fig. 12.3).

The hemisphere is in contact (under a controlled normal force) with the glass plate which is moved in translation at a controlled speed (only one single passage performed). The tangential force (friction force) is measured and the evolution of the contact area during friction is followed with a video camera. Surface energy of PDMS films was determined by wettability measurements. Equilibrium contact angles of different liquids drops (water, a polar liquid, and diiodomethane, a non-polar liquid) were measured with an automated Kruss apparatus. All experiments are performed in ambient air, at room temperature (20 °C).

12.3
Nano-Adhesion Investigation

Table 12.1 reports the elastomers characteristics: Sol fraction (weight proportion of free chains in %), molecular weight between crosslinks Mc, and Young's modulus E for PDMS E1 and E2 (before extraction of free chains). Sol fraction and

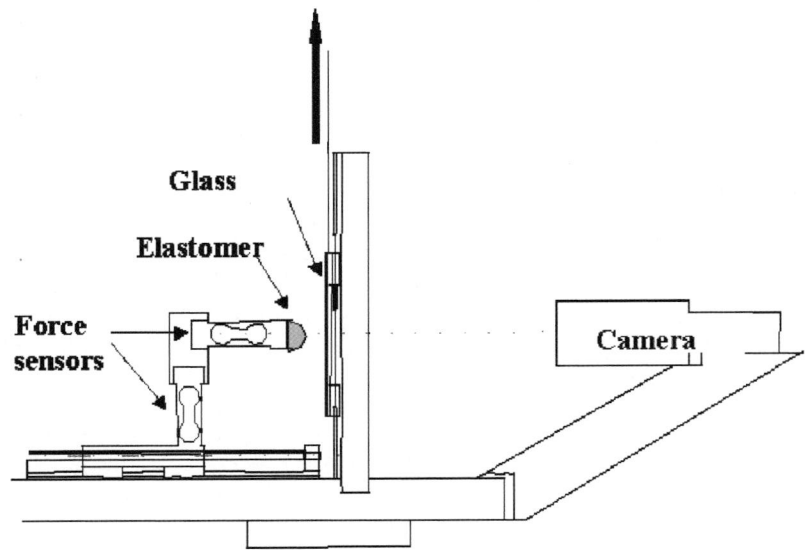

Fig. 12.3. Translation tribometer

Table 12.1. PDMS characteristics: sol fraction, molecular weight between crosslinks Mc, and Young's modulus E

PDMS	Sol fraction (%)	Mc (g/mol)	E (MPa)
E1	4	7500	1.40
E2	14	18500	0.42

molecular weight between crosslinks have been determined by using the swelling method (immersion in toluene).

Sol fraction values indicate that the quantity of free chains is higher for PDMS E2. These free chains are also longer for E2 compared to E1. After extraction, the modulus of PDMS is decreased from 1.40 MPa (E1) to 1.20 MPa (E1') and from 0.42 MPa (E2) to 0.24 MPa (E2').

Similar surface energies are obtained for all PDMS (with and without free chains), close to $27 \, \text{mJ/m}^2$, with a dispersive component equal to $27 \, \text{mJ/m}^2$ and a non-dispersive component equal to zero.

Nanoscale measurements have been performed with AFM in contact mode [48, 49]. Nano-adhesion is proportional to the maximum cantilever deflection D during separation (force-distance curves). Adhesion force (expressed in Newton) is calculated by multiplying the maximum deflection D by the spring constant (0.58 N/m). Nano-adhesion measurements are not possible for sample E2 because the AFM tip can not be separated from the PDMS surface within the measurable cantilever deflection range. Also the adhesion value of E2 can not be determined, it has to be considered as important. After extraction of free chains (PDMS E2'), measurements are

Table 12.2. Nano-adhesion force of PDMS E1, E1′, and E2′, for two applied loads

Sample	E1	E1′	E2	E2′
Adhesion force (nN) (normal force = 0 nN)	37 ± 2	36 ± 2	>> 261	261 ± 15
Adhesion force (nN) (normal force = 70 nN)	39 ± 2	39 ± 2	>> 383	383 ± 15

possible. Nano-adhesion results are reported in Table 12.2 for samples E1, E1′, and E2′, and for two applied normal forces (0 and 70 nN).

For PDMS E1 and E1′, the adhesion force is quite constant before and after extraction, and is not significantly dependant on the normal force, taking into account experimental errors, even if a slightly higher adhesion level is measured when the applied load is increased. The effect of applied load is more pronounced for E2′, with a great increase of adherence for a higher normal load. A higher contact area will then favor chain adsorption onto the tip in the case of sample E2′ which has a lower modulus compared to E1 and E1′ and also more numerous and longer pendant chains. For PDMS E2, the tip can not be separated from the elastomer surface, indicating a much greater adhesion for E2 compared to E2′. Elimination of free chains decreases the nano-adhesion, for samples E2 and E2′. Important adsorption phenomena of numerous and long free chains of E2 on the AFM tip could explain the higher nano-adhesion before extraction. The mobility of these free chains (greater than pendant chains) allows a greater adsorption onto the tip, avoiding the separation (with the same experimental device, i.e. cantilever stiffness). The effect of free chains was quite negligible for E1 and E1′, probably due to the fact that the length and quantity of free chains is lower for E1 compared to E2.

The comparison between E1 and E2 shows that the nano-adhesion increases when the crosslinking density is decreased. A lower stiffness and a higher content of free and/or pendant chains are able to firstly induce a greater chain adsorption onto the tip and secondly more dissipative chain extensions and movements during separation. Both effects are able to explain the higher adhesion of E2 and E2′ compared to E1 or E1′.

12.4
Nano-Friction Investigation

For nano-friction determination, the TMR value, which is directly proportional to the friction force, is measured for different normal loads (deflection set point value in volts) and friction speeds (tip velocity, obtained by varying the scan frequency), for PDMS E1, E1′, and E2′. Nano-friction measurements are indeed impossible for sample E2 (before extraction) because the AFM tip is still "trapped" in the surface layer: friction between the tip and the surface of PDMS E2 can therefore be considered as very high. Figures 12.4 and 12.5 illustrate the evolution of the TMR as a function of the applied load for different friction speeds, for PDMS E1 and E1′, respectively.

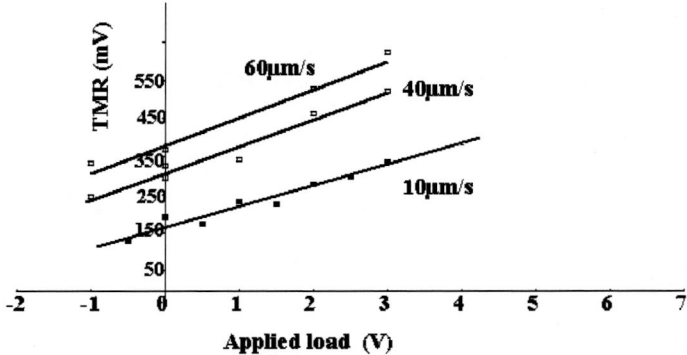

Fig. 12.4. Evolution of the TMR (nano-friction force) as a function of applied load; for different friction speeds, measured for PDMS E1

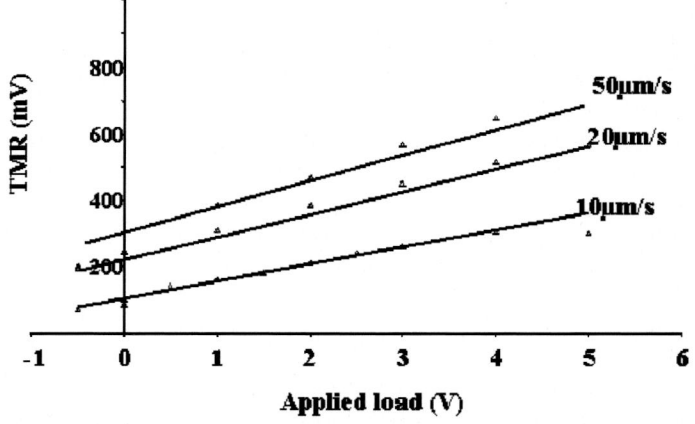

Fig. 12.5. Evolution of the TMR (nano-friction force) as a function of applied load; for different friction speeds, measured for PDMS E1′

A linear evolution is observed for the TMR in the given normal load range indicating a quite constant friction coefficient whatever the applied load. The evolution of the friction force versus normal load is also linear for E2′ (Fig. 12.6).

For all PDMS, a significant TMR value is measured for an applied load equal to zero, corresponding to a significant adhesive contribution. An adhesive contact (chain adsorption onto the tip) is then created even at zero load. TMR values at zero load are similar for samples E1 and E1′, but are much greater for E2′, indicating a higher adhesive contribution. Table 12.3 presents the friction force (TMR, in volts), measured for PDMS E1 and E1′ for various speeds and applied loads (in volt).

The presence of free chains has a major effect on nano-friction, with a higher friction for E1 compared to E1′. This effect is also more pronounced for lower normal load and speed.

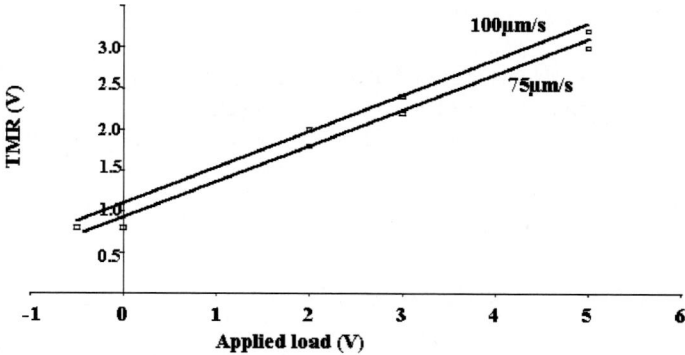

Fig. 12.6. Evolution of the TMR (nano-friction force) as a function of applied load; for different friction speeds, measured for PDMS E2′

Table 12.3. Friction force (TMR values in volt, ΔTMR $= 0.01$ volt) of PDMS E1 and E1′ for different friction speeds and applied loads

	Speed (μm/s)	Applied load (volt)		
		-0.5	0	1
E1	5	0.11	0.13	0.16
	50	0.28	0.33	0.39
E1′	5	0.06	0.09	0.13
	50	0.21	0.28	0.40

The influence of speed is also pronounced with an increase of friction with speed for both PDMS. Chain motions during nano-friction and pull-out mechanisms, which are rate-sensitive (viscoelastic dissipation), are able to explain this behavior.

Nano-friction measurements were possible for PDMS E2′, indicating that the friction of E2′ is lower compared to E2, for which the tip could not be moved from the polymer surface (free chains of E2 were adsorbed onto the AFM tip, avoiding its sliding). Table 12.4 reports nano-friction results measured for PDMS E2′ for different normal loads and friction speeds.

Table 12.4. Friction force (TMR values in volt, ΔTMR $= 0.1$ volt) of PDMS E2′ for different friction speeds and applied loads

Speed (μm/s)	Applied load (volt)		
	2	3	5
75	1.8	2.2	3.0
100	2.0	2.4	3.2

Table 12.4 shows a slight effect of friction speed, taking into account the experimental errors. Friction values obtained for lower speeds allow us to confirm the slight effect of speed. Nano-friction values obtained for E2′ (and consequently E2) are much higher compared to PDMS E1 and E1′. The longer and more numerous free and/or pendant chains present in E2 or E2′ will induce a great adsorption on the AFM tip, increasing the interfacial shear resistance.

12.5
Relation Between Adhesion and Friction at Nanoscale

The first similarity between nano-adhesion and nano-friction is that both measurements were impossible for PDMS E2. Long and numerous free chains of E2 avoid the tip separation for both cases. The correlation between adhesion and friction is also based on the adhesive contribution measured during friction for a zero normal load. For all PDMS, a significant TMR value was measured at zero load, with a much greater value for E2′, compared to E1 and E1′ whose values were very close. Nano-adhesion also underlined the greater adhesive properties of PDMS E2′ compared to E1 and E1′. Moreover, the friction forces measured for E2′ are also much greater than E1 and E1′. The lower stiffness and the longer and more numerous pendant chains present in E2′ induce a great adsorption on the AFM tip, increasing the friction. The same effect (adsorption of longer chains) was proposed to explain the higher nano-adhesion of E2′. An interesting correlation between nano-adhesion and nano-friction can then found: both behaviors are governed by chain adsorption phenomena on the tip, favored by mobility of free or pendant chains [39, 49]. A schematic illustration of the molecular mechanisms involved in nano-adhesion and friction is proposed on Fig. 12.7.

The interactions of the AFM tip and PDMS surface show that short free or pendant chains of E1 or E1′ have a lower adsorption onto the tip compared to long and more numerous free or pendant chains of E2 or E2′.

To resume nanoscale behavior, when the degree of crosslinking decreases (longer chains), nanoscale adhesion and friction increase, due to chain adsorption onto the tip. And when free chains are present, nano-adhesion and nano-friction increase (greater mobility which favors the wetting of the tip), especially for E2, which contains more numerous and longer free chains than E1.

AFM tip

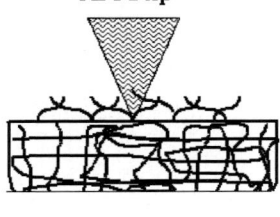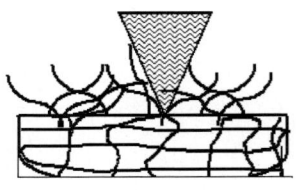

Fig. 12.7. Schematic representation of the interactions between AFM tip and PDMS surface

PDMS E1 and E1'   **PDMS E2 and E2'**

12.6
Comparison with Macroscale Results

Macro-adhesion is quantified by using a tack test. A polymer hemisphere is put into contact (with an approach speed of 10 mm/min) with a cleaned glass plate, during a given contact time and under a controlled normal force. Elastomer and glass are then separated at a given speed, and the separation force is measured. The apparatus also measures the apparent contact area between the substrate and the elastomer, which allows us to calculate the adherence energy, in J/m^2. Table 12.5 presents the adherence energy of elastomer samples, before and after extraction of free chains [50].

Tack results show a higher adherence energy for PDMS E2 and E2′ compared to E1 and E1′. The higher chain mobility of samples E2 and E2′ (lower stiffness, more free and/or pendant chains) will induce a better contact with the substrate, at a molecular scale, and consequently a greater number of interactions per area with the glass substrate.

Moreover, the mechanical separation will activate viscoelastic dissipation: chains are indeed extended and these motions and disentanglements (and probably breaking) are more dissipative for E2 and E2′, due to the higher chain length. A higher number of interfacial interactions, which will transfer stress and consequently activate dissipative chain motions, is able to explain the higher adherence of PDMS E2 and E2′ compared to E1 and E1′. This result is in good correlation with nanoscale adhesion results, which have shown a decrease of adherence when the crosslinking density was increased.

Table 12.5 also underlines a significant effect of separation speed with an increase of the adherence energy with speed, explained by the rate-sensitivity (viscoelasticity) of large-scale interfacial chain motions [39, 50].

Additionally, elimination of free chains induces a higher adherence energy, both for PDMS E1′ and E2′ (especially for a high separation speed), contrary to nanoscale results where adhesion was higher before extraction, especially for E2 compared to E2′. Free chains exhibit a high mobility (which favors adsorption onto the substrate) but they are not chemically bonded to the network, avoiding an efficient stress transfer. Free chains, due their low cohesion, could indeed act like a weak boundary layer: their adsorption onto the glass is important, but the strength of the created interface is low. At nanoscale, adsorbed chains governed the friction resistance, and the low strength of this kind of "viscous layer" does not affect friction, due to the low size of

Table 12.5. Adherence energy (in J/m^2, ± 0.1) of PDMS for two separation speeds (normal force $= 1\,N$, initial contact time $= 300\,s$)

Adherence energy (J/m^2)	$V = 1\,mm/min$	$V = 100\,mm/min$
E1	0.5	1.0
E1′	1.2	2.0
E2	1.7	3.7
E2′	1.7	7.0

the contact. On the contrary, pendant chains, which are still present after extraction, seem to play a major role in macro-adhesion: they have a free extremity (mobility and consequently adsorption ability) and the other extremity is chemically bonded to the network (efficient stress transfer).

Macro-friction properties have been quantified with a translation tribometer, which measures the friction force, and the contact area evolution between the PDMS hemispheres and the glass substrate. Friction stress (in N/mm^2) is calculated by dividing each force value by the corresponding contact area. Table 12.6 reports macroscale friction results for PDMS samples [39, 48].

The most surprising result is that a higher friction stress is measured for E1 compared to E2, despite the greater adherence of E2. During friction, the permanent contact (induced by the applied normal force) could act like a "forced wetting" and compensate the lower adsorption and adhesive contact of E1. Moreover, the lower content of free and pendant chains of E1 (compared to E2) is able to induce a direct and efficient contact between the crosslinked network and the glass substrate. The network will be then directly constrained and deformed during friction, thus increasing the bulk dissipation.

The longer chains of E2 (especially free and pendant chains) can be oriented in the friction direction by the interfacial shear stress. Previous study has indeed evidenced an anisotropy of chains after friction (orientation of transferred chains onto the substrate) [51]. This alignment could favor the sliding (self-lubrication), by decreasing the bulk viscoelastic dissipation and consequently the friction resistance. Friction speed has a negligible effect taking into account experimental errors.

The effect of extraction on macro-friction is globally slight, with quite identical values before and after extraction. Macro-adhesion was higher after extraction of free chains, especially for E2' compared to E2, and nanoscale adhesion and friction were lower after extraction.

To summarize the macroscale results, a higher adherence is measured for PDMS E2 compared to E1. The effect of free chain elimination is great, with an increase of adherence after extraction. A higher separation speed induces an increase of adherence energy. Macroscale friction results indicate a higher friction resistance for PDMS E1 compared to E2, and a slight effect of extraction and friction speed.

Nanoscale behavior is globally different from macroscopic properties. Both scales are very sensitive to molecular parameters like crosslinking degree (or chain

Table 12.6. Friction stress (in N/mm^2, $\pm 0.01\,N/mm^2$) for PDMS in contact with glass, for different friction speeds (normal force $= 1\,N$)

PDMS	Friction stress (N/mm^2)	
	25 mm/min	125 mm/min
E1	0.22	0.19
E1'	0.22	0.21
E2	0.07	0.07
E2'	0.07	0.07

length) and the presence of free chains. However, their effect is different at macro and nanoscale. Mobility of free chains induces a great adsorption onto the AFM tip, increasing nano-adhesion and friction. But their low cohesion generates a decrease of macro-adhesion, like a weak boundary layer, their effect on macro-friction being lower. A lower crosslinking degree allows us to increase macro-adhesion (longer chains are better adsorbed and their motions are more dissipative) but this decreases macro-friction (due to chain orientation phenomena). A higher crosslinking degree increases nanoscale adhesion and friction, due to the better adsorption of longer chains onto the tip. The experimental results underline the major role of molecular parameters such as free and pendant chains. Hence, nanoscale behavior appears to be more sensitive to the adsorption ability due to molecular mobility.

The same molecular parameter is then able to increase or decrease adhesion and friction, depending on the solicitation scale. In that specific case, both macro and nanoscale are sensitive to the same molecular parameter, even if its consequence is different at both scales.

However, AFM is sometimes able to detect a change in nano-friction, despite a similar macro-friction. Sensitivity of AFM to molecular structure has for example allowed us to underline the influence of chain length of polystyrene on nano-friction [52]. Chemically modified AFM tips were used in order to investigate nano-friction of different atactic polystyrenes varying by their molecular weight. Friction between the polystyrene films and a hydrophobic tip (methyl-terminated grafted layer) or hydrophilic tip (hydroxyl-terminated tip) was quantified by using AFM as a function of sliding velocity and normal force. Self-assembled monolayers (SAMs) formed by adsorption of organic molecules on a solid surface were used to modify the surface chemistry of AFM tips. This method allows us to modify the surface chemistry without varying the roughness, geometry, and mechanical properties. Hydrophilic tips were obtained by immersion in a piranha solution (H_2SO_4/H_2O_2) and hydrophobic tips were prepared by a chemical grafting of hexadecyltrichlorosilane onto a hydroxylated tip.

Experimental results show that the friction coefficients measured with a hydroxylated tip were always larger than those obtained with a hydrophobic tip, indicating a relationship between nano-friction and interfacial interactions. Elsewhere, a higher friction was obtained for the higher molecular weight polystyrene in contact with hydroxylated tips. However, differences between polystyrenes become negligible in the case of hydrophobic tips. Nano-adhesion was also quantified and corresponding results also underline higher adhesion for the hydroxylated tips in contact with polystyrenes compared to hydrophobic tips. Moreover, a greater adhesion is measured for the higher molecular weight polystyrene in the case of hydrophilic tips, the effect of polystyrene chain length being negligible in the case of hydrophobic tips.

The effect of tip chemistry was in good correlation with macro-friction results, performed on polystyrene cylinders sliding against hydrophilic and hydrophobic silicon wafers: a higher friction is obtained, at both scales, for the hydrophilic surface (tip or wafer) [52,53]. However, differences between polystyrenes were negligible at macroscale, whatever the substrate (hydrophilic or hydrophobic). That point underlines the sensitivity of AFM to detect nanoscale adhesion or friction differences. Nanoscale measurements are indeed more sensitive to molecular variation such as chain length. Polystyrene is a glassy polymer at room temperature. The low mobility state could explain the fact that macroscale friction measurements are not able to detect dissipation difference. The AFM tip is able to probe the local mobility, even if a minimum level of adhesion is required to detect a difference, as shown by the necessity to use hydrophilic tips to discriminate polystyrenes.

The influence of the nature of interfacial interactions on the sliding friction of polymers can be unfortunately more complex. A previous study has investigated the macro-friction behavior of PDMS elastomers in contact with a hydrophilic substrate (hydroxylated silicon wafer) and a hydrophobic substrate (silicon wafer grafted with a CH_3-terminated silane) [54]. Experimental results indicated that the friction coefficients of PDMS against hydrophilic and hydrophobic substrates are different for low friction speed, with a significantly greater friction coefficient for hydrophilic wafer. But, for higher speeds, friction coefficients obtained with both types of substrates become identical, as illustrated on Fig. 12.8, which represents the evolution of the friction coefficient with speed for a PDMS elastomer in contact with hydrophilic and hydrophobic substrates [54].

The friction coefficient is then increased as a function of speed in the case of the hydrophobic substrate and is decreased for the hydrophilic substrate: the effect of speed is strongly dependent on the substrate chemistry, and vice versa. At high speed, the friction coefficients of both hydrophobic and hydrophilic systems are identical: the effect of the interface becoming negligible. Specific rheological behavior of this confined interfacial layer has been proposed to explain these complex speed dependences. At low speeds, interfacial interactions control the friction: the role of the substrate surface chemistry is then significant, with a higher friction against hydrophilic surfaces. At higher speeds, the friction becomes governed by the rheological behavior of the constrained polymer surface and the influence of the substrate surface becomes negligible. Higher shear rate and stress are indeed able to induce a chain orientation at the PDMS surface [51], which will probably modify the rheological behavior of the polymer interface, with a lower shear resistance. At low speed, adhesion is dominant and stronger interactions between PDMS and hydrophilic substrate will activate a viscoelastic dissipation mechanism, leading to a high friction. When the speed is increased, chains are oriented in the friction direction. The shear resistance of this oriented layer could be lower (or less dissipative)

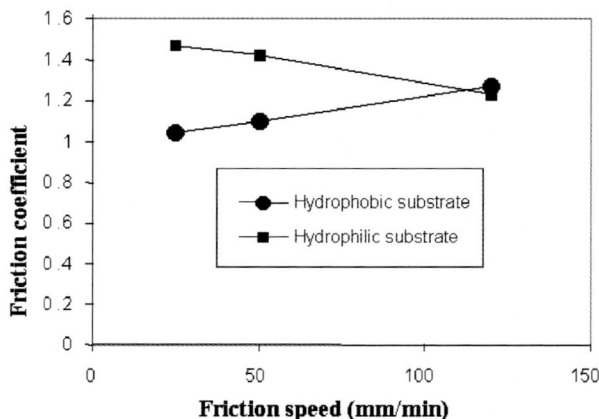

Fig. 12.8. Evolution of the friction coefficient with speed for a PDMS elastomer in contact with hydrophilic and hydrophobic substrates

than hydrophilic wafer/PDMS interactions, inducing a decrease of friction with speed for the hydrophilic substrate. However, the shear (or sliding) resistance of this oriented layer could be greater (more dissipative) than the weak hydrophobic wafer/PDMS interactions, explaining the increase of the friction with speed for the hydrophilic substrate. At high speed, shear probably occurs within this confined layer (not exactly at the polymer–substrate interface), leading to similar friction for both substrates. Analysis of the transfer layer (chains deposited onto the substrate during friction) could allow a better understanding of the involved mechanisms [51]. A complex competition between interfacial interactions and rheological behavior of the confined polymer surface could therefore explain the increase or decrease of the friction coefficient, depending on the substrate chemistry. This complex behavior can probably not be generalized to other polymers. Elastomers are soft materials and a balanced competition between interfacial interaction and shear resistance of the confined surface chains is then possible.

The effect of interface chemistry on friction is thus not so evident: polymer friction against hydrophilic and hydrophobic surfaces can be identical or different, depending on the experimental conditions. Friction speed is also able to play a major role, through its influence on polymer rheology and viscoelasticity.

To conclude, investigation of nano and macroscale adhesion and friction of model elastomers has underlined the major role of molecular parameters such as free and pendant chains. However, their effect is subtle, the same structural parameter being able to increase nanoscale adhesion and friction but decrease macro-adhesion.

In addition, chemically modified AFM tips can be used to magnify the effect of interface and discriminate close molecular structures, for which macroscale behaviors are identical. The role of interfacial interactions is then intricate and model tips but also controlled polymer chemistries and structures would allow us to better understand the molecular mechanisms responsible for polymer friction. Finally, as each polymer exhibits a specific behavior, multi-scale approaches are essential to better understand friction and adhesion of these complex materials.

References

1. Liu H, Bhushan B (2002) Ultramicroscopy 91:185
2. Liu H, Bhushan B (2003) Ultramicroscopy 97:321
3. Kaijiyama T, Tanaka K, Takahara A (1997) Macromolecules 30:280
4. Hammerschmidt JA, Gladfelter WL, Haugstad G (1999) Macromolecules 32:360
5. Hammerschmidt JA, Moasser B, Gladfelter WL, Haugstad G (1996) Macromolecules 29:8996
6. Wang D, Ishida H (2006) C. R. Chimie 9:90
7. Scott WW, Bhushan B (2003) Ultramicroscopy 97:151
8. Carpick RW, Ogletree DF, Salmeron M (1997) Appl Phys Lett 12:70
9. Tanaka K, Takahara A, Kaijiyama T (1997) Macromolecules 30:6626
10. Tanaka K, Taura A, Takahara A, Kaijiyama T (1996) Macromolecules 29:3040
11. Lee W (1999) Polymer 40:5631
12. Paiva A, Sheller N, Foster MD, Crosby AJ, Shull KR (2000) Macromolecules 33:1878
13. Gauthier S, Aimé JP, Bouhacina T, Attias AJ, Desbat B (1996) Langmuir 12:5126
14. Feldman K, Fritz M, Hahner G, Marti A, Spencer ND (1998) Tribol Int 31:99

15. Bhushan B, Israelachvili JN, Landman U (1995) Nature 374:607
16. Gao C, Vo T, Weiss J (1998) J Tribol 120:369
17. Bhushan B (1995) Tribol Int 28:85
18. Bhushan B (1998) Tribology issues and opportunities in MEMS, Kluwer Academic Pub-
    lishers, Dordrecht
19. Bhushan B, Sundararajan S (1998) Acta Mater 11:793
20. Paiva A, Sheller N, Foster MD (2001) Macromolecules 34:2269
21. Bhushan B, Dandavate C (2000) J Appl Phys 87:1201
22. Tallal J, Gordon M, Berton K, Charley AL, Peyrade D (2006) Microelectron Eng 83:851
23. Stifter T, Marti O, Bhushan B (2000) Phys Rev B 62:13667
24. Bhushan B, Tokachichu DR, Keener MT, Lee SC (2006) Acta Biomater 2:39
25. Ta NS (2005) Ultramicroscopy 105:238
26. Zhang Newby BM, Chaudhury MK (1997) Langmuir 13:1805
27. Amouroux N, Petit J, Léger L (2001) Langmuir 17:6510
28. Israelachvili JN (1991) Intermolecular and surface forces, Academic Press, London
29. Yoshizawa H, Israelachvili JN (1994) Thin Solid Films 246:71
30. Heuberger M, Drummond C, Israelachvili JN (1998) J Phys Chem B 102:5038
31. Yamada S, Israelachvili JN (1998) J Phys Chem B 102:234
32. Chen YL, Helm CA, Israelachvili JN (1991) J Phys Chem 95:10736
33. Heuberger M, Luengo G, Israelachvili JN (1999) J Phys Chem B 103:10127
34. Ghatak A, Vorvolakos C, She H, Malotky DL, Chaudhury MK (2000) J Phys Chem B
    104:4018
35. She H, Malotky D, Chaudhury MK (1998) Langmuir 14:3090
36. Chaudury MK, Owen MJ (1993) J Phys Chem 97:5722
37. Deruelle M, Léger L, Tirrell M (1995) Macromolecules 28:7419
38. Amouroux A, Léger L (2003) Langmuir 19:1396
39. Galliano A, Bistac S, Schultz J (2003) J Colloid Interf Sci 265:372
40. Lee SW, Yoon J, Kim HC, Lee B, Chang T, Ree M (2003) Macromolecules 36:9905
41. Gasco MC, Rodriguez F, Long T (1998) J Appl Polym Sci 67:831
42. Zhang S, Lan H (2002) Tribol Int 35:321
43. Liu Y, Evans DF, Song Q, Grainger DW (1996) Langmuir 12:1235
44. Wallace WE, Fischer DA, Efimenko K, Wu WL, Genzer J (2001) Macromolecules 34:5081
45. Luengo G, Schmitt FJ, Hill R, Israelachvili J (1997) Macromolecules 30:2482
46. Luengo G, Israelachvili JN, Granick S (1996) Wear 200:328
47. Israelachvili JN, Kott SJ (1989) J Colloid Interf Sci 129:461
48. Bistac S, Galliano S (2005) In: Possart W (ed) Adhesion – current research and application,
    Wiley-VCH, Weinheim, p 59
49. Bistac S, Galliano A (2005) Tribol Lett 18:21
50. Galliano A, Bistac S, Schultz J (2003) J Adhes 79:973
51. Elzein T, Galliano A, Bistac S (2004) J Polym Sci, part B 42:2348
52. Bistac S, Ghorbal A, Schmitt M (2007) J Phys: Conf Ser 61:130
53. Bistac S, Ghorbal A, Schmitt M (2006) Prog Org Coat 55:345
54. Bistac S, Ghorbal A, Schmitt M (2007) Fundamental of friction and wear on the nanoscale,
    Springer-Verlag, p 647

13 Mechanical Characterization of Materials by Micro-Indentation and AFM Scanning

Gabriella Bolzon · Massimiliano Bocciarelli · Enzo J. Chiarullo

Abstract. As the realism and the interpretative capacity of constitutive models employed to describe the material response in different mechanical systems grow, the number of parameters to be experimentally determined becomes more and more numerous and less amenable to direct separate measurement. The combination of experiments, their simulation (by computer or, sometimes, by analytical solutions) and inverse analysis represents then the envisaged approach to material characterization, which has been applied to indentation tests mainly exploiting the traditional indentation curves as the source of experimental data. Recent literature shows that geometrical data concerning the imprint generated by indentation represents important supplementary information to be exploited for mechanical characterization purposes, with satisfactory results in terms of accuracy, robustness, and of the number of identifiable material characteristics. The present contribution intends to illustrate this novel methodology with the aid of some selected applications.

Key words: Materials characterization, Indentation, AFM scanning

13.1
Introduction

The main purpose of experimental mechanical tests on materials is the quantitative assessment of the parameters included in the constitutive models to be used for structural analysis and design in many engineering situations. Indentation tests represent a practical methodology for material characterization since not only hardness, like in the origin of this experimental methodology [1,2], but also material and surface properties such as elastic modulus, yield strength, and scratch resistance can be inferred from the indentation curves (imposed force versus penetration depth) derived from laboratory tests on various materials at different scale.

At present, indentation procedures are widely used in industries dealing with micro and nano technology and with thin-film coatings [3,4] as well as for the characterization of local constitutive behavior in a large and still growing number of application contexts: in MEMS, where thin gold layers are deposited on silicon to allow electrical connections with other devices; in metal-working tools, precision ball bearings and computer magnetic storage systems, where surfaces are coated by hard ceramic and diamond-like materials in order to improve wear and corrosive resistance; in lubrication systems, where friction and wear can be lowered by the surface deposition of soft, low shear-strength materials such as graphite; in turbines or solar engineering, where multi-layered or functionally graded coatings act as thermal or control barriers [5, 6]; in the automotive industry, where high-pressure die-cast magnesium lightweight alloy and TRIP steel components exhibit performances depending on the local stress–strain response of the skin/interior microstructure and of the retained austenite, respectively [7–9]; in medical, control, seismic-resistant design and retrofitting applications, where shape memory alloys have strain-recovery properties depending on thermal processing [10].

As the most commonly used coating technique consists of vapor deposition in a variety of modes, parameters like deposition rate, temperature, gas pressure, and impurity level affect the characteristics and the mechanical properties of the coating and of the interface with the underlying material. The mechanical properties of thin films can be very different from those of the corresponding bulk materials also because of their unique microstructure, large surface to volume ratio, and reduced dimensions. Indentation is therefore exploited, in order to perform material mechanical characterization, for its capability of deforming materials at a very low scale. A comprehensive review of mechanical applications of this technique can be found in the book by Bhushan [11] and in the review paper by Bhushan and Li [12].

In order to obtain accurate estimates of the material properties in layered systems, it is crucial to distinguish between film and substrate contributions to the overall specimen response. In the approach most widely used, penetration depth is kept smaller than some critical value that depends on film-substrate properties (see [13, 14] and references therein) and approaches developed for homogeneous specimens are followed. However, it could be difficult to obtain reliable parameter estimates for very thin films, since the suggested indentation depth could be too small to consider the indentation response representative of the macroscopic material behavior. Bocciarelli and Bolzon [15] have found that, in the case of perfect adhesion between the layers, the consideration of the residual imprint geometry is beneficial for the identification of material properties of film–substrate systems since the presence of the substrate can be accounted for and its mechanical properties identified, if unknown, together with those of the coating.

Thermal, electrical, and mechanical characteristics of layer systems are of great interest for their applications, but interface resistance is a further important macroscopic property to be verified, see for example [16]. Existing methods for measuring adhesion are mostly qualitative and can be used for comparative purposes only, see [17,18] and references therein. For instance, interfacial properties are often identified by means of the scratch test, which consists of moving a diamond conical tip across the external film surface at constant velocity, with increasing vertical load leading to the coating detachment, observed by either optical microscopy or scanning electron microscopy. The maximum applied force is referred to as the critical load and is assumed to be a measure of the adhesive strength.

While comparative tests are appropriate for some applications, a quantitative estimation of the interface strength and of the fracture energy resulting from the film–substrate coupling is required in several situations, for example in view of numerical simulations aimed at life-time predictions. These properties can be evaluated through Vickers indentation performed at the interface between the layers on a specimen cross section. The length of the cracks which can be observed after unloading, measured from the center of the imprint, is in fact assumed to reflect the material toughness in the linear elastic fracture mechanics (LEFM) context, according to a theory developed by Lawn et al. [19] for bulk materials and subsequently improved [20]. Experimental investigations have, however, shown a systematic dependence of toughness value estimated in this way on the indentation load and as the corresponding crack length also in rather brittle situations. This result, which is inconsistent with LEFM assumptions, is likely to originate from various error sources: the experimental difficulties related to a reliable measurement of the actual crack length; the existence of a cohesive zone at the crack tip; the possible influence of residual stresses; see, for example [6, 16, 21]. The prediction of interfacial toughness by this method can be likely improved by the AFM mapping of the displacement field in the vicinity of the crack tips, as recently proposed, for example in [22, 23].

Delamination induced by instrumented indentation performed on the external coating surface of film–substrate systems has been recently considered to evaluate adhesion properties [24–29]. A kink in the indentation curve is often observed at the initiation of through-thickness cracks, for example in carbon layers deposited on a silicon substrate, information that can be exploited to estimate fracture energy. Data gathered from the indentation curve and from the measurement of the horizontal displacement field around the residual imprint can also be combined to this purpose, see for example [30–32].

As the realism and the interpretative capacity of constitutive models grow, together with the complexity of the investigated systems, material parameters become more numerous and less amenable to direct separate measurement. Therefore, the combination of experiments, their simulation (by computer or, sometimes, by analytical solutions) and inverse analysis represents an approach to material characterization with roots in elastic–plastic structural mechanics that is increasing in popularity [33–37]. Such methodology has been applied to indentation tests in the recent literature, mainly exploiting the traditional indentation curves as a source of experimental data; see, among others: [38–46].

Capehart and Cheng [47] have, however, shown that even low noise level may preclude accurate identification when the only available experimental information consists of loading–unloading curves in conical indentation. Their results agree with the remarks by Venkatesh et al. [39] and by Dao et al. [42], according to whom plastic properties extracted from indentation curves can be strongly influenced by small variations of input data. Cheng and Cheng [48] showed that different parameter sets corresponding to the same indentation curves may exist, which return rather different uniaxial behavior. Recently, Chen et al. [49] have further demonstrated that multiple indentations by different tip geometry, as proposed in [50–52], do not always provide a remedy to this limitation.

Fairly satisfactory identification results were reported for various material properties in the case of spherical indentation, following the analytical method proposed in [53], which however requires that the yield stress or, at least, the ratio of yield stress to Young's modulus be known to determine the strain hardening exponent. Further, friction, between indenter tip and specimen surface, should be carefully accounted for, even if this quantity is usually difficult to evaluate, since it can significantly affect the applied load versus penetration depth curves resulting from instrumented indentation [54, 55].

Different studies (e.g., [56, 57]) showed that Young's modulus, initial yield strength, and the strain-hardening exponent, besides friction, influence the amount of piling-up or sinking-in around the indenter tip. Therefore, the material characterization methods proposed in [58–62] consider the imprint profile or the digital reconstruction of the indenter tip and of the residual imprint, mapped by scanning interference microscopy, to accurately determine the contact area or for validating inverse analysis results. Bolzon et al. [63] have further shown that geometrical data concerning the imprint generated by indentation tests represent an important experimental information, supplementary to the usual one consisting of the indentation curves, to be exploited for parameter identification of isotropic elastic–plastic-hardening constitutive models, with satisfactory results in terms of accuracy, robustness and of the number of identifiable parameters, including coefficient of friction between the specimen surface and the indentation tool.

Indentation of anisotropic solids (because of material micro-structure or production processes like extrusion or lamination) has also been the subject of research carried out from both theoretical and applicative standpoints; see, for example: [64–70]. In this case, the shape of the imprint plays a fundamental role in the determination of the material symmetry axes, while indentation curves can only reflect average constitutive properties [71, 72].

The shape of the residual imprint represents an essential information also to estimate the directionality of stress states of biaxially strained specimens, since its profile at the contact boundary is clearly sensitive to the presence of initial stresses [73, 74].

Suresh and Giannakopoulos [75] proposed a method for the evaluation of average surface stresses based on the measurement of the difference in contact areas between loaded and reference specimens, a practical method for stress values close to the yield limit since the influence on the contact area can be relatively small [76, 77]. Lee et al. [78] have shown that the amount and the observed anisotropy in pile-up morphologies around the impressions left by an axi-symmetric indenter can be related to the ratio of the applied principal stresses. Bocciarelli and Maier [79] have further demonstrated that the consideration of the whole geometry of the residual imprint represents an improvement with respect to the methods available in the literature for initial stress recovery.

At present, instruments apt to measure the local profile of an indented specimen are fairly frequently available in laboratories for different purposes. In particular, atomic force microscopes (AFM) can provide an accurate mapping of the geometrical consequences of micro indentation tests [80–83], besides being sometimes used as loading tools (see, e.g., [84–89]), to evaluate the local mechanical properties of metals, polymers, and composites (e.g. [85, 90–92]), to understand the local deformation mechanisms under load (e.g. [6, 83, 85, 93–98]), to evidence and investigate the complex size effects arising at the nanoscale (see, e.g., [93, 99–107]).

Scanning force microscopes have been profitably used to map the actual geometry of indenter tips, one of the major sources of uncertainty in micro- and nanoindentation tests, to calibrate area functions to be inserted in analysis methods developed for perfect model shapes or in finite element simulations of the indentation test, to return more accurate predictions of the material response [108–110].

The material characterization approach to be illustrated in the present Chapter through various applications gathered from the Authors' experience, exploits all available information from load-penetration depth curves returned by instrumented micro-indenters, as well as from the mapping of the residual imprint and of displacement field in the surrounding surface of the specimen, by combined AFM/DIC techniques [31,94,111–113]. These experimental data are used, in combination with finite element simulations of the indentation tests and with inverse analysis, to calibrate material parameters in popular constitutive models developed for bulk materials and interfaces [15,64,73,80]. Emphasis will be given to the relatively novel data achievable by imprint mapping.

The considered experimental techniques will be introduced first. A brief description is then provided of the conventional finite element (FE) models generated and of the constitutive laws adopted in the simulations intended to evaluate the consequences of indentation in terms of residual (irreversible) displacements and deformations.

Available measurements can be better exploited by a proper design of the laboratory test and of the identification procedure, based on the outcome of some sensitivity analyses, which will be outlined as well; details can be found, for example in [114].

13.2
Experimental Techniques

13.2.1
Micro-Indentation

Indentation tests at various scales are routinely exploited in many laboratories and industrial environments. Instrumented indentation returns the reaction force exerted on the tip versus its penetration depth. The maximum load for commercial micro-indenters is in the range 0.5–20 N; the penetration depth commonly varies from about 0.1 nm to about 100 μm; these thresholds can be overcome by special devices. A non-exhaustive list of available instrumentation can be found in [11,12].

Figures 13.1 and 13.2 reproduce two representative examples of the material response to Vickers (pyramidal) indentation, relevant to zirconia and copper specimens. In many cases, the observed experimental scatter represents the prevailing uncertainty source, much higher than that given by the nominal equipment resolution, which is of the order of 1 μN or less for the load, less than 0.01 nm for the displacements.

Traditional semi-empirical formulae [38, 46, 115] relate Young's modulus E to the initial slope of the unloading indentation curve, S, and to the contact area at maximum penetration depth, A, as follows:

$$\frac{1 - \nu^2}{E} + \frac{1 - \nu_{in}^2}{E_{in}} = \frac{\gamma \sqrt{A}}{S}. \tag{13.1}$$

In the above equation, E_{in} and ν_{in} denote the elastic moduli of the indenter material while γ represents a correction factor which depends on the indenter shape. This calibration parameter plays an important role on the elastic modulus resulting from

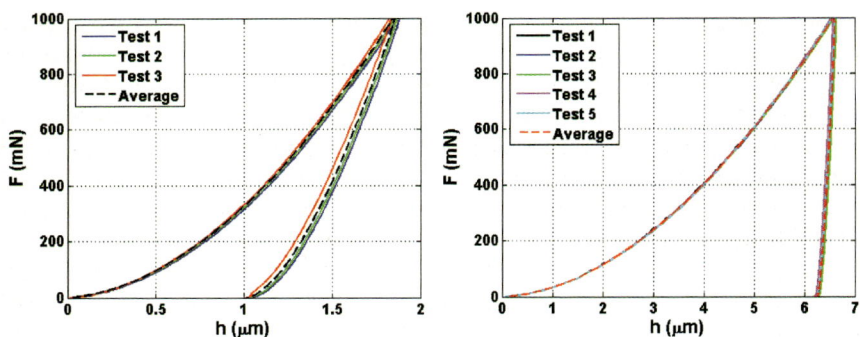

Fig. 13.1. Vickers indentation of zirconia (*left*) and copper (*right*) specimens

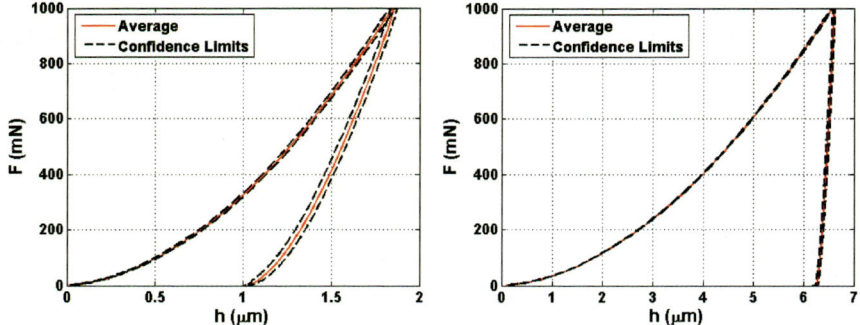

Fig. 13.2. Average indentation curves and the corresponding scatter deduced from the experimental data in Fig. 13.1

Eq. (13.1), which is often automatically returned by the software implemented in available experimental equipment for given Poisson's ratio ν [116].

An accurate estimation of E is subordinated to a precise evaluation of the slope S and of the area A, which influences also hardness. This occurrence has motivated the most-recent investigations concerning the geometry of the imprint [58–60] and, for a small penetration depth, of the indenter tip [108–110].

Several other material parameters can be inferred from the indentation curves, especially through combined numerical-experimental techniques. Some of them have been briefly reviewed in the introduction.

13.2.2
AFM Scanning

Atomic force microscopes are instruments suitable for describing the geometrical consequences of indentation at the micro-scale, in terms of the deformed configuration in and around the residual imprint.

Principles and challenges of AFM are reviewed, for example in a recent paper by Giessibi [117]. The variations in signal related to the load-displacement response of the elastic cantilever supporting the AFM tip is used to generate images of relative contrast of the topography of the scanned surface and to return a possible digital reconstruction of it. Typical monitored areas are in the range 0.5×0.5 to $200 \times 200\,\mu m^2$, depending on the equipment characteristics, with corresponding orthogonal displacements varying between 0.4 and $10\,\mu m$ and nominal resolution which can be less than 0.1 Å. Clearly, it is not trivial to obtain an absolute calibration of the corresponding measuring systems [85].

Examples of the output returned by AFM scanning are shown in Fig. 13.3, concerning the topography maps of the residual imprints left on the metal phase of a copper–alumina composite and on yittria stabilized zirconia (YSZ) by Vickers (pyramidal) and Rockwell (conical) indentation, respectively. Data about the originally mapped areas ($15 \times 15\,\mu m^2$ and $80 \times 80\,\mu m^2$, respectively) are returned in matrices of 512×512 elements.

The discrepancy between nominally identical profiles, see Fig. 13.4, provide information about material anisotropy and/or tip imperfection, account taken of the different data acquisition mode: accuracy is usually higher in the fast scanning direction, Fig. 13.5. The instrumental scatters evidenced in Fig. 13.4 usually represent the major error source in the input data for material characterization procedures.

The expected symmetries of the residual imprints can be exploited to check the accuracy of the data returned by the instrumentation. In the case of Vickers inden-

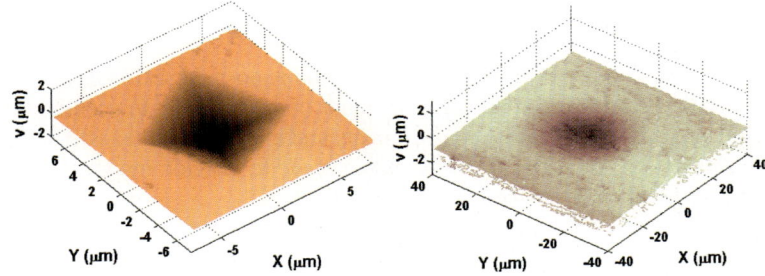

Fig. 13.3. AFM topography map of the residual imprint of: (*left*) Vickers indentation on the copper phase of a metal–ceramic composite (maximum applied load 250 mN); (*right*) Rockwell C indentation of zirconia (maximum applied load 25 N)

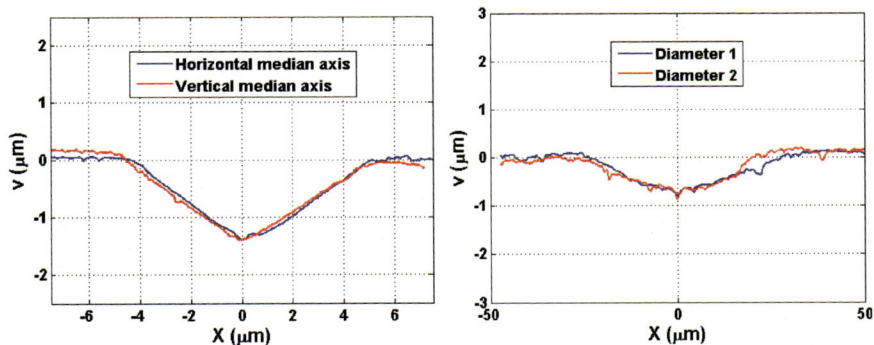

Fig. 13.4. Profiles of the residual imprint shown in Fig. 13.3, as returned by AFM scanning

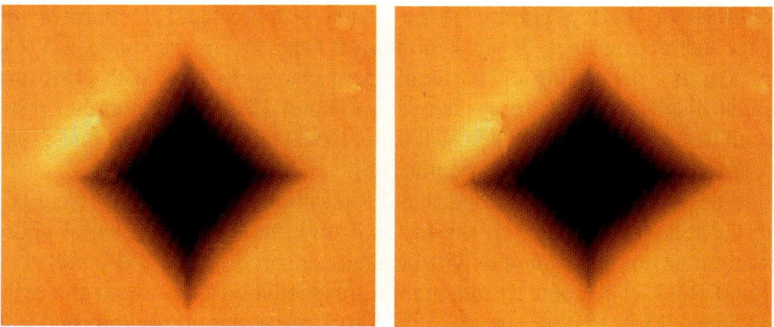

Fig. 13.5. AFM images of the same residual imprint left by Vickers indentation on zirconia: fast scanning (0.5 Hz) along the horizontal (*left*) and the vertical (*right*) direction

tation, geometrical information can be reduced to the average quarter (for instance, with respect to median axes) of the original mapped surface, combined with its standard deviation, see Fig. 13.6. The average and the confidence limits of the diagonals and of a median axis are shown in Fig. 13.7. The confidence interval is here defined by adding and subtracting the square root of the variance to the average. Similarly, the conical geometry of the residual imprint left on an isotropic material by the Rockwell indenter can be reduced to an axial-symmetric profile with a confidence interval, as shown in Fig. 13.8.

AFM images of the consequences of micro- and nano-indentation tests have been used to examine plastic deformation mechanisms under load and to accurately determine residual areas for hardness measurements and elastic modulus estimation; see, for example [80–83]. Several efforts are being made to provide integrated scanning capabilities at AFM scale to indentation instruments.

The high-resolution micrographs obtained by AFM constitute an ideal input for a digital-image correlation (DIC) technique, a recently developed robust methodology to return full-field displacement measurements in real-time on the surface of

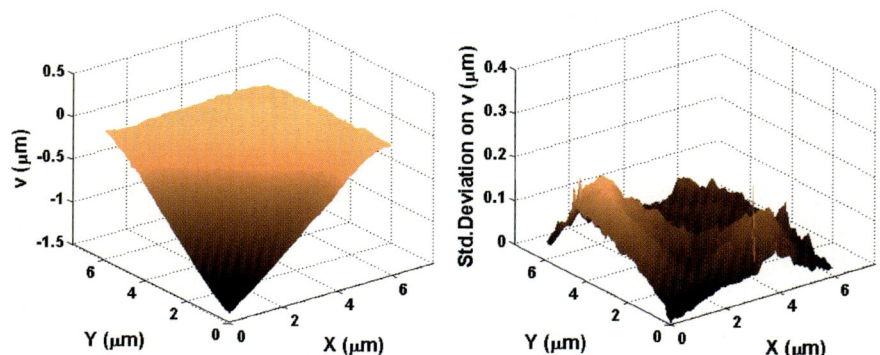

Fig. 13.6. The average quarter (*left*) and the relevant standard deviation (*right*) of the imprint in Fig. 13.3 (*left*)

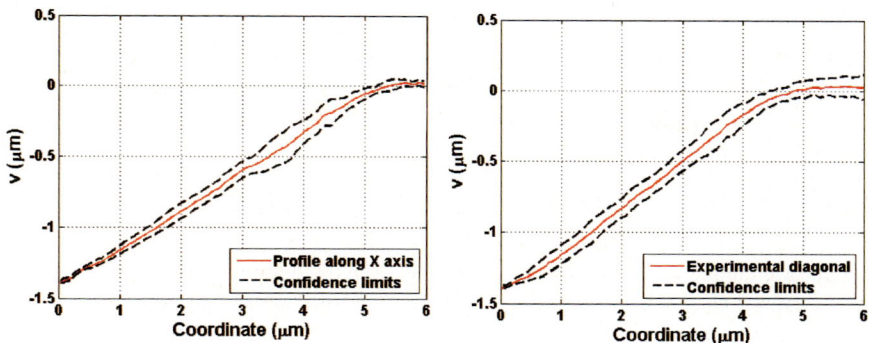

Fig. 13.7. The semi-profiles, along the diagonal and along the *x*-axis, of the average quarter in Fig. 13.6 (*left*) with the corresponding confidence limits

a deforming body. DIC is based on the comparison of different digital records of the same physical surface before and after deformation [118]. Its implementation requires high contrast random patterns to be created on the specimen surface, naturally provided on AFM images by the surface roughness that acts as distributed markers [113, 119].

Since displacements are recovered by the comparison of digital images, their accuracy is controlled by the hardware resolution, which usually amounts to integer multiples of an image pixel. Precision can be enhanced, for example by the use of non-linear spline interpolations [120]. Among the error sources of image acquisition by scanning techniques (SEM-AFM), the different drift between the scanner and the object should be further considered, as it can lead to micrograph distortion [31]. In most cases, spatially resolved displacements and strains can be evaluated with an accuracy of about 0.1 image pixel and 0. 1%, respectively, also depending on the scanning direction. According to the experience gathered by Knaus et al. [112], strains smaller that 0.02% can be resolved with current AFM instrumentation with 1024×1024 pixel resolution.

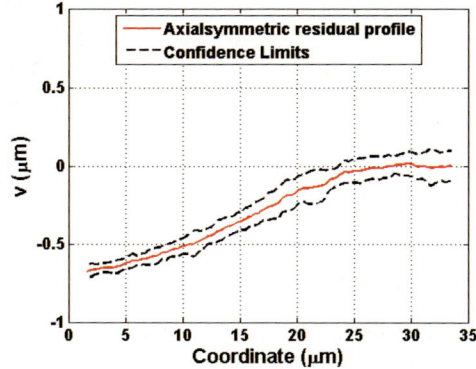

Fig. 13.8. The axial-symmetric profiles and the confidence interval deduced from the median axes and two diagonals of the imprint in Fig. 13.3 (*right*)

Information gathered by the application of DIC techniques to AFM images can be profitably combined to the topography maps directly provided by AFM scanning, to return the full 3D displacement field with one measurement tool only, see [112].

13.3
Inverse Analysis for Materials Characterization

Indentation curves and imprint geometry reflect the mechanical properties of the considered material in a rather involved manner. Therefore, the interpretation of these results and the recovery of constitutive parameters benefit significantly from sensitivity analyses and identification procedures resting on the combination of the test with its (often numerical) simulation, and from the comparison between the available experimental information (e.g., the indentation curve and the displacement field in and around the residual imprint) and the corresponding quantities returned by computations; see, for example: [7, 35, 38–46, 64, 73, 80].

A brief description is here provided of the conventional finite element discretizations generated and of the material models more frequently adopted in the simulation of indentation tests.

13.3.1
Simulation of the Test

The consequences of an indentation test, in terms of load-displacement curve and of residual (irreversible) displacements and deformations produced in different material systems, are often analyzed by means of widely used computing codes based on the presently popular finite element method; see, for example [42, 47, 48, 53, 121, 122].

Figures 13.9 and 13.10 represent possible FE meshes for the simulation of Vickers (square pyramidal tip with 136° opening angle) and Rockwell (conical indenter with 120° opening angle and 200-μm spherical tip radius) indentation of isotropic materials; the assumed discretization and the boundary conditions are selected in view of the geometric and physical symmetries and of the hypothesis that outside

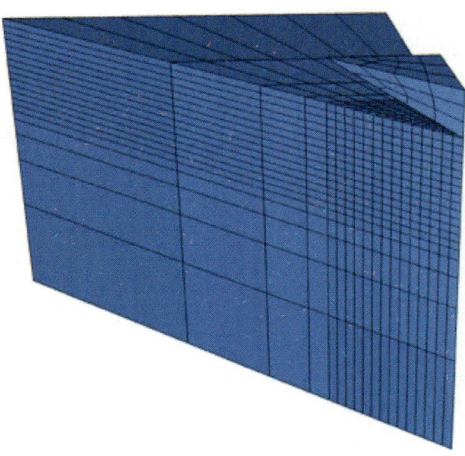

Fig. 13.9. Finite element model of Vickers indentation test on isotropic materials exploiting the symmetries of the problem

Fig. 13.10. Finite element model of Rockwell indentation test on isotropic materials exploiting the symmetries of the problem

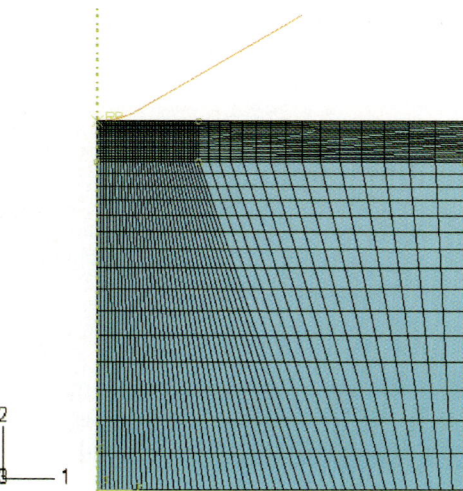

the visualized region the material response will remain in the linear elastic regime. This condition, realistic when the modeled domain dimensions are suitably chosen, permits us to preliminarily condense the response of the specimen portion exterior to the visualized one in a set of elastic constraints that are attributed to the boundary.

The contact interface between the tool and the specimen can be characterized by Coulomb friction without dilatancy, i.e. by a non-associative rigid–plastic interface model. Although this contact has been assumed frictionless in most instances in the available literature [42, 48, 100, 122], the friction coefficient can be regarded as a parameter with a role in the material response, particularly in the extension of the piling-up portion of the indented material around the imprint [53, 55, 63].

An essential ingredient of these computer simulations is the constitutive model embedding the material parameters to be identified on the basis of the experimental data. The classical Drucker–Prager (DP) model (see, e.g., [123, 124]) is suited to describing the mechanical response of ceramics. It can also be adopted for metals and metal–ceramic composites, respectively, through its particularization into the also classic model by Hencky–Huber–Mises (HHM) and its generalization introduced by Bocciarelli et al. [125]; see also [126–128].

DP yield criterion defines the elastic domain of isotropic pressure-sensitive materials in the stress space defined by components σ_{ij}. Its formulation for linear isotropic hardening, where constitutive parameters are functions of the material point position x, reads:

$$f = \sqrt{\frac{1}{2}\sigma'_{ij}\sigma'_{ij}} + \alpha(x)I_1 - K(x) - h(x)\lambda \leq 0 \qquad (13.2)$$

where: σ'_{ij} $(i, j = 1, 2, 3)$ represent the components of the deviatoric stress tensor and I_1 is the first stress invariant. Constitutive parameters $\alpha(x)$ and $K(x)$ at the material point x can be correlated to the compressive and tensile overall material strength

$\sigma_{0c}(\boldsymbol{x})$ and $\sigma_{0t}(\boldsymbol{x})$, respectively, according to:

$$\alpha(\boldsymbol{x}) = \frac{1}{\sqrt{3}} \frac{\sigma_{0c}(\boldsymbol{x}) - \sigma_{0t}(\boldsymbol{x})}{\sigma_{0c}(\boldsymbol{x}) + \sigma_{0t}(\boldsymbol{x})} \tag{13.3}$$

$$K(\boldsymbol{x}) = \frac{2}{\sqrt{3}} \frac{\sigma_{0c}(\boldsymbol{x})\sigma_{0t}(\boldsymbol{x})}{\sigma_{0c}(\boldsymbol{x}) + \sigma_{0t}(\boldsymbol{x})}. \tag{13.4}$$

Parameter $h(\boldsymbol{x})$ governs material hardening, while $\lambda(\,> 0)$ is the cumulative multiplier of the plastic deformations, which evolves according to the classical (complementarity) rules:

$$f \leq 0 \quad \dot{\lambda} \geq 0 \quad f\dot{\lambda} = 0. \tag{13.5}$$

A superimposed dot denotes a rate quantity.

In the hypothesis of small strains, or in the case of large plastic deformations, strain rate components $\dot{\varepsilon}_{ij}$ can be additively decomposed into their elastic (reversible) $\dot{\varepsilon}_{ij}^{e}$ and plastic (irreversible) $\dot{\varepsilon}_{ij}^{p}$ parts:

$$\dot{\varepsilon}_{ij} = \dot{\varepsilon}_{ij}^{e} + \dot{\varepsilon}_{ij}^{p}. \tag{13.6}$$

Irreversible strains are usually assumed to develop orthogonally to a potential surface, here $g(\sigma_{ij}, \boldsymbol{x})$, such that:

$$\dot{\varepsilon}_{ij}^{p} = \dot{\lambda} \frac{\partial g}{\partial \sigma_{ij}} \tag{13.7}$$

$$g = \sqrt{\frac{1}{2}\sigma_{ij}'\sigma_{ij}'} + \beta(\boldsymbol{x})I_1 \tag{13.8}$$

where β represents the dilatancy coefficient. HHM theory is recovered as the internal friction angle α is set equal to zero, and associative plasticity results as $f = g$.

The elastic strain components are related to stresses through the elastic compliances C_{ijrs}:

$$\dot{\varepsilon}_{ij}^{e} = C_{ijrs}\dot{\sigma}_{rs}. \tag{13.9}$$

Overall (including the elastic behavior as well) the model depends on six parameters in the case of isotropic homogeneous materials, namely: elasticity (Young's) modulus E; Poisson's ratio ν and the already introduced coefficients governing the inelastic behavior (α, β, K, h).

For metal–matrix composites, the value of the above-listed macroscopic constitutive parameters can be computed at each material point according to suitable homogenization rules, as a combination of the local counterparts for each component phase and of the volume fraction distribution; see for example [5,15,40,45,46,65,129–131].

Alternative models of material behavior can be simply defined within the same framework by modifying the entries C_{ijrs} of the elastic compliance tensor and the expression of the yield and potential functions, f and g, respectively.

For instance, linear elasticity in an orthotropic material is characterized by nine engineering constants, associated with the material principal directions x_1, x_2, and x_3, defined by the intersections of the three planes of material symmetry: elastic moduli E_1, E_2, and E_3; transversal ratios ν_{12}, ν_{23}, and ν_{31} (taking into account $\nu_{ij}/E_j = \nu_{ji}/E_i$); shear moduli G_{12}, G_{23}, and G_{31}.

The inelastic behavior of orthotropic ductile materials can be represented by the above relationships (13.5), (13.6), (13.7), (13.9), by associative plasticity $(f=g)$ and

$$f = k_{11}(\sigma_{22} - \sigma_{33})^2 + k_{22}(\sigma_{33} - \sigma_{11})^2 + k_{33}(\sigma_{11} - \sigma_{22})^2$$
$$+ k_{23}\sigma_{23}^2 + k_{31}\sigma_{31}^2 + k_{12}\sigma_{12}^2 - 1. \tag{13.10}$$

Relation (13.10) has been proposed by Hill [132] as a straightforward generalization of HHM criterion (13.2).

The six material parameters k_{ij} in Eq. (13.10), which define the elastic domain as $f \le 0$, are related to the yield limits (same in tension and compression) for uniaxial stress along the three axes of orthotropy ($\sigma_{110}, \sigma_{220}, \sigma_{330}$) and to the shear yield stress in the three planes of symmetry ($\sigma_{120}, \sigma_{230}, \sigma_{310}$) according to the following relationships:

$$k_{11} = \frac{1}{2}\left(\frac{1}{\sigma_{220}^2} + \frac{1}{\sigma_{330}^2} - \frac{1}{\sigma_{110}^2}\right); \quad k_{22} = \frac{1}{2}\left(\frac{1}{\sigma_{110}^2} + \frac{1}{\sigma_{330}^2} - \frac{1}{\sigma_{220}^2}\right);$$

$$k_{33} = \frac{1}{2}\left(\frac{1}{\sigma_{220}^2} + \frac{1}{\sigma_{110}^2} - \frac{1}{\sigma_{330}^2}\right) \tag{13.11}$$

$$k_{23} = \frac{1}{\sigma_{230}^2}; \quad k_{31} = \frac{1}{\sigma_{310}^2}; \quad k_{12} = \frac{1}{\sigma_{120}^2}$$

Thus, Eq. (13.10) reduces to the classical HHM expression as $\sigma_{110} = \sigma_{220} = \sigma_{330} = \sigma_{0c} = \sigma_{0t} = \sigma_0$ and $\sigma_{120} = \sigma_{230} = \sigma_{310} = \sigma_0/\sqrt{3}$.

Fracture processes in quasi-brittle materials and delamination in film–substrate systems are easily induced by either pyramidal (Vickers) or conical (Rockwell) indenters and can effectively be simulated using interface and cohesive crack models, see Figs. 13.11 and 13.12. In these situations, fracture is regarded as a phenomenon of gradual separation between two initially bonded surfaces, resisted by cohesive tractions which depend on the opening (w_n) and sliding (w_{t1}, w_{t2}) relative displacements and which vanish when the kinematics discontinuities grow beyond some critical values.

The relationship between tractions and displacement discontinuities at the interface is often based on the assumption of the existence of a free energy density function φ, which represents a potential function for normal and tangential stresses (σ_n, σ_{t1}, and σ_{t2}, respectively), such that:

$$\sigma_n = \frac{\partial \varphi}{\partial w_n}; \quad \sigma_{ti} = \frac{\partial \varphi}{\partial w_{ti}} \quad (i = 1, 2) \tag{13.12}$$

see, for example [133] and references therein.

A frequently adopted formulation is that originally proposed in [134] for mode I fracture of metals and bimetallic interfaces, and extended to two and three-dimensional situations in [135–137] through the introduction of a scalar measure

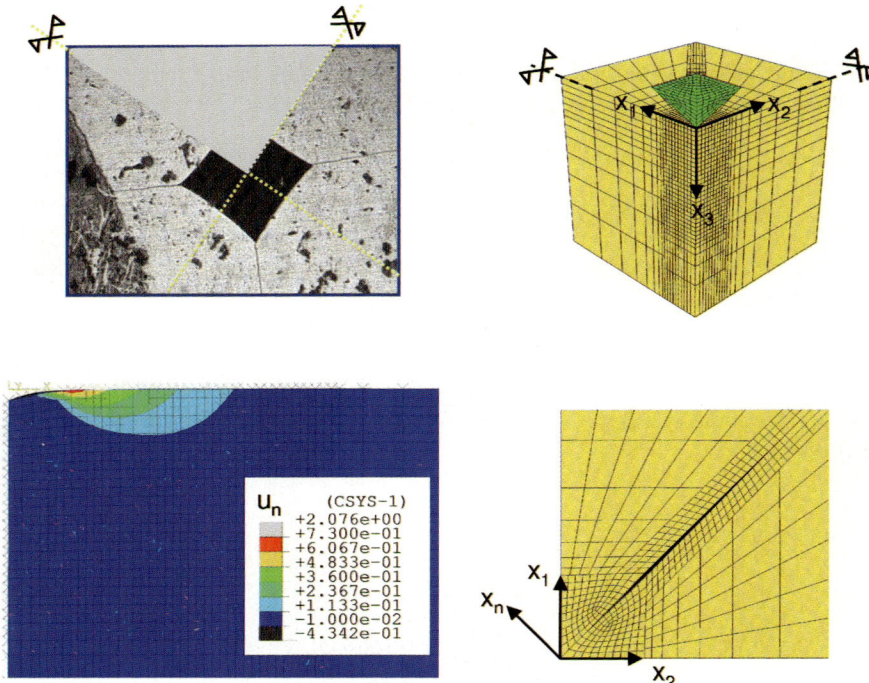

Fig. 13.11. Indentation induced fracture and the corresponding simulation by a finite element model

of the displacement jump across the interface:

$$\delta = \sqrt{\chi^2 \left(w_{t1}^2 + w_{t2}^2\right) + w_n^2} \tag{13.13}$$

where the parameter χ assigns different weights to opening (w_n) and sliding (w_{t1}, w_{t2}) displacements.

The potential φ is then defined as a function of δ, for example as:

$$\varphi = e\sigma_c\delta_c \left[1 - \left(1 + \frac{\delta}{\delta_c}\right) \exp\left(-\frac{\delta}{\delta_c}\right)\right] \tag{13.14}$$

where: e is the basis of natural logarithms (Neper's number); σ_c indicates the maximum cohesive normal traction and δ_c represents a characteristic relative displacement.

The traction versus relative displacement law resulting from relationships (13.12), (13.13), and (13.14) is represented in Fig. 13.13 for pure opening and sliding mode. Irreversible effects can be governed by the maximum attained effective displacement jump $\delta_{\max}$, as the only internal variable in the model; its evolution is governed by:

$$\dot{\delta}_{\max} = \begin{cases} \dot{\delta} & \text{if } \delta = \delta_{\max} \\ 0 & \text{otherwise} \end{cases} \tag{13.15}$$

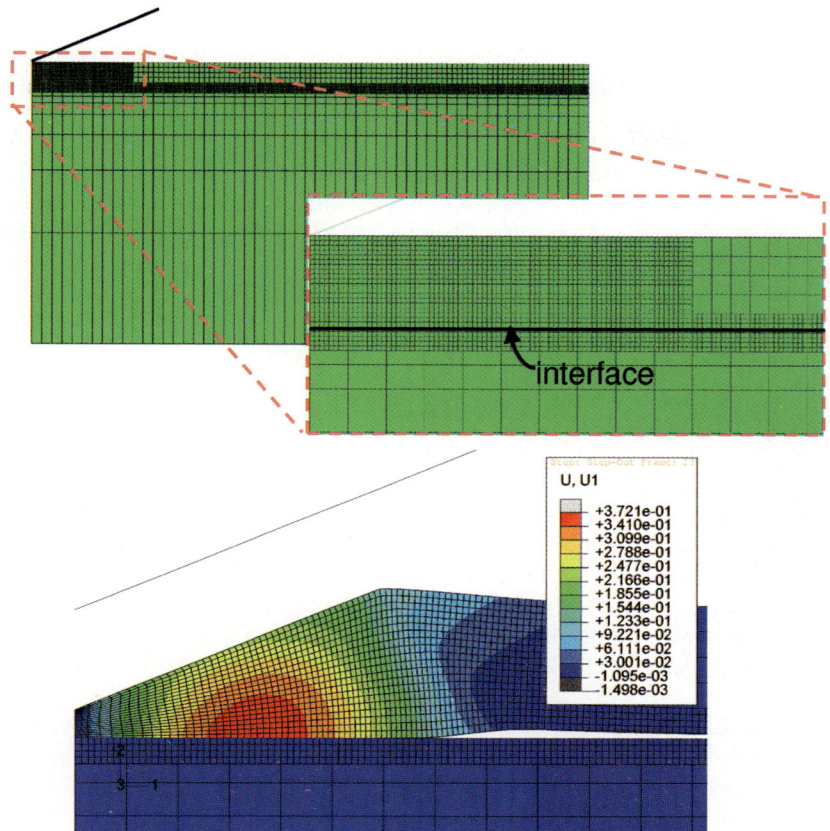

Fig. 13.12. Finite element model of delamination induced by indentation of film-substrate systems

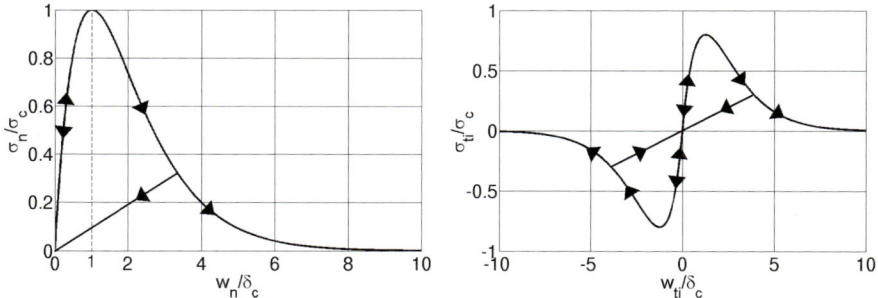

Fig. 13.13. Interface fracture model in pure opening (*left*) and sliding (*right*) mode, for $\chi = 0.8$

It can be easily shown that, in the model defined by Eq. (13.14), the quantity $e\sigma_c\delta_c$ plays the role of the fracture energy G_c required to generate a new free unit surface. G_c is graphically represented by the area enclosed by the cohesive curve defined for $w_n > 0$ as $w_{t1} = w_{t2} = 0$.

The present interface law depends on three independent material parameters, which can be chosen as either σ_c, δ_c and χ or, with a more direct physical meaning, as: mode I (opening) resistance σ_c, mode II (sliding) resistance $\tau_c = \chi \sigma_c$ and fracture energy G_c.

This or similar formulations have been often exploited to describe the delamination effects of indentation on brittle and quasi-brittle material systems [25–29,32,38] as well as to simulate crack propagation in functionally graded metal–ceramic composites, as an alternative approach to linear elastic fracture mechanics; see, for example [130, 138] and references therein.

13.3.2
Optimization Procedures

The inverse analysis problem of recovering the parameters embedded in constitutive models (like the ones presented in the above section) can be formulated as the minimization, with respect to the unknown material parameters, of a norm that quantifies the overall discrepancy between quantities measured in a laboratory or in situ experiment and the corresponding values computed through the simulation of the test.

Both deterministic batch or sequential stochastic approaches such as Kalman filtering (see, for example [36,40,46, 129]) can be employed for parametric identification purposes.

In conventional deterministic batch approaches, the available experimental information are exploited all together without processing uncertainties, which may affect both the measurements and the system modeling, which are taken into account by stochastic approaches.

In the case of indentation tests, input data for material characterization can be gathered from the curves representing the reaction force on the instrument versus the penetration depth, for example sampling M couples of force F_{mi} and corresponding displacement h_{mi}. Here subscript m means measured, index $i = 1, 2, \ldots, M$, see for example Fig. 13.14.

At the end of the process, horizontal and/or vertical displacement components u_{mj} are recovered in N selected points on the surface of the indented material specimen, for example by the above-mentioned AFM mapping and DIC techniques, in order to describe the geometry of the residual imprint, including piling-up and sinking-in deformations in its neighborhood, see for example Fig. 13.14.

The measurable quantities $F_{ci}(\mathbf{z})$ and $u_{cj}(\mathbf{z})$, corresponding to the experimental data F_{mi} and u_{mj}, are returned by the computational model of the tests (here subscript c means computed), as a function of the material parameters supplied with the adopted constitutive relationships, here collected by vector $\mathbf{z}$.

In a classical least-square context, it is then possible to define a discrepancy norm, for example as follows:

$$\omega(\mathbf{z}) = \sum_{i=1}^{M} \left(\frac{F_{ci}(\mathbf{z}) - F_{mi}}{w_{Fi}} \right)^2 + \sum_{j=1}^{N} \left(\frac{u_{cj}(\mathbf{z}) - u_{mj}}{w_{Uj}} \right)^2 \qquad (13.16)$$

Fig. 13.14. Possible sampling of available information for material characterization purposes

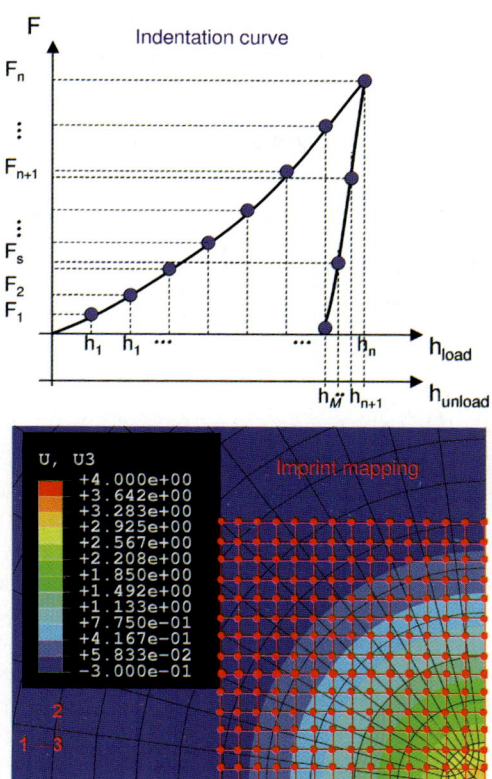

where w_{Fi} and w_{Ui} represent weights on force and displacement components, which can be assumed to be inversely proportional to the variances which quantify the measurement uncertainty.

The optimum value of the sought parameters is represented by the entries of vector $\mathbf{z}$ which returns the minimum discrepancy. The minimization of the objective function $\omega(\mathbf{z})$, defined by relation (13.16) can be performed by a number of numerical methods implemented in widely available optimization tools, for example see [138]. The iterative solution procedure starts from a suitably chosen set of initial parameters, usually those expected on the basis of handbooks, previous experience or expert's judgment. The unknown vector $\mathbf{z}$ is then updated following directions, in the parameter space, which minimize the discrepancy $\omega(\mathbf{z})$ or some convenient approximation of it (often quadratic). The usually highly non-linear and non-convex function $\omega(\mathbf{z})$ can present multiple minimum points. To avoid the problem of being stuck in one local minimum, the algorithm is usually restarted several times from quite different initialization points.

Sequential procedures, like the classical Kalman filtering, basically consist of a sequence of estimations, which starts from an a priori guess and exploits, one by one, the flow of the available experimental measures, accompanied by the statistical characterization of their uncertainties. At the i-th step, the estimates $\hat{\mathbf{z}}_{i-1}$ obtained at the previous instant are updated (and generally improved) together with their covariance matrix $\hat{\mathbf{C}}_{i-1}$, until convergence is achieved. The iterative algorithm, which here considers the case of experimental data from the indentation curves only, as in [40, 46, 129], usually relies on the updating scheme:

$$\hat{\mathbf{z}}_i = \hat{\mathbf{z}}_{i-1} + \mathbf{K}_i \left(F_{mi} - F_{ci}(\hat{\mathbf{z}}_{i-1}) \right), \quad \mathbf{K}_i = \hat{\mathbf{C}}_{i-1} \mathbf{L}_{i-1}^T \left(\mathbf{L}_{i-1} \hat{\mathbf{C}}_{i-1} \mathbf{L}_{i-1}^T + C_{Fi} \right)^{-1} \tag{13.17}$$

$$\hat{\mathbf{C}}_i = \hat{\mathbf{C}}_{i-1} + \mathbf{K}_i \mathbf{L}_{i-1} \hat{\mathbf{C}}_{i-1} \tag{13.18}$$

A key ingredient of this procedure is the so-called gain-matrix $\mathbf{K}_i$, which depends on the variance of the measurement C_{Fi} and on the sensitivity matrix $\mathbf{L}_{i-1}$, evaluated from the model at the previous step. Sensitivity matrix gathers the derivatives of the measurable quantities with respect to the parameters to identify, namely:

$$\mathbf{L}_{i-1} = \left. \frac{\partial F_{c\mathbf{K}}(\mathbf{z})}{\partial \mathbf{z}} \right|_{\mathbf{z}=\hat{\mathbf{z}}_{i-1}} \tag{13.19}$$

The repeated evaluation of sensitivities represents the most-significant burden of this otherwise rather effective identification tool, especially heavy when the computational model does not admit a closed-form analytical formulation and derivatives have to be replaced by finite-difference approximations. For a small number of unknown variables, computational costs can be significantly reduced by interpolation techniques introduced in [140].

Preliminary sensitivity analyses intended to quantify the influence of each sought parameter on measurable quantities are usually performed, independently of the implemented identification scheme, for the selection of the most effective quantities to measure during the experimental test and for corroborating conjectures on parameter identifiability, see for example [36, 114]. These aspects will be exemplified in the next section, devoted to applications.

13.4
Applications

Some examples are presented in this Section in order to elucidate, in computational terms, potentialities and limitations of imprint mapping as a supplement to the indentation curves for the calibration of material parameters for several different application fields.

As a common rule, the performances of any considered inverse analysis procedure is checked first by using "exact" pseudo-experimental input data, computed on the basis of a certain set of material parameters, which should be found as output of the inverse analysis. The robustness with respect to experimental errors of the envisaged methodology is then tested by means of "noisy" information, obtained by adding random perturbations to the pseudo-experimental data to be input in the identification process, while modeling errors are ruled out. Finally, the consideration of truly experimental information is used to fully validate the implemented procedure, including the selection of the most appropriate constitutive law.

The following applications, taken from the available literature, consider all these alternative inputs; some selected results are presented and discussed.

13.4.1
On the Role of Friction

Friction between the indenter and the specimen is often neglected in the literature on hardness tests since the tip angle of conical or pyramidal indenters can be selected in order to minimize frictional effects, which usually have weak influence on traditional experimental data consisting of force versus indentation depth.

In most cases, the friction coefficient between the most-common diamond indenter tips and metal surfaces has been assumed as known, equal to either 0 or 0.15 [42, 48, 100, 122]. However, friction is worth being investigated in many situations [53–55] in view of its intrinsic uncertainty and of its expected influence on the residual displacements, which might be considered for material characterization purposes. Parametric computer studies concerning conical indentation of elastic–plastic materials characterized by parameters typical of steel, show that the friction angle affects significantly the specimen residual deformations, particularly on the piling-up portion of the indented material around the imprint when the friction coefficient varies in the range 0–0.3, see for example the graphs in Fig. 13.15 (details in [63]). The effect of friction becomes less and less important as the ratio between the yield stress and the Young's modulus E of the material increases and, hence, piling-up tends to disappear; see also [141].

Figure 13.16 (left) visualizes the sensitivity (13.19) of the indentation force with respect to the friction coefficient c, over the set of the considered indentation depths: the sensitivity of the indentation force at a given penetration depth is almost negligible all along the loading process; it increases in the unloading process but remains quite limited. On the contrary, the sensitivity with respect to friction exhibited by the residual vertical displacements on the specimen surface, Fig. 13.16 (right), is high not only at points around the boundary of the contact region, but also in the contact zone due to the plastic flow of the material along the indentation tool.

Data about the geometry of the residual imprint can then be exploited to identify the friction coefficient between the indenter tool and the specimen surface, besides the constitutive parameters of isotropic and anisotropic materials as done, for example in [15, 63, 72].

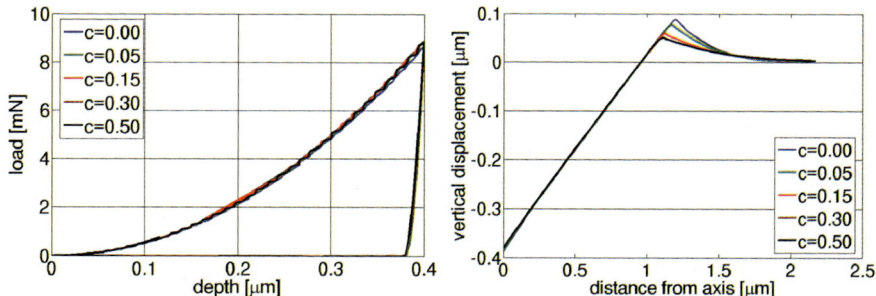

Fig. 13.15. Indentation curves and imprint profiles for different friction coefficients generated by simulation, considering the penetration of a conical tip in steel [63]

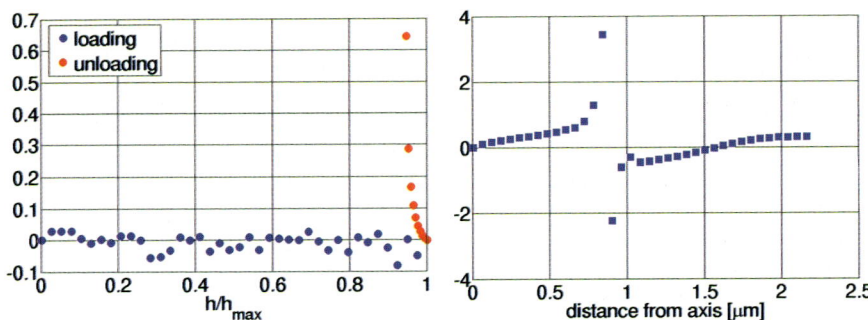

Fig. 13.16. Sensitivity with respect to the friction coefficient c (*left*) of the indentation force over the set of indentation depths; (*right*) of the residual vertical displacements on and near the imprint, at points along the radius from the vertical symmetry axis of the indenter [63]

13.4.2
Tests Concerning HHM and DP Models

The classical HHM elastic-perfectly-plastic constitutive law is widely used in engineering practice to describe the response of ductile metals, characterized by elastic modulus E, Poisson ration ν, and elastic limit σ_0, equal in tension and compression. This relatively simple three-parameter model has been conveniently used for the comparative assessment of procedures, which combine inverse analysis and indentation with and without imprint mapping, for parameter calibration purposes. As an example, Fig. 13.17 shows the convergence pattern for the batch deterministic identification procedure outlined in the previous Section, for sought material parameters $E = 200$ GPa; $\nu = 0.3$, $\sigma_0 = 1.25$ GPa. Pseudo experimental data F_{mi}, u_{mj} entering the discrepancy function (13.16), are generated through the computational model of the conical indentation test, see [63]. Output results of this numerical analysis are corrupted by random noise before being considered as given input information for the inverse analysis. Results in Fig. 13.17 (normalized on the expected value of each parameter, in this case a priori known) evidence the significant benefits arising from the additional experimental information relevant to the residual displacement field.

This feature has been further verified in the case of truly experimental data gathered from Vickers indentation of yittria stabilised zirconia (YSZ), visualized in Figs. 13.1, 13.2, and 13.18. The homogeneous material behavior is in this case represented by associative DP model. Parameters to be identified consist of: elastic modulus E, elastic limit in compression σ_{0c}, hardening parameter h, and internal friction angle α, see relations (13.2), (13.3), and (13.4). Table 13.1 collects the results of the proposed inverse analysis procedure and shows how the use of the imprint geometry, besides the traditional indentation curve, can improve the accuracy (in terms of reduced dispersion) of the estimated material parameters and the robustness of the procedure, which returns results almost independent of the initialization vector even if this has been selected in a rather wide parameter space. On the contrary, the use of information relevant to the indentation curves only cannot get rid of almost equivalent local minimum points, corresponding to rather different identified material parameter sets, collected by vector $\mathbf{z}_{opt}$, approximating the available

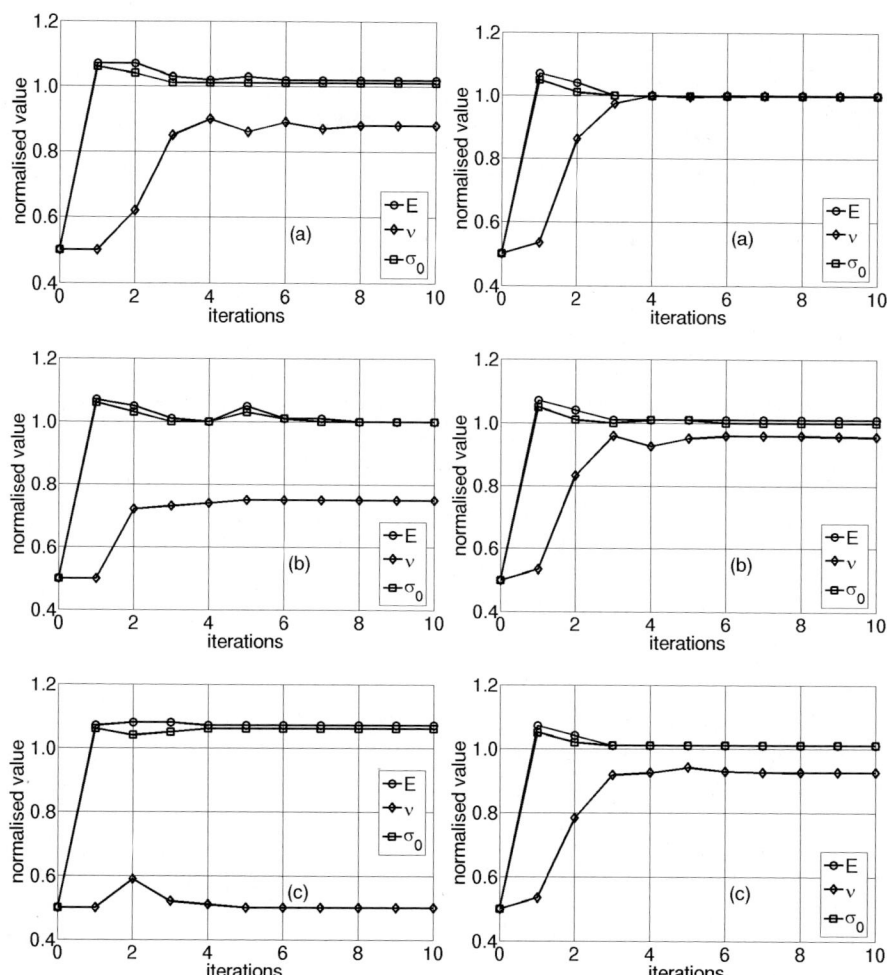

Fig. 13.17. HHM parameters estimated by inverse analyses based on indentation curves alone (plots on the *left*) or on imprint mapping as well (on the *right*) with: (**a**) zero noise; (**b**) ±2.5% noise and (**c**) ±5% noise [63]

experimental information almost to the same accuracy, measured by the discrepancy value $\omega(\mathbf{z}_{opt})$. Figure 13.18 compares the available experimental information with the corresponding quantities computed on the basis of the identified parameters, listed on the right columns of Table 13.1: the good agreement between these different data confirms the suitability of the constitutive model selected for this application.

13.4.3
Anisotropic Materials

Some numerical tests have been performed in [72] for the preliminary validation of the above-proposed parameter calibration methodology in the case of transversally

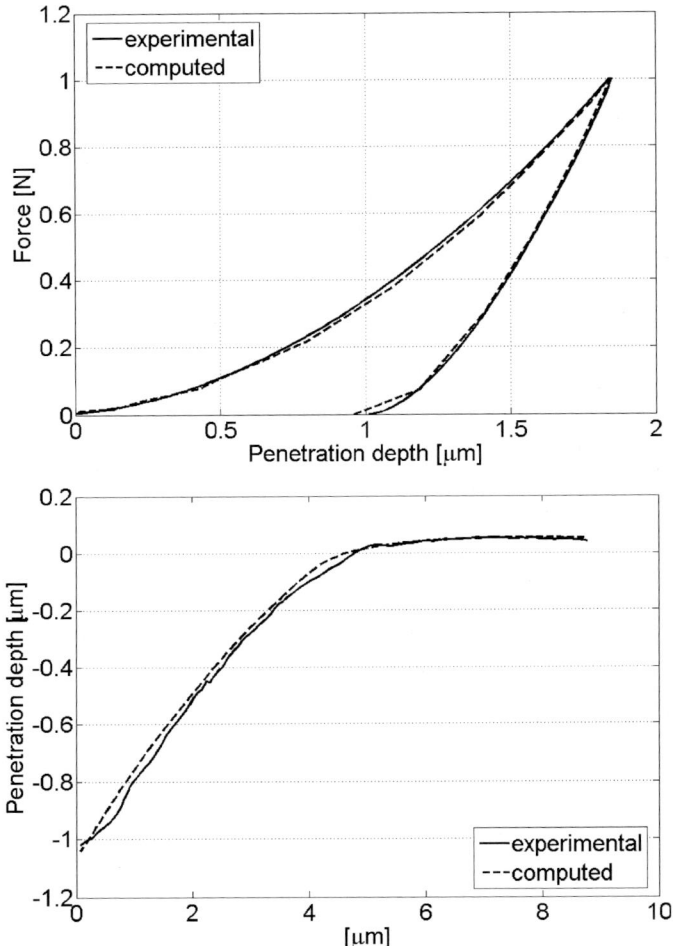

Fig. 13.18. Comparison between experimental and numerical results, the latter calculated on the basis of the identified parameters listed in Table 13.1, concerning the average indentation curve and imprint diagonal resulting from Vickers indentation of a YSZ specimen

isotropic materials. The analyzed situation considers the isotropy plane orthogonal to the x_1 axis while conical indentation is performed along the material directions x_3 and x_1, see Fig. 13.19. Hill's model for anisotropic perfect elasto-plasticity has been adopted to describe the material behavior, see relations (13.10) and (13.11). The five independent parameters that characterize the elastic response of the considered material were assumed as: $E_1 = E = 42$ GPa; $E_2 = E_3 = E_t = 36$ GPa, $v_{12} = v_{23} = v_{31} = 0.3$; $G_{12} = G_{31} = 16$ GPa. The independent plastic parameters are reduced to three, namely: $\sigma_{220} = \sigma_{330} = \sigma_{T0} = 30$ MPa, $\sigma_{110} = \sigma_0 = 35$ MPa, $\sigma_{120} = \sigma_{130} = \tau_0 = 18$ MPa. These values are typical of thin aluminum foils employed in laminates for food containers.

The considered pseudo-experimental data consisted of the indentation curves and of the corresponding residual vertical displacements, generated by computation

Table 13.1. Identification of DP constitutive parameters for YSZ specimen based on experimental information gathered from indentation curves only or from indentation curves and imprint mapping

Indentation curve					Indentation curve and imprint mapping				
	E [GPa]	σ_{0c} [MPa]	h [MPa]	α		E [GPa]	σ_{0c} [MPa]	h [MPa]	α
Initial	150	1,500	129,326	40	Initial	100	700	11,547	10
Estimate	**201**	**1,995**	**89,726**	**18.5**	Estimate	**209.14**	**2,715**	**22,885**	**31.9**
$\omega(z_{opt})$	0.00652				$\omega(z_{opt})$	0.09180			
Initial	300	2,000	86,602	30	Initial	160	1,500	40,415	10
Estimate	**211**	**991**	**83,840**	**27**	Estimate	**209.47**	**2,713**	**22,898**	31.9
$\omega(z_{opt})$	0.00543				$\omega(z_{opt})$	0.09178			
Initial	180	450	86,602	20	Initial	250	5,000	86,602	27
Estimate	**192.5**	**481**	**59,536**	**30**	estimate	**209.95**	**2,716**	**22,824**	31.8
$\omega(z_{opt})$	0.00598				$\omega(z_{opt})$	0.09188			

Fig. 13.19. Finite element model developed for the calibration of a transversally isotropic constitutive law: the isotropy plane is defined by the axes x_2 and x_3

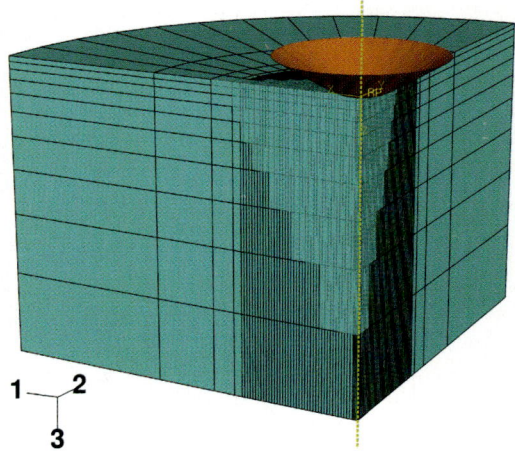

and corrupted by random noise, distributed with uniform probability density over the interval $\pm 2.5\%$ of each assumed measurable quantity. Contour mapping in Fig. 13.20 represents the geometry of the residual imprint, left after indentation in the $x_1 - x_2$ plane, which clearly reflects material anisotropy, while the indentation curve represents an average material response, only, quite insensitive to the directions associated to the different mechanical characteristics.

The inverse problem of finding material parameters back from the assumed available information has been tackled in [72] by considering a two-stage procedure, which exploits, in a sequence, data obtained by indenting the specimen along the material directions x_3 and x_1, respectively, the latter in the isotropy plane. Already at the end of the first stage the plastic parameters and the friction coefficient are identified almost exactly, while larger errors characterize the identification of the elastic moduli. At the end of the second stage, a maximum 15% residual error on E is found. Clearly, the dominance of plastic deformation during the considered indentation pro-

Fig. 13.20. The geometry of the residual imprint left after indentation in the $x_1 - x_2$ plane of a transversely isotropic elastic-perfectly-plastic material (isotropy plane orthogonal to x_1 axis)

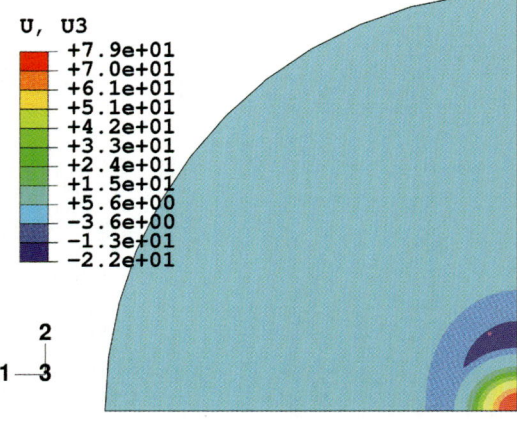

cess fosters the identifiability of the inelastic parameters, as also pointed out by the sensitivity analyses illustrated in [72].

13.4.4
Self Stresses

A stressed specimen subjected to indentation is known to exhibit different mechanical response in terms of indentation curve and imprint geometry if compared to an unstressed one. In particular, if the initial self-stresses are predominantly tensile, the material starts yielding at a smaller applied load as compared to the case of an unstressed specimen; vice-versa if it is compressive. In other terms, for a given indentation force, the maximum penetration depth will be larger in the case of initial tensile stresses. This information can provide an average value of the stress applied in the plane orthogonal to the indenter axis, but cannot return stress directionality [73–75, 142, 143]. On the contrary, the shape of the imprint geometry is directly influenced by both the magnitude and the directionality of a state of surface stresses. As an example, Fig. 13.21 represents the contour map of the residual vertical displacement for an initial stress state characterized by the principal values $\sigma_I = -500$ MPa and $\sigma_{II} = 0$, along directions rotated counter-clockwise of the angle $\phi = 56.25°$; these results refer to an ideal conical indentation experiment carried out on an elastic-perfectly-plastic metallic material, details can be found in [79]. Notice that the residual imprint presents an ellipsoidal shape, with major and minor axes of the iso-displacement contour lines, which coincide with the principal directions of the initial stress state. The imprint profile is also affected by the stress state in the material; see for example [144].

The usually non-symmetrical morphology of the imprint left on pre-stressed samples has been experimentally verified by Lee and co-workers, who performed three-sided pyramidal indentation on the surface of artificially strained tungsten single crystals [76] and conical indentation tests on small-grained API X65 steel [77], in different states of biaxial stresses generated at prescribed and controlled load level. An empirical relationship was found between the ratio of the two stress components

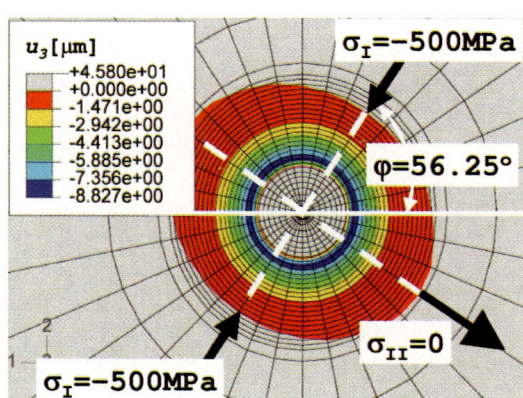

Fig. 13.21. Contour map of the residual vertical displacement of a pre-stressed metallic specimen after the indentation

and the ratio of the pile-up height shifts along the loading axes, compared to the unstressed specimen.

The identification of pre-existing stresses on the basis of their effect on the imprint geometry was proposed first in [145]; superior performances were, however, shown by the procedure proposed by Bocciarelli and Maier [79], which exploits all the information that can be recovered from the AFM mapping of the residual imprint.

13.4.5
Coatings and Layered Systems

Indentation procedures are widely used in industries dealing with thin film and coatings due to their capability of deforming materials at a very small scale, in situations where cutting specimens for different characterization techniques is prohibitive.

In order to obtain accurate estimates of the material properties from the indentation of layered systems it is crucial to distinguish between film and substrate contributions to the overall specimen response, see for example Figs. 13.22 and 13.23, concerning the simulated response to conical indentation of a thin gold film on a silicon substrate and of a zirconia-coated steel specimen, compared with the case of pure gold and pure silicon, respectively (details in [15]).

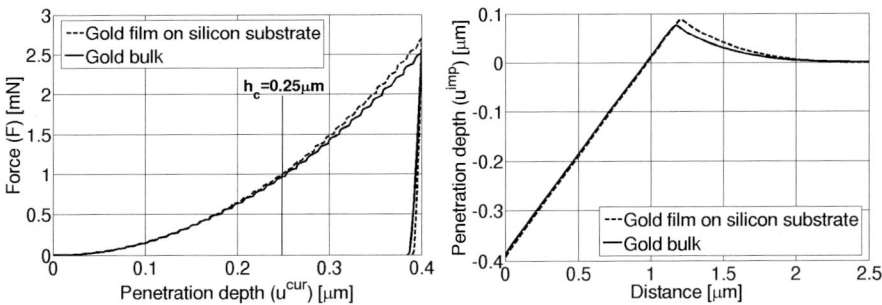

Fig. 13.22. Simulated response to conical indentation of a gold film (1-μm thick) deposited on a silicon substrate: indentation curve (*left*) and residual imprint (*right*) compared with the corresponding curves from indentation of bulk gold

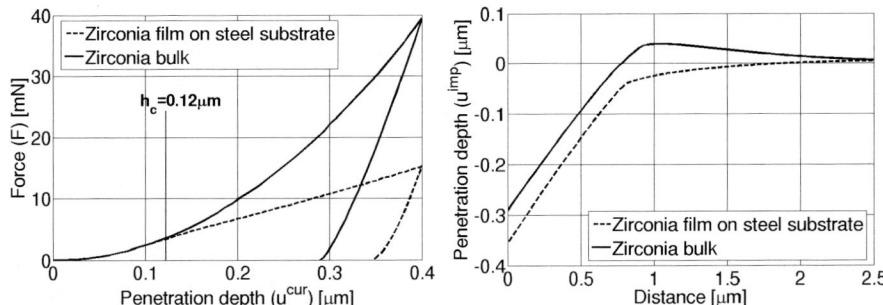

Fig. 13.23. Simulated response to conical indentation of a zirconia film (1-μm thick) deposited on steel: indentation curve (*left*) and residual imprint (*right*) compared with the corresponding curves from indentation of bulk zirconia

In the approach most widely used, penetration depth is kept smaller than some critical value (h_c in Figs. 13.22 and 13.23) that depends on the film–substrate properties. The presence of the substrate and its influence on the test results are hence neglected, and formulae or methods developed for homogenous media are applied. Alternatively, the mechanical parameters characterizing the film can be derived, once the substrate properties are known, through empirical and analytical models developed for layered composites, see for example [146, 147]. Beyond the elastic range, the film properties can be estimated through techniques based on FE analyses and either the rule of mixture [148, 149] or dimensional analysis [59, 150].

Huber and co-workers [44, 151] developed a neural network methodology for the identification of material parameters entering elastic–plastic constitutive laws of film–substrate systems, considering indentation depth up to twice the film thickness. Their procedures, really effective in terms of computing costs, basically rest on the assumption that the same constitutive description can be ascribed to both the coating material and its support.

The use of an inverse approach, based on a numerical model of the entire film–substrate system, permits us to overcome this problem in a consistent way since the presence of the substrate is explicitly accounted for and its mechanical properties, if unknown, can be identified together with those of the coating material, independently of the envisaged constitutive models, see [15]. The procedure, which considers indentation curves and imprint geometry as available sources of experimental information, has been validated for the identification of material properties of ductile films on hard substrates (as found, for example in MEMS technology) and of hard protective coatings (mainly ceramics) on metals, in the case of perfect adhesion between isotropic layers.

Tables 13.2 and 13.3 report the results of parameter identification exercises exploiting the pseudo-experimental data represented in Figs. 13.22 and 13.23. Modeling errors are ruled out but data from the indentation curves and from the imprint geometry are affected, as in other applications, by a randomly distributed numerical noise defined within the range $\pm 2.5\%$ of each corresponding measurable quantity.

In the considered cases, gold and steel have been described by the elastic-perfectly-plastic HHM model, zirconia by associative DP constitutive law, relations (13.2), (13.3), (13.4), (13.5), (13.6), (13.7), (13.8), while the silicon substrate has been conceived as linear elastic. The number of unknown parameters to be identified varies between four, for the case of gold film on silicon substrate (namely, E_f, σ_{0f}, E_s and c, subscripts f and s referring to the film and its substrate, respectively) and six, for the case of a zirconia–steel layered system (namely, E_f, σ_{c0f}, α_f, E_s, σ_{0f}, and c) including the friction coefficient c between the coating and the indentation tool. Poisson's ratios ν_f and ν_s are assumed to be a priori known due to their relatively low influence on the indentation results.

In the case of hard film on ductile substrate, most material parameters are identified with a maximum error of the same order of magnitude as the added noise. Sensitivity analyses showed that the only parameter which is badly identified in the case of noisy data, namely the friction coefficient c, is in fact poorly reflected by all the considered measurable quantities in this case of sinking-in response. Because of this low sensitivity, recourse should be made to some other experiment to get a reliable estimate of this surface characteristic whenever needed; for instance, the

Table 13.2. Inverse analysis results of the identification exercises based on both indentation curve and imprint geometry data, affected by a 2.5% numerical noise, in the case of a hard film on a ductile substrate; three different initialization vectors are considered

	c	E_f [GPa]	σ_{0cf} [MPa]	α_f [°]	E_s [GPa]	σ_{0s} [MPa]
Correct	**0.15**	**210**	**2,700**	**32**	**210**	**510**
Initial	0.075	105	1,350	16	105	255
Estimate	**0.2**	**207.1**	**2,662.8**	**33.4**	**217.1**	**507.4**
Error (%)	33.2	1.4	1.4	4.3	3.4	0.5
$\omega(\mathbf{z}_{opt})$	5.321×10^{-3}					
Initial	0.225	315	4,050	48	315	765
Estimate	**0.24**	**211.9**	**2,664.3**	**33.1**	**208.2**	**510**
Error (%)	63.6	0.9	1.3	3.5	0.9	0.0
$\omega(\mathbf{z}_{opt})$	5.122×10^{-3}					
Initial	0.225	105	4,050	16	315	255
Estimate	**0.33**	**216.2**	**2,726.7**	**31.7**	**200.0**	**510.6**
Error (%)	122.6	2.9	1.0	1.0	4.7	0.1
$\omega(\mathbf{z}_{opt})$	5.044×10^{-3}					

indenter could be used in scratch mode [11]. On the other hand, in the considered inverse analysis procedure, one can fix this parameter to any reasonable value since the simulation results and, hence, the estimation of the remaining parameters are not significantly affected by an even larger error on c.

Table 13.3. Inverse analysis results of the identification exercises based on both indentation curve and imprint geometry data, affected by a 2.5% numerical noise, in the case of a ductile film on an elastic substrate; three different initialization vectors are considered

	E_f [GPa]	σ_{0f} [MPa]	E_s [GPa]	c
Correct	**80**	**200**	**200**	**0.15**
Initial	40	100	100	0.075
Estimate	**803**	**200.4**	**202.5**	**0.15**
Error (%)	0.4	0.2	1.3	0.0
$\omega(\mathbf{z}_{opt})$	3.314×10^{-3}			
Initial	160	300	300	0.225
Estimate	**70.4**	**201.2**	**247.4**	**0.147**
Error (%)	12	0.6	23.71	2.0
$\omega(\mathbf{z}_{opt})$	3.613×10^{-3}			
Initial	40	300	100	0.225
Estimate	**80.1**	**200.5**	**204.1**	**0.149**
Error (%)	0.09	0.3	2.0	0.5
$\omega(\mathbf{z}_{opt})$	3.428×10^{-3}			

Also in the considered case of ductile film on a hard substrate, material parameters are mostly identified with a maximum error of the same order as the added noise. The largest discrepancy is on the elastic moduli: after three (rather severely chosen) initializations, the average value obtained for E_f is affected by about 4% residual error only, while the error for E_s is about 9%. In many engineering situations, such uncertainty levels are still well acceptable.

These results are corroborated by the outcome of sensitivity analyses showing that, with the exception of the unloading branch, the indentation curves are practically insensible to the variation of the elastic moduli E_f, E_s: the specimen response during increasing load and the residual imprint are dominated by a large amount of plastic deformation while the elastic contribution is by far less significant in this process. The friction coefficient c is moderately reflected by the indentation curves, but it affects quite strongly the geometry of the residual imprint, especially in the piling-up area due to the plastic material flow along the contact surface toward the upper boundary of the specimen, which enhances friction effects.

Thermal, electrical, and mechanical characteristics of layered systems are of great interest for many applications, but a further important macroscopic property to be verified is adhesion across the interface region, strictly related to the coating deposition technique and to the mechanical/physical properties of the film and of its substrate, see for example [16].

Delamination induced by instrumented indentation performed on the coated surface of film–substrate systems can be exploited for the characterization of fracture properties. Information gathered from the indentation curves usually permits us to discriminate situations where material separation does or does not occur, see Fig. 13.24, but it is not rich enough to return reliable estimates of the interfacial strength and toughness [32].

Work in progress concerns the case of strong interfaces, where shear delamination dominates the film-coating detaching phenomenon, promoted by conical indentation, without any apparent film buckling. Inverse analysis procedures are exploited to characterize the fracture characteristic properties of the interface, in terms of mode II resistance τ_c and fracture energy G_c entering the cohesive crack model described by relations (13.12), (13.13), (13.14), and (13.15).

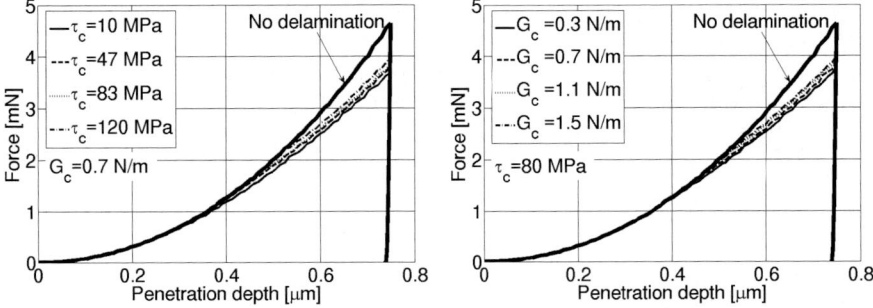

Fig. 13.24. Influence of parameters G_c and τ_c on indentation curves. Comparison is made with the response in the case where delamination does not occur (*thick line*)

The test is simulated in a finite element context by the use of interface elements, see Fig. 13.12. Axis-symmetric response is hypothesized as reasonable response for isotropic interfacial properties, as experimentally observed by Kim et al. [21].

The bulk material parameters and the friction coefficient between the indenter tip and the sample are supposed to be a priori known. In fact, these parameters can be determined by the inverse analysis procedure presented above, applying forces which do not induce delamination, an event usually reflected by the indentation curves (see Fig. 13.24) and which can be further verified by impact-echo techniques [152] or by mechanical impedance [153].

Data about the horizontal displacement field, which can be acquired by the coupling of AFM scanning and DIC techniques [30, 31, 94, 111–113, 118, 119] are considered part of the available experimental information.

Table 13.4 compares the identification results obtained in a preliminary validation exercise, for increasing noise level, when different sources of experimental information are considered, namely: (i) only the indentation curve; (ii) indentation curve and horizontal displacement field; (iii) indentation curve and vertical displacement field; (iv) all available data. Each inverse problem has been solved with 500 different random noise extractions and with different initialization vectors. The main conclusion that can be drawn from the computational experience gathered so far is that accurate values of the maximum shear traction τ_c and of the fracture energy G_c can be returned in a quite robust manner for rather noisy input data by exploiting the horizontal displacement field around the indentation area, while data from the vertical profile of the residual imprint are much less meaningful for this application.

A foreseen future solution for overcoming the interface problems experienced, for example by metal–ceramic coupling, consists of functionally graded materials (FGMs) [154]. The continuously varying homogenized mechanical properties of these composites can be related to the spatial distribution of their material phases by a modified rule of mixture proposed by Tamura, Tomota, and Ozawa (and henceforth named TTO model) for metal alloys [155], but applied to FGMs of different composition by the proper calibration of a characteristic "stress transfer" parameter, say q ($0 \leq q \leq \infty$), which measures the ratio between the difference of average stresses and strains in the material phases; see for example [40] and [156].

Table 13.4. Results in terms of identification error for an interface characterized by the fracture parameters $\tau_c = 80$ MPa, $G_c = 0.544$ N/mm and $\sigma_c = 100$ MPa (the latter assumed a priori known). Sources of input data for the identification procedure consist of: (i) only the indentation curve; (ii) indentation curve and horizontal displacement field; (iii) indentation curve and vertical displacement field; (iv) all the available data

Noise	Case (i)		Case (ii)		Case (iii)		Case (iv)	
	$Err(\tau_c^{id})$	$Err(G_c^{id})$	$Err(\tau_c^{id})$	$Err(G_c^{id})$	$Err(\tau_c^{id})$	$Err(G_c^{id})$	$Err(\tau_c^{id})$	$Err(G_c^{id})$
2.5%	9.6%	10.7%	3.3%	3.4%	8.0%	8.6%	4.0%	3.4%
5.0%	15.0%	18.5%	9.9%	9.5%	12.9%	15.4%	8.7%	8.3%
7.5%	21.2%	25.5%	16.1%	14.8%	20.6%	22.9%	17.0%	15.1%
10.0%	27.9%	37.2%	22.7%	20.1%	27.8%	30.1%	22.5%	19.9%

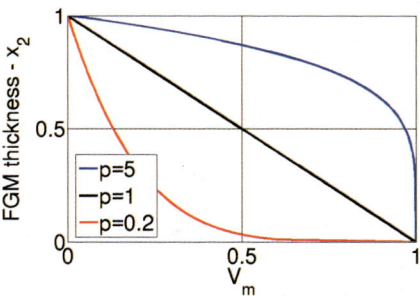

Fig. 13.25. Exponential volume fraction distribution of the metal phase in a FGM coating depending on the exponent p

In the material model proposed by Bocciarelli and co-workers [125], the TTO homogenization rule governs the transition from HHM theory, typical of metals, toward a DP constitutive model more suitable to describe the mechanical response of ceramics.

In the case of FGMs, the volume distribution of the components plays also a significant role on the overall material characteristics and therefore needs careful verification. An often assumed design profile for the volume fraction of metal and ceramic constituents is described by an exponential law, governed by parameter p as in Fig. 13.25. The simulated response to conical indentation of FGM Al-Al$_2$O$_3$ coating (1μm thick, on Al substrate) having different volume fraction distribution,

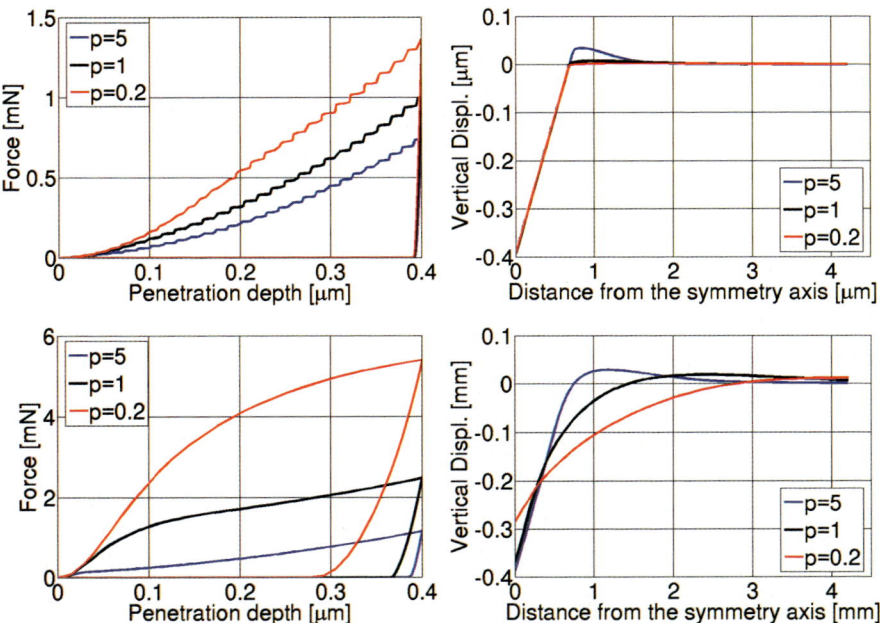

Fig. 13.26. Computed response to conical indentation of the Al-Al$_2$O$_3$ FGM coating on Al for different metal/ceramic volume fraction distribution and for the alternative choices $q = 0$ (*above*) and $q = 400$ (*below*)

is represented in Fig. 13.26 for different values of p and q; the assumed properties for Al and Al_2O_3 are listed in Table 13.5.

As can be seen, both p and q affect significantly the material response. Their identification based on inverse analysis procedures is therefore possible and particularly desirable, since they are not amenable to direct measurement, while the small thickness of typical FGM coatings makes the indentation test particularly appropriate, see [40, 46, 125, 126].

Results collected in Table 13.6 confirm, also for this application, the good performances of the envisaged parameter identification procedure based on the exploitation of experimental information gathered from indentation curves and imprint mapping.

Table 13.5. Material properties of the single phases of an alumina-aluminum FGM (taken from the literature)

Material properties	Al (metal)	Al_2O_3 (ceramic)
E [GPa]	69	340
ν	0.33	0.22
α	0.0	0.476
K [MPa]	57.73	252.85

Table 13.6. Inverse analysis results, in terms of mean value and standard deviation, for the identification of parameters p and q in five different combination, starting from pseudo-experimental data corrupted by 500 random noise extractions

Case	p_{exp}	q_{exp}	p_{id}	q_{id}
1	1.000	0.288	0.999 ± 0.012	0.287 ± 0.009
2	0.439	0.016	0.439 ± 0.009	0.016 ± 0.001
3	0.439	1.451	0.439 ± 0.014	1.443 ± 0.125
4	2.279	1.451	2.280 ± 0.023	1.451 ± 0.062
5	2.279	0.016	2.275 ± 0.049	0.016 ± 0.002

References

1. Tabor D (1951) The hardness of metals, Clarendon Press, Oxford
2. Mott BW (1957) Microindentation hardness testing, Butterworths, London
3. Alexopoulos PS, O'Sullivan TC (1990) Annu Rev Mater Sci 20:391
4. Spearing SM (2000) Acta Mater 48:179
5. Cetinel H, Uyulgan B, Tekmen C, Ozdemir I, Celik E (2003) Surf Coat Technol 174:1089
6. Chen J, Bull SJ (2006) Thin Solid Films 494:1
7. Shan Z, Gokhale AM (2003) Mater Sci Eng A 361:267
8. Furnémont Q, Kempf M, Jacques PJ, Göken M, Delannay F (2002) Mater Sci Eng A 328:26
9. Wang XD, Huang BX, Rong YH, Wang L (2006) Mater Sci Eng A 438:300
10. Frick CP, Ortega AM, Tyber J, Maksound AE, Maier HJ, Liu Y, Gall K (2005) Mater Sci Eng A 405:34
11. Bhushan B (1999) Handbook of micro/nano tribology. CRC Press, Boca Raton
12. Bhushan B, Li X (2003) Int Mater Rev 48:125

13. Saha R, Nix WD (2002) Acta Mater 50:23
14. Gamonpilas C, Busso EP (2004) Mater Sci Eng A 380:52
15. Bocciarelli M, Bolzon G (2007) Mater Sci Eng A 448:303
16. Hivart P, Crampon J (2007) Mech Mater 39:998
17. Volinsky AA, Moody NR, Gerberich WW (2002) Acta Mater 50:441
18. Ohring M (1991) The materials science of thin films, Academic Press, New York
19. Lawn BR, Evans AG, Marshall DB (1980) J Am Ceram Soc 63:574
20. Lawn BR (1993) Fracture of brittle solids, Cambridge University Press, Cambridge
21. Kim JJ, Jeong JH, Lee KR, Kwon D (2003) Thin Solid Films 441:172
22. Keller J, Vogel D, Schubert A, Michel B (2004) Displacement and strain field measurements from SPM images. In: Bhushan B, Fuchs H, Hosaka S (eds) Applied scanning probe methods. Springer, pp 253–276
23. Keller J, Gollhardt A, Vogel D, Michel B (2006) In: Gdoutos EE (eds) Proceedings of the 16th European conference of fracture, Springer, p 518
24. Xiaodong L, Dongfeng D, Bhushan B (1997) Acta Mater 45:4453
25. Abdul-Baqi A, Van der Giessen E (2001) Thin Solid Films 381:143
26. Abdul-Baqi A, Van der Giessen E (2002) Int J Sol Struct 39:1427
27. Li W, Siegmund T (2003) Scripta Mater 49:497
28. Li W, Siegmund T (2004) Acta Mater 52:2989
29. Liu P, Zhang YW, Zeng KY, Lu C, Lam KY (2007) Eng Fract Mech 74:1118
30. Vogel D, Grosser V, Schubert A, Michel B (2001) Opt Las Eng 36:195
31. Vogel D, Gollhardt A, Michel B (2002) Sens Actuat A 99:165
32. Maier G, Bocciarelli M, Bolzon G, Fedele R (2006) Int J Fract 138:47
33. Maier G, Giannessi F, Nappi A (1982) Eng Struct 4:89
34. Bittanti S, Maier G, Nappi A (1983) Inverse problems in structural elasto-plasticity: a Kalman filter approach. In: Sawczuk A, Bianchi G (eds) Plasticity today. Elsevier Applied Science, London, UK, p 311
35. Bui HD (1994) Inverse problems in the mechanics of materials: an introduction, CRC Press, Boca Raton
36. Stavroulakis G, Bolzon G, Waszczyszyn Z, Ziemianski L, (2003) Inverse analysis. In: Karihaloo B, Ritchie RO, Milne I (eds) Comprehensive structural integrity, chap 13. 3 Elsevier Science Ltd, Kidlington
37. Maier G, Bocciarelli M, Fedele R (2004) Some innovative industrial prospect on inverse analysis. In: Mroz Z, Stavroulakis G (eds) Parameter identification of materials and structures, Springer-Verlag
38. Giannakopoulos AE, Suresh S (1999) Scripta Mater 40:1191
39. Venkatesh TA, Van Vliet KJ, Giannakopoulos AE, Suresh S (2000) Scripta Mater 42:833
40. Nakamura T, Wang T Sampath S (2000) Acta Mater 48:4293
41. Tardieu N, Constantinescu A (2000) Inv Prob 16:577
42. Dao M, Chollacoop N, Van Vliet KJ, Venkatesh TA, Suresh S (2001) Acta Mater 49:3899
43. Kucharski S, Mróz Z (2001) Mater Sci Eng A 318:65
44. Huber N, Nix WD, Gao H (2002) Proc R Soc Lond A 458:1593
45. Fischer-Cripps AC (2003) Surf Coat Technol 168:136
46. Gu Y, Nakamura T, Prchlik L, Sampath S, Fallace J (2003) Mater Sci Eng A 345:223
47. Capehart TW, Cheng YT (2003) J Mater Res 18:827
48. Cheng YT, Cheng CM (1999) J Mater Res 14:3493
49. Chen X, Ogasawara N, Zhao M, Chiba NJ (2007) J Mech Phys Sol 55:1618
50. Chollacoop N, Dao M, Suresh S (2003) Acta Mater 51:3713
51. Cao YP, Lu J (2004) Acta Mater 52:1143
52. Cao YP, Qian XQ, Lu J, Yao ZH (2005) J Mater Res 20:1194
53. Taljat B, Zacharia T, Kosel F (1998) Int J Sol Struct 35:4411

54. Di Carlo A, Yang HTY, Chandrasekar S (2004) Int J Num Meth Eng 60:661
55. Mata M, Alcalá J (2004) J Mech Phys Sol 52:145
56. Cheng YT, Cheng CM (1998) Philos Mag A 78:115
57. Alcala J, Barone AC, Anglada M (2000) Acta Mater 48:3451
58. Matsuda K (2002) Philos Mag A 82:1941
59. Tunvisut K, Busso EP, O'Dowd NP, Brantner HP (2002) Philos Mag A 82:2013
60. Mulford R, Asaro RJ, Sebring RJ (2004) J Mater Res 19:2641
61. Stauss S, Schwaller P, Bucaille JL, Rabe R, Rohr L, Michler J, Blank E (2003) Microelectron Eng 67:818
62. Liu Y, Wang B, Yoshino M, Roy S, Lu H, Komanduri R (2005) J Mech Phys Sol 53:2718
63. Bolzon G, Maier G, Panico M (2004) Int J Sol Struct 41:2957
64. Vlassak J, Nix WD (1994) J Mech Phys Sol 42:1223
65. Jørgensen O, Giannakopoulos AE, Suresh S (1998) Int J Sol Struct 35:5097
66. Mahajan P (1998) Compos Sci Technol 58:505
67. Yu HY (2001) Int J Sol Struct 38:2213
68. Swadener JG, Pharr GM (2001) Philos Mag A 81:447
69. Vlassak JJ, Ciavarella M, Barber JR, Wang X (2003) J Mech Phys Sol 51:1701
70. Hengsberger S, Enstroem J, Peyrin F, Zysset Ph (2003) J Biomech 36:1503
71. Swadener JG, Rho JY, Pharr GM (2001) J Biom Mat Res 57:108
72. Bocciarelli M, Bolzon G, Maier G (2005) Mech Mater 37:855
73. Tsui TY, Pharr GM (1999) J Mater Res 14:292
74. Swadener JG, Taljat B, Pharr GM (2001) J Mater Res 16:2091
75. Suresh S, Giannakopoulos AE (1998) Acta Mater 46:5755
76. Lee YH, Kwon KD (2003) Scripta Mater 49:459
77. Lepienski CM, Pharr GM, Park YJ, Watkins TR, Misra A, Zhang X (2004) Thin Solid Films 447:251
78. Lee YH, Takashima K, Higo Y, Kwon D (2004) Scripta Mater 51:887
79. Bocciarelli M, Maier G (2007) Comput Mater Sci 39:381
80. Harvey S, Huang H, Venkataraman S, Gerberich WW (1993) J Mater Res 8:1291
81. Ma Q, Clarke DR (1995) J Mater Res 10:853
82. Petzold M, Landgraf J, Füting M, Olaf JM (1995) Thin Solid Films 264:153
83. Ma LW, Cairney JM, Hoffman M, Munroe PR (2005) Surf Coat Technol 192:11
84. Bhushan B, Kulkarni AV, Bonin W, Wyrobek JT (1996) Philos Mag A 74:1117
85. Baker SP (1997) Thin Solid Films 308:289
86. Serre C, Gorostiza P, Pérez-Rodríguez A, Sanz F, Morante JR (1998) Sens Actuat A 67:215
87. Serre C, Pérez-Rodríguez A, Morante JR, Gorostiza P, Esteve J (1999) Sens Actuat A 74:134
88. Sundararajan S, Bhushan B, (2002) Sens Actuat A 101:338
89. Sundararajan S, Bhushan B, Namazu T, Isono Y (2002) Ultramicroscopy 91:111
90. Khanna SK, Ranganathan P, Yedla SB, Winter RM (2003) J Eng Mater Technol 125:90
91. Wang M, Liechti KM, White JM, Winter RM (2004) J Mech Phys Sol 52:2329
92. Liu Y, Wang B, Yoshino M, Roy S, Lu H, Komanduri R (2005) J Mech Phys Sol 53:2005
93. Thome F, Göken M, Vehoff H (1999) Intermetallics 7:491
94. Chasiotis I, Knaus WG (2002) Exp Mech 42:51
95. Espinosa HD, Prorok BC, Fischer M (2003) J Mech Phys Sol 51:47
96. Krupička A, Johansson M, Hult A (2003) Prog Org Coat 46:32
97. Stauss S, Schwaller P, Bucaille JL, Rabe R, Rohr L, Michler J, Blank E (2003) Microelectron Eng 67:818
98. Xie ZH, Hoffman M, Moon RJ, Munroe PR (2006) J Mater Res 21:437
99. Bhushan B, Kulkarni AV, Bonin W, Wyrobek JT (1996) Philos Mag A 74:1117
100. Begley MR, Hutchinson JW (1998) J Mech Phys Sol 46:2049

101. Shu JY, Fleck NA (1998) Int J Sol Struct 35:1363
102. Bobji MS, Bhushan B (2001) Scripta Mater 44:37
103. Guo Y, Huang Y, Gao H, Zhuang Z, Hwang KC (2001) Int J Sol Struct 38:7447
104. Yuan H, Chen J (2001) Comp Mater Sci 38:8171
105. Bazant ZP, Guo Z (2002) Int J Sol Struct 39:5633
106. Swadener JG, George EP, Pharr GM (2002) J Mech Phys Sol 50:681
107. Vodenitcharova T, Zhang LC (2003) Int J Sol Struct 40:2989
108. Herrmann K, Hasche K, Pohlenz F, Seemann R (2001) Measurement 29:201
109. Herrmann K, Jennett NM, Wegener W, Meneve J, Hasche K, Seemann R (2000) Thin Solid Films 377:394
110. Tyulyukovskiy E, Huber N (2007) J Mech Phys Sol 55:391
111. Vendroux G, Schmidt N, Knauss WG (1998) Exp Mech 38:154
112. Knaus WG, Chasiotis I, Huang Y (2003) Mech Mater 35:217
113. Cho SW, Chasiotis I (2007) Exp Mech 47:37
114. Kleiber M, Antúnez H, Hien TD, Kowalczyk P (1997) Parameter sensitivity in nonlinear mechanics. Theory and finite element computations, John Wiley & Sons, Chichester
115. Oliver WC, Pharr GM (1992) J Mater Res 7:1564
116. Prchlik L (2004) J Mater Sci 39:1185
117. Giessibi FJ (2003) Rev Mod Phys 75:949
118. Berfield TA, Patel JK, Shimmin RG, Braun PV, Lambros J, Sottos NR (2007) Exp Mech 47:51
119. Scrivens WA, Luo Y, Sutton MA, Collette SA, Myrick ML, Miney P, Colavita PE, Reynolds AP, Li X (2007) Exp Mech 47:63
120. Cuitino AM, Wang Y (2002) Int J Sol Struct 39:3777
121. Bhattacharya AK, Nix WD (1988) Int J Sol Struct 24:1287
122. Jayaraman S, Hahn GT, Oliver WC, Rubin CA, Bastias PC (1998) Int J Sol Struct 35:365
123. Lubliner J (1990) Plasticity theory, Macmillan Publishing Company
124. Lemaitre J, Chaboche JL (1990) Mechanics of solid materials, Cambridge University Press
125. Bocciarelli M, Bolzon G, Maier G (2008) Comput Mater Sci 43:46
126. Giannakopoulos AE, Larsson PL (1997) Mech Mat 25:1
127. Giannakopoulos AE, Suresh S (1997) Int J Sol Struct 34:2393
128. Giannakopoulos AE (2002) Int J Sol Struct 39:2495
129. Nakamura T, Yu G, Prchlik L, Sampath S, Wallace J (2003) Mater Sci Eng A 345:223
130. Tilbrook MT, Moon RJ, Hoffman M (2005) Mater Sci Eng A 393:170
131. Moon RJ, Tilbrook M, Hoffman M, Neubrand A (2005) J Am Ceram Soc 88:666
132. Hill R (1948) A theory of yielding and plastic flow of anisotropic metals. Proc R Soc Lond A 193:281
133. Bolzon G, Corigliano A (1997) Comp Meth Applied Mech Eng 140:329
134. Rose JH, Ferrante J, Smith JR (1981) Phys Rev 9:675
135. Xu XP, Needleman A (1993) Mod Sim Mat Sci Eng1:111
136. Ortiz M, Suresh S (1993) J Appl Mech 60:77
137. Ortiz M, Pandolfi A (1999) Int J Num Meth Eng 44:1267
138. Zhi-He J, Glaucio H, Paulino RH, Dodds J (2003) Eng Fract Mech 70:1885
139. The MathWorks Inc. (2004) Matlab 2004, User's guide and optimization toolbox, Release 6.13, USA
140. Aoki S, Amaya K, Sahashi M, Nakamura T (1997) Comp Mech 19:501
141. Rodríguez J, Garrido Maniero MA (2007) Mech Mater 39:987
142. Bolshakov A, Oliver WC, Pharr GM (1996) J Mater Res 11:760
143. Carlsson S, Larsson PL (2001) Acta Mater 49:2179
144. Xu ZH, Li X (2005) Acta Mater 53:1913
145. Bijak-Zochowski M (1993) Arch Mech Eng 40:29

146. Yu HY, Sanday BB, Rath J (1990) J Mech Phys Solids 38:745
147. Chechenin NG (1997) Thin Solid Films 304:78
148. Chechenin NG, Bottiger J, Krog JP (1995) Thin Solid Films 261:219
149. Korsunsky AM, McGurk MR, Bull SJ, Page TF (1998) Surf Coat Technol 99:171
150. Tunvisut K, O'Dowd NP, Busso EP (2001) Int J Sol Struct 38:335–351
151. Huber N, Tsagrakis I, Tsakmakis C (2000) Int J Solids Struct 37:6499
152. Stavroulakis GE (1999) Eng Fract Mech 62:165
153. Bamnios Y, Douka E, Trochidis A (2002) J Sound Vibr 256:287
154. Suresh S, Mortensen A (1998) Fundamentals of functionally graded materials, The University Press, Cambridge, UK
155. Tamura I, Tomota Y, Ozawa H (1973) Inst Met (London) Monogr Rep Ser 1:611
156. Jin ZH, Paulino GH, Dodds RH (2003) Eng Fract Mech 70:1885

14 Mechanical Properties of Metallic Nanocontacts

G. Rubio-Bollinger · J.J. Riquelme · S.Vieira · N. Agraït

Abstract. The mechanical properties of the reduced number of atoms forming the apex of a tip are interesting both from a fundamental point of view and for the interpretation of experiments related to scanning local probe methods. These mechanical properties can be studied by establishing a very small contact, a nanocontact, between a tip and a surface. The elasticity and fracture events during the controlled breaking of a nanocontact as the tip is separated from the surface provide information about the mechanical properties of the tip apex. In the case of metallic tips, electron transport through the nanocontact also provides information on its mechanical properties, because at the scale of a few atoms forming the nanocontact the mechanical and electron transport properties are strongly related.

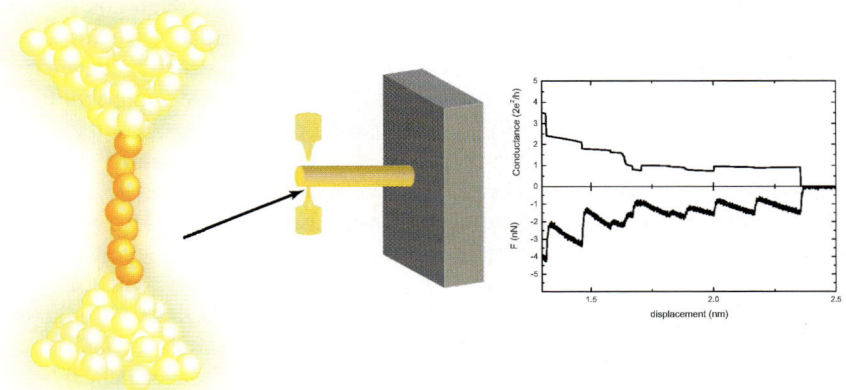

Key words: Nanomechanics, Metallic nanocontacts, Quantum point contacts

Abbreviations

AFM Atomic force microscope
DFT Density functional theory
DOS Density of states
MCBJ Mechanically controlled break-junction
MD   Molecular dynamics
MFM Magnetic force microscopy

PC   Point contact
PCS   Point contact spectroscopy
SNOM   Scanning near-field optical microscopy
SPM   Scanning probe microscopy
STM   Scanning tunneling microscope
TF   Tuning fork
UHV   Ultra high vacuum

14.1
Introduction

Local probe techniques are often based on the interaction of a tip in close proximity with a surface. The highest imaging resolution is achieved when the tip apex is brought as close as possible to the surface but avoiding direct contact, so that neither the tip apex nor the surface are modified or damaged during scanning. The transition from the non-contact to the contact regime plays therefore a crucial role in the achievement of high-resolution images and their interpretation. In most of the cases this transition is not smooth but there is a mechanical instability which results in a jump to contact process [1, 2]. The instability occurs at a probe to sample distance at which the elasticity of the probe becomes lower than the force gradient of the interaction between the tip and the surface. There are two main contributions to the elasticity of the probe. The first, which could be named extrinsic, has to be considered if the tip is mounted on an elastic device, such as the cantilever beam of an atomic force microscope. Second, a non-negligible intrinsic contribution has its origin in the finite elasticity of the apex of the tip and the spot on the surface close to the tip. While the extrinsic elasticity is macroscopic in nature, the intrinsic elasticity is at the nanoscale because it involves very few atoms located at the tip apex and its surroundings. These nanoscale mechanical properties are the subject of this chapter. Scanning probe methods are used to study friction and wear between a tip and a surface. In such a situation, an atomic scale contact is established between the tip and the surface. The mechanical properties of this kind of structure, a nanocontact, are different from that of macroscopic contacts. The detailed arrangement of the atoms in the contact can result in a variety of mechanical behaviors and quantum effects become relevant [3, 4]. At the very smallest contact, a one-atom contact, electron transport and mechanical properties have been shown to be intimately related. Theoretically, molecular dynamics [5] and density functional theory calculations have provided deeper insight into the behavior of matter at the nanoscale. Despite the apparent simplicity of structures composed of a small number of atoms, theoretical modeling still does not provide a full quantitative way of finding the mechanical properties of atomic-size structures, and further research is still necessary.

In addition, there is an increasing interest in techniques that allow establishment of a reliable interface between macroscopic or mesoscopic electric circuits, and atomic-sized structures tailored to have functional and practical mechanical or electron transport properties. That is the case of functional molecules and carbon-based electronic devices, such as carbon nanotubes or graphene nanoribbons [6]. These tiny, but subtle, structures will probably be interfaced with (scanning or static) atomic-sized tips whose mechanical and electron transport properties will play a significant role in the functionality and performance of these nanoscale devices.

14.2
Experimental Tools

We will focus in this section on devices that are especially well suited for the study of metallic nanocontacts at low temperatures. Despite many successful low temperature scanning tunneling microscopy experiments the number of results obtained by low-temperature scanning force microscopy is still small. Moreover, only a few commercial manufacturers have presented a low-temperature scanning force microscope so far [7–9] because of the difficulties in adapting the usual force-detection techniques to cryogenic environments.

Binnig and Rohrer [10] developed the Scanning Tunneling Microscope (STM) more than 20 years ago. The STM has allowed study of the topography and electronic properties of metallic surfaces with atomic resolution. Moreover, the STM is the first of a family of local probes, the Atomic Force Microscope (AFM) being the most important [11]. Soon after its invention, Gimzewski and Möller employed the STM to fabricate a nano-sized contact [2]. They were measuring the electrical resistance while they were approaching from the tunneling regime and finally touched gently the surface with the STM tip. After this, a nano-sized protuberance was observed. Subsequently, Dürig et al. [1] investigated the adhesion forces using an AFM.

Local elastic properties can be studied by both static and dynamic AFM methods. The static methods include force-distance curves, while the dynamic techniques track the frequency shift of a resonator vs. the distance. In order to measure both mechanical and electrical properties the microscope has to be supplemented with a force sensor. Conventional AFMs with conductive tips are usually not well suited for these experiments because the cantilever elastic constant is too low ($\leq 1\,\mathrm{N\,m^{-1}}$) to be able to fabricate and manipulate metallic nanocontacts, whose effective elastic constant is above $10\,\mathrm{N\,m^{-1}}$ [12–14]. Therefore, modified versions of combined AFM/STM techniques are required [4], using bending beams with elastic constants above $100\,\mathrm{N\,m^{-1}}$.

There are different devices to detect the deflection of the force sensor. The most commonly used method in commercial AFMs measures the deflection of a laser beam on the backside of the cantilever using a photodiode [15]. Interferometric detection is a very precise technique that detects the motion of the cantilever using the interference of light reflected from the cantilever with that reflected from the end of a fiber, which is mounted very close to the backside of the cantilever [16, 17]. Piezoresistive methods measure the change of the resistance path on the backside of the cantilever and extract its deflection [18]. Furthermore, a STM can also be used not only as a tunneling current probe but also for displacement sensing applications, see Fig. 14.1. It was used by Binnig, Quate, and Gerber in the first AFM [11]. A tunneling tip on the conductive backside of a cantilever was used to measure deflection. Such

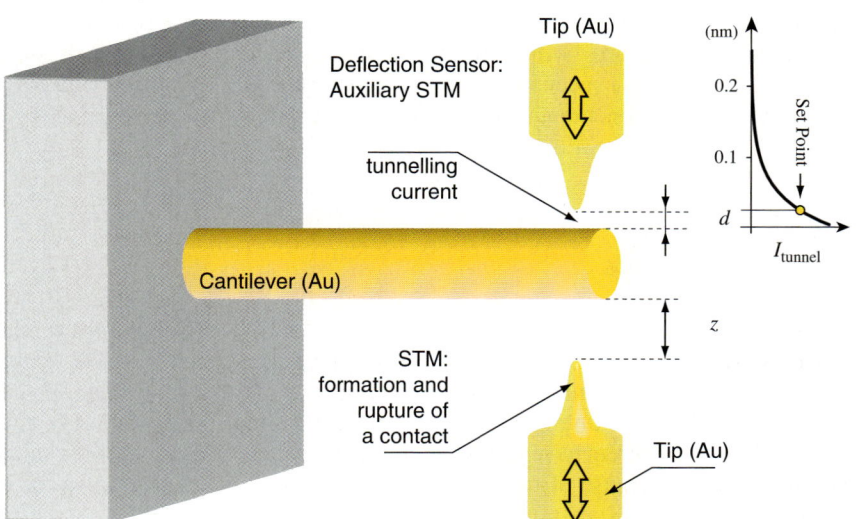

Fig. 14.1. Sketch of a STM supplemented with a force sensor. The cantilever deflection is followed by an auxiliary STM working in constant current mode, providing an extremely high-resolution measurement of the deflection of the cantilever

a sensor features extreme sensitivity for the detection of displacement, given that the current decays exponentially over a decay length of 44 pm for an apparent tunneling barrier height of 5 eV. Among the possible force-detection methods cited above, we present an implementation that uses the STM as a force sensor, because it enables us to study the mechanical properties of metallic nanocontacts at low temperatures with extremely high resolution in the cantilever deflection measurement.

In addition to the static cantilever deflection measurement, it is possible to measure the force gradient if the cantilever is forced to oscillate. The dynamic methods use some signal derived from the force between tip and sample as a feedback parameter to track the topography of a surface (for a review, see for example [19, 20]). The detection devices listed above are also used in these dynamic methods. Besides, piezoelectric detectors based on quartz tuning forks (TF) are also used as sensors [21, 22]. The TFs feature high stiffness and a high quality factor. The first feature maintains the stability of the microscope, while the latter allows precise tracking of the resonance frequency. This high quality factor is a consequence of low coupling between the preferred vibrational mode of the tuning fork where both legs move in the same plane and in opposite phase, and other modes, leading to a high quality factor. In contrast to micromachined silicon cantilevers, the quartz forks are large enough to permit attachment to one prong of a wide variety of conductors, thus enabling the use of a bulk metallic tip. An additional advantage of piezoelectric sensors when used for low-temperature applications is that the electric dissipation is negligible when compared with laser beam bounce, interferometric or piezoresistive methods. All this goes to show that such a sensor is a very suitable tool to explore the mechanical properties of metallic nanocontacts at low temperature.

14.2.1
The Scanning Tunneling Microscope Supplemented with a Force Sensor

Although very accurate, there are some important issues to be considered when using the STM as a deflection sensor, since the position of the tunneling tip has to be maintained within fractions of an Angstrom during operation. First, in order to measure the deflection of the cantilever with subatomic accuracy, one needs to keep the tip positioned over the same atom. Thus, it is necessary to consider the diffusion of atoms over surfaces since it may mislead the measurement of the cantilever deflection. Second, the STM must be carefully designed to avoid external disturbances in the tunnel current signal. This is best accomplished in designs that result in a rigid, compact STM. A rigid STM with a high resonance frequency helps to attenuate the external low frequency excitations while a compact design enables one to suspend the STM from springs. Third, another issue in SPM comes from the rather complicated relation between the mechanical deformation of the piezoelectric ceramic material used for moving the tip over the sample and the applied voltage. One may consider that the piezoelectric actuator deformation is proportional to the applied voltage. However, a hysteretic behavior may considerably modify this proportional dependence. This behavior arises from piezoelectric creep. Fourth, differential thermal contractions of the mechanical parts of the microscope can result in small but unavoidable displacement drift. Modern SPMs overcome these problems by using different correction strategies implemented as closed-loop feedback systems.

These issues are naturally overcome when using a setup in a cryogenic environment, because at liquid helium temperature (4.2 K) all thermally activated processes are hampered. Additionally, as the sensor uses the tunneling current as the control signal for the feedback loop, it is crucial to prevent the presence of adsorbates, contamination, and oxides both in the tip and in the backside of the cantilever. When operated at low temperature the STM is in a cryogenic vacuum environment that minimizes the exposition of fresh surfaces to contaminants. Furthermore, the piezoelectric actuator's non-linear behavior is strongly temperature-dependent. As a result, these non-linearities are strongly reduced when the microscope is operated at low temperatures.

However, there are some remarkable technical difficulties at low-temperature. Usually, materials with similar thermal expansion coefficients are chosen in order to minimize undesired stress that may lead to fracture of a STM part. The coarse positioning system of the STM needs some care if designed to work at low temperatures. There are several designs of piezoelectric inertial motors [23–25] or friction motors [26–28] in STMs at low temperatures. A design that follows the one by Pan [28] is used here because it has been found to be highly reliable at low temperature while allowing a compact STM design.

In order to reduce the generation of mechanical noise it is advisable to use a superinsulated cryostat that does not require liquid N_2 for thermal shielding. A compact, low-mass STM design and a low-heat loss setup minimize the liquid He consumption and thus the bubbling noise. Moreover, the high mechanical stiffness of the Dewar vessel and the insert does not amplify mechanic and acoustic vibrations. In addition to that, acoustic disturbances are amenable to being attenuated using foam adhered to its enclosure. These cautions are important because at low temperatures there are few possibilities to use a viscoelastic damping system.

14.2.2
The Mechanically Controllable Break-Junction Technique

The mechanically controllable break-junction technique (MCBJ) has its origin in the work done by Moreland and Ekin [29], but the configuration outlined below resembles more closely the one developed by Muller et al. [30]. It has been widely used during the last decade to study electron transport through metallic nanocontacts, and has been recently supplemented with a tuning-fork force sensor in order to simultaneously study their mechanical properties [12, 31].

The MCBJ is depicted in Fig. 14.2. It consists of a conducting thin wire (diameter 0.01–0.25 mm) of the material that one wants to investigate, which is fixed at two closely spaced spots on top of a flexible insulated substrate. In order to take care that the wire will break it has a notch at the position between the two spots. The notch will decrease the force needed to break the wire and can be fabricated straightforwardly in the case of ductile materials using a knife. The substrate is mounted in a three-point bending configuration, where either the central or the counter supports can be displaced resulting in controlled bending of the substrate. This bending causes the top surface of the substrate to expand, resulting in the elongation of the weakest part of the wire at the notch until rupture. In this way, two clean fracture surfaces are exposed. A contact between the fractured surfaces can be re-established by relaxing the bending force on the substrate, hence the name "break junction." A combination of micrometric screws and piezoelectric actuators are commonly used for fine control of the opening at the notch.

One main advantage of the MCBJ technique is that the freshly exposed surfaces are free of contamination, and can be kept clean if the device is kept in UHV or cryogenic vacuum. A second advantage of the MCBJ is the mechanical stability of the two electrodes with respect to each other, which results from the short mechanical loop between the electrodes. That is, the distance between the two fixed points of the wire at both sides of the notch can be as small as 0.1 mm. Lithographically fabricated MCBJs present an outstanding stability since the interelectrode distance is reduced

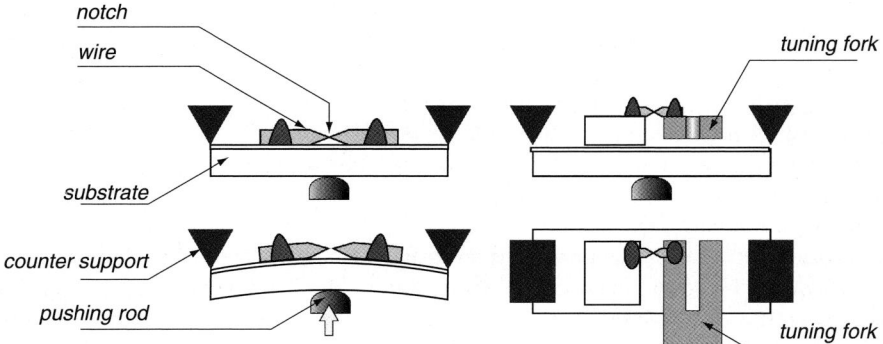

Fig. 14.2. Schematic top and side view of a MCBJ (*left*) and a MCBJ supplemented with a tuning fork (TF) resonator. The TF resonance frequency is sensitive to the force gradient between the tips

to $\sim 1\,\mu$m [32]. On the contrary, STM has scanning capabilities and permits the fabrication of contacts between two different metals, two features that are not easily incorporated in MCBJ implementations.

The MCBJ supplemented with a force sensor, based on a microfabricated quartz tuning fork resonator, enables simultaneous current and force-gradient measurements [12, 31]. The frequency shift of the TF can be accurately tracked using a frequency modulation technique [33]. Tuning forks have recently been introduced in different scanning probe microscopy setups, such as AFM [21, 34–36], Magnetic Force Microscopy (MFM) [37], and Scanning Near-field Optical Microscopy (SNOM) [38, 39]. It has very good properties as a resonator because its high quality factor leads to high frequency resolution, while its high stiffness avoids jumps to contact [40] and anharmonic vibration. A description of a MCBJ supplemented with a tuning fork follows, as shown in Fig. 14.2.

Commercially available tuning forks are inside a metal cap. This cap is partially removed leaving only a little ring around the base. The ring is then soldered onto a substrate, so that the prongs of the fork are free standing with the preferred oscillation mode parallel to the substrate (see Fig. 14.2). Next to the tuning fork, a little metal block is mounted on the substrate, such that the top of the block and the top of the tuning fork are at the same height. A thin metal wire is soldered on both one prong and the block with a notch in between the prong and the block.

In the experiments by Rubio-Bollinger et al. [12] the TF electrodes were used for both the excitation and the resonance detection. Valkering et al. [31] have used an external excitation source based on a magnetic actuator. This resonant frequency is tracked by implementing a phase-locked loop oscillator [12, 20, 39, 41, 42]. The shift of the resonance frequency Δf of the TF excited with vanishing amplitude is proportional to the force gradient or stiffness k of the interaction between the electrodes: the measured resonance frequency $f_0 + \Delta f$ is given by

$$f_0 + \Delta f = \frac{1}{2\pi} \sqrt{\frac{k_{\text{lever}} + k}{m_{\text{eff}}}} \tag{14.1}$$

with m_{eff} the effective mass in resonance. Since the factor $\alpha = \Delta f / k$ depends on the mechanical properties of the TF, a high spring constant k_{lever} reduces the sensitivity of the force derivative measurement.

Compared with previous measurements of atomic contact forces using atomic force microscope cantilevers, the use of a TF has a series of advantages. First, the stiffness (10^3 N m^{-1} [21, 42]) of the lever supporting the contact is high enough to prevent early jump to contact. Second, TF has a very high quality factor [21] (larger than 10^4 in the experimental configuration in a vacuum at low temperature [36, 42]), permitting a precise measurement of the frequency shift. Moreover, it allows for a low power electrical measurement of the resonance, which is particularly convenient for low-temperature experiments. In addition, it is possible to use a TF to form and break the contact. This avoids the usual uncertainties in the calibration of the interelectrode distance in MCBJ experiments. One can control the interelectrode distance using the piezoelectric effect of the TF itself by applying a dc voltage in addition to the ac drive. This piezoelectric displacement can be calibrated at room temperature.

This calibration does not change at low temperature because the TF is fabricated from a single crystal of quartz.

To sum up, this setup allows us to make use of the high mechanical stability of the MCBJ while the tuning fork has a high spring constant of $10,000 \, \mathrm{N \, m^{-1}}$, which in combination with a high quality factor, permits detection of very small frequency shifts (smaller than 100 mHz) and at small vibration amplitudes (smaller than 10 pm).

14.3
Electron Transport Through Metallic Nanocontacts

It is possible to extract information on the contact mechanical properties from magnitudes related to the electronic transport, because at the scale of the few atoms forming the nanocontact these mechanical and electronic transport properties are strongly related.

Mostly, these electronic transport properties are determined by only a few atoms and thus provide rich information related to the size, shape, state of strain, and elastic deformation of metallic nanocontacts [3].

In metallic contacts the electronic transport description changes as their dimensions are reduced down to the nanoscale. Macroscopic conductors are characterized by Ohm's law, which establishes that the conductance G of a given sample is directly proportional to its transverse area S and inversely proportional to its length D, i.e.

$$G = \sigma S/D, \tag{14.2}$$

where σ is the electrical conductivity of the sample.

As the typical length scale of our sample L is reduced, one can identify different transport regimes. These regimes are defined by the relative size of various length scales. The differences between these regimes will become clear as we move through them one by one. In metallic point contacts, an important length scale is related to the phase-coherence length L_ϕ, which measures the distance over which quantum coherence is preserved. As long as $L >> L_\phi$ the contact is in the macroscopic regime, whereas if $L << L_\phi$ the contact is in the so-called mesoscopic regime. Another relevant scale is associated with the elastic mean free path l, which defines the change from the diffusive regime if $L >> l$ to the ballistic regime if $L << l$ (see Fig. 14.3).

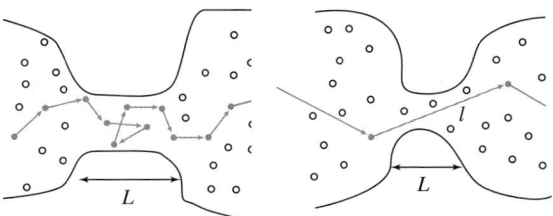

Fig. 14.3. Scheme of the electronic transport in metallic conductors: diffusive (*left*) and ballistic regimes

In the classical limit, the conductance of a constriction was calculated by Maxwell for a constriction of hyperbolic geometry [43], and is proportional to the constriction radius a and the conductivity $G \propto a\sigma$. When the dimensions of a contact are much smaller than their mean free path, the electrons will pass through ballistically. In such contacts, the large potential gradient near the contact accelerates the electrons within a short distance. The conduction through this type of contact resembles the effusion of a dilute gas [44]. Sharvin first solved this problem using a semiclassical approximation [45], providing an expression of the conductance for a pure ballistic contact, the so-called Sharvin's conductance

$$G = \frac{2e^2}{h}\left(\frac{k_F a}{2}\right)^2,\tag{14.3}$$

where h is the Planck constant, e is the electron charge, k_F is the Fermi wave vector, and a is the contact radius. Sharvin's conductance depends on the material only through k_F, and it is completely independent of conductivity and the mean free path, contrary to the conductance in the diffusive regime. Sharvin's formula may be used in ballistic contacts to estimate the area of minimal cross-section from the conductance. The quantity $2e^2/h$ is called the conductance quantum G_0. Corrections to this formula where given by Torres, Pascual, and Saenz using an exact quantum calculation for a circular cross-section [46].

The ultimate atomic contact has one atom between two electrodes. In this case, a fully quantum description is needed because the quantum effects are increasingly important as the size of the constriction is comparable to the Fermi wavelength of the conduction band electrons. It has been shown that the conductance of a one atom contact depends dramatically on its chemical nature [47].

14.4
Mechanical Properties of Metallic Nanocontacts

Nanocontacts are structures where the number of atoms in the minimum cross-section of the constriction ranges from tens of atoms to a single atom. Although there are several ways of preparing such structures [48, 49], the experiments at low temperature usually involve the use of probes related to STM or MCBJ. This section is dedicated to the fabrication procedure of nanocontacts between gold electrodes, whose electronic and mechanical properties have been extensively studied in the past decade.

14.4.1
Fabrication of Metallic Nanocontacts

Single atomic-sized contacts between metals can be produced with a scanning tunneling microscope. In its constant current operation mode the tip is scanned over the sample surface without making contact. The tip–sample separation is kept constant by controlling the current that flows between tip and sample, due to the tunneling effect, when a fixed bias voltage is applied between them. Typical currents are $\sim$1 nA, corresponding to a tunneling junction resistance in the GΩ range, for tip–sample voltages $\sim$1 V and the tip–sample distances $\sim$0.1 nm.

However, the STM was soon used to modify the sample surface on a nanometer scale. In the experiment by Gimzewski and Möller [2] the surface was gently touched with the tip and the transition from the tunneling regime to metallic contact was observed as an abrupt jump in the current. This jump in current was due to a change in resistance of about 13 kΩ. According to Sharvin's formula, a conductance of 1 G_0 corresponds to a contact diameter of 0.25 nm, suggesting a mechanical jump to contact forming a single atom bridge between the tip and the surface.

After the jump to contact, the indentation continues and the conductance increases showing a characteristic staircase pattern. In the case of gold subsequent retraction of the tip results in breaking the contact: the neck between the two electrodes gets thinner as the sense of the tip motion is reversed. Again, the conductance has intervals in which it is relatively constant (plateaus) separated by jumps. The last plateau, in the case of gold, shows a quite well-defined conductance with a value of approximately of 1 G_0. For other metals the conductance during the indentation cycle will look somewhat different, depending on the electronic structure of the metal [47] but the stepwise behavior of the conductance is still observed. However, it has been reported that in some cases the jump does not occur. Instead, there is a continuous increase of current from tunneling to contact: it occurs for some metals, in particular for Ni, W, and Ir [3].

Figure 14.4 shows an indentation cycle in gold without breaking the contact. The bias voltage is 10 mV using a STM supplemented with a force sensor. As the contact

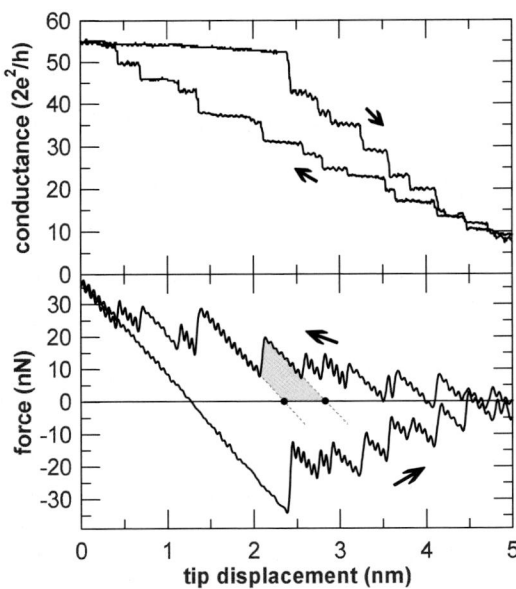

Fig. 14.4. Simultaneous measurement of conductance and force during an indentation cycle, for Au at 300 K. The *arrows* indicate compression, (positive forces) and elongation of the nanocontact. Reprinted with permission from [52]. Copyright (1997) by the American Physical Society

size increases, the conductance increases showing a step-like behavior. Nevertheless, the measured curve shows a force dependence on the distance that looks like a saw-tooth signal. There is a sequence of linear stages separated by sudden relaxations. The simultaneously measured conductance remains almost constant in the elastic stages, and there is an abrupt conductance change when the force relaxes. Landman et al. [5] predicted this behavior in early molecular dynamics (MD) simulations. This yielding results in a sudden change of the minimal cross-section due to the atomic rearrangement. These processes result in a sharp jump in the contact conductance. The relaxation in the tensile force is the result of a change in the length of the constriction. This change of length is the distance between successive relaxed configurations in the nanocontact separated by a yielding event, marked as two black dots in Fig. 14.4. In gold nanocontacts, this plastic relaxation length ranges from 0.2 to 1 nm, which implies that only a few atomic layers participate in the plastic deformation process. This dependence of both conductance and force vs. the electrode separation is unique since the arrangements are different for every indentation cycle.

14.4.2
Elasticity and Fracture of Metallic Nanocontacts

The simultaneous measurement of conductance and tensile force during the indentation cycle described above provides information about the different mechanical behavior of macroscopic specimens and metallic nanocontacts. Macroscopic objects change their shape when they are deformed by a stress. As long as they regain their original shape when the load that causes the stress is removed, they are subjected to an elastic deformation. In the elastic regime, the stress applied to an isotropic material is linearly related to the strain by Young's modulus and Poisson's ratio by the generalized Hooke's law (see for example [50]). Consider the simple case of a rod of length L_z and diameter b subjected to a stress through its longitudinal axis z. The stress s is proportional to the stress uniaxial strain $\delta L_z/L_z$, being the proportionality constant Young's modulus. At the nanoscale, elastic strain is associated to a change of the interatomic distance. Thus, Young's modulus measures the resistance of the bonds between atoms to deform. The effective stiffness of the nanocontacts can be extracted from the slope of the elastic stages of the force vs. separation curve.

When a macroscopic object subjected to a load does not recover its original shape after the load is removed, but is deformed in a permanent, non-recoverable fashion, the deformation lies in the plastic regime. The elastic regime holds for most metallic materials when the strain is smaller than 0.005. Higher strain values cause a transition from elastic to plastic regime on yielding. In macroscopic samples, it is difficult to fix the lower limiting stress below which no plastic deformation is appreciable because the transition from plastic to elastic regime takes place gradually. The yield strength is conventionally defined as the stress necessary to produce a plastic strain of 0.002 under uniaxial stress. This yield strength depends on heat treatment, sample purity, and prior deformation of the sample. On the contrary, Young's modulus is not sensitive to these factors.

Plastic deformation is equivalent at the nanoscale to permanent change in the positions of the atoms: they do not recover their original positions after the load is removed. Hence, when subjected to a plastic deformation, the atoms break the bonds with their neighbors and form new bonds with new neighbors. The simplest model of plastic deformation of a perfect crystal, that is, one with no defects, considers the sliding of two compact planes with respect to each other. Frenkel [51] calculated the maximum shear stress required for this process to occur and found a value of $\tau_{max} \approx G_{shear}/30$, where G_{shear} is the shear modulus of the material. This value of the shear modulus is much larger than observed in macroscopic metal specimens. This discrepancy is explained by the presence of dislocations, which can glide at low stress values. Experiments on whiskers find high values of the shear stress, and even higher values are expected in nanocontacts, where dislocations are expelled from nm-volume regions [5].

The maximum pressure on nanocontacts can be extracted from the combined force and conductance measurements before the relaxation, see Fig. 14.4. In ballistic contacts, the area of minimum cross-section and the conductance are related through the Sharvin formula. The apparent pressure found in gold nanocontacts ranges from 3 to 6 GPa for the compression and elongation parts of the indentation cycle. This value is much larger than the maximum pressure that a macroscopic sample can sustain. Moreover, it is of the same order of magnitude as the theoretical value in the absence of dislocations and is consistent with the theoretical maximum shear stress value for gold. A comparison of the elastic constant of nanocontacts with continuum mechanics models shows good agreement [4].

The energy dissipated in each force relaxation, that is, the energy necessary to produce a configurational change, can be directly obtained from the force cycle. In Fig. 14.4, the energy to pass from one configuration to the next is shown by the area in gray. The value of this energy is of the order of 0.1-eV per atom in the minimal cross-section of the contact. If we compare that with the heat of fusion (0.13 eV/atom) we find that configurational changes take place only at the zone around the narrowest part of the nanocontact.

14.4.3
The Shape of Metallic Nanocontacts

The shape of nanocontacts has been studied using a STM. Local modifications on a surface have been produced by approaching a Pt–Ir tip to a Ag substrate in UHV, touching the surface and imaging after tip indentation [2]. The surface after a gentle indentation cycle of the tip into substrate exhibits a nanometer-sized protrusion. This is attributed to the formation of a small neck while pushing the tip that subsequently is stretched and broken.

In order to estimate the shape of the constriction Untiedt et al. [52] used a slab model for the constriction. Experimentally, the conductance vs. displacement curves are recorded during an indentation cycle (Fig. 14.5) [52]. When the constriction is subjected to a force, the stresses are highest in the narrowest part, with cross-sectional area A_i. Therefore, one may assume that plastic deformation takes place in that narrowest slab, in a zone of depth λ, leaving the rest of the neck unmodified. In this model, a new slab is formed from a piece of length λ_i and area A_i. The new slab cross-sectional area A_{i+1} is given by volume conservation, being $\lambda_i + \Delta l$ the length of the new slab, where Δl is the deformation. Given that only the narrowest

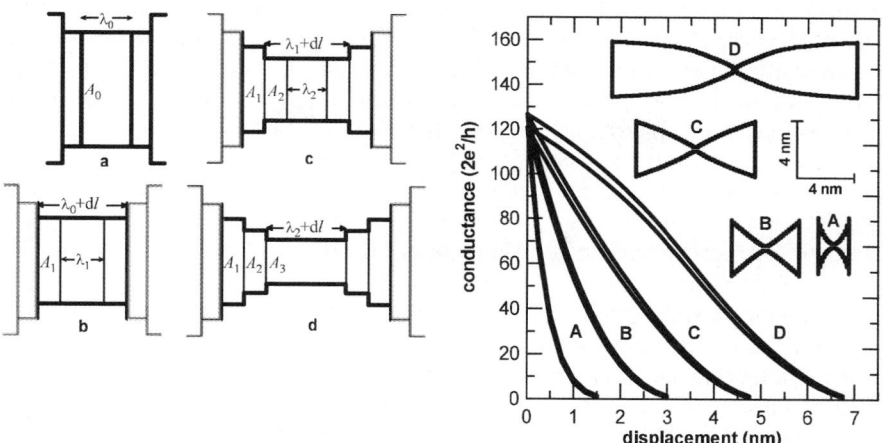

Fig. 14.5. The *left panel* shows the plastic deformation of a nanocontact using the slab model. From one configuration to the next only the central slab elongates, while the rest of the neck does not change. The *right panel* shows calculated shapes and conductance for the contacts of Fig. 14.6 using the slab model. Reprinted with permission from [52]. Copyright (1997) by the American Physical Society

slab is modified, the shape of the constriction after a number of plastic deformation processes results in a sequence of values of A_i and λ_i. The plastic deformation length λ_i can be obtained from the experimental $G(z)$ curve noting that for the limit $\Delta l \rightarrow 0$, $\lambda = -(d \ln A/dl)^{-1}$, where A is the cross-section of the narrowest portion of the constriction.

As is shown in Fig. 14.6, gold nanocontacts may exhibit different behaviors in the conductance curve corresponding to different indentation cycles. They have markedly different slopes in the conductance curves. From the slopes of the

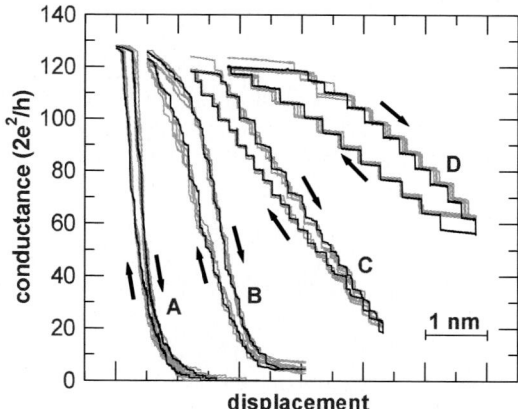

Fig. 14.6. Conductance curves $G(z)$ for four different sets of indentation cycles for Au at 4.2 K. Each set consists of five indentations. Reprinted with permission from [52]. Copyright (1997) by the American Physical Society

conductance curves Untiedt et al. obtained the constriction shape (Fig. 14.5, right panel) using the described slab model. The analysis indicates that the steeper $G(z)$ curves correspond to constrictions with larger opening angle and involve shorter plastic deformation lengths. Hence, only one atomic layer is involved in plastic deformation at yielding events in constrictions with this shape.

14.4.4
Inelastic Scattering by Phonons in Nanocontacts

Up to this point, only elastic scattering has been considered, but when applying a bias voltage in order to measure conductance, there are also effects derived from inelastic scattering that can be related to mechanical properties. In ballistic point contacts such as nanocontacts, the dominant inelastic scattering mechanism is electron-phonon scattering. This interaction has been shown to be a useful tool when studying nanocontacts. The derivative of the differential conductance of a point contact contains information about the inelastic electron backscattering. The Eliashberg function for the electron-phonon interaction in the point-contact situation, or point-contact spectroscopic (PCS) curve, is proportional to the density of states of phonons at a given energy $F(eV)$ times a factor related to the electron-phonon interaction α^2, as shown in [53], by

$$\alpha^2 F = -\frac{3}{32\sqrt{2}} \frac{h^{3/2} k_F^2}{4\pi^2 m_e} G^{-3/2} \frac{d^2 I}{dV^2}, \qquad (14.4)$$

where $G = (dI/dV)$ is the differential resistance and m_e is the electron mass.

The amplitude of the phonon-induced peaks is reduced if there is elastic scattering, for example, due to impurities or defects. Consequently, large PCS amplitude indicates that the constriction is indeed ballistic. Hence, it is possible to obtain experimental information on the degree of disorder in the constriction using point-contact spectroscopy [54].

Untiedt et al. [52] found that in a point contact at low temperature the PCS curve does not change with the conductance as the derivative of the conductance scales with $G^{-3/2}$ as given by Eq. (14.4). Figure 14.7 shows the PCS curves corresponding to contacts obtained from indentation cycles similar to D, C, and B in Fig. 14.6. The PCS curve does not change with the conductance since d^2I/dV^2 scales with $G^{-3/2}$, but the amplitudes of the phonon peaks are different for different realizations of the contact.

For a contact such as D, the amplitude of the phonon peaks is maximum (the black curve in Fig. 14.7) and similar to that of previously reported spectra [55] for ballistic point contacts. This is an indication of the degree of order in these contacts that can be attributed to a crystalline contact [52]. In contacts such as B and C, the amplitude of the phonon peaks in the PCS curve is reduced (the gray curves in Fig. 14.7) due to elastic scattering in the neck region, indicating the presence of defects, but still the necks are far from being disordered. From this evidence, they concluded that nanocontacts are ballistic, and are not disordered. This idea is supported by the results of the MD simulations [5, 56].

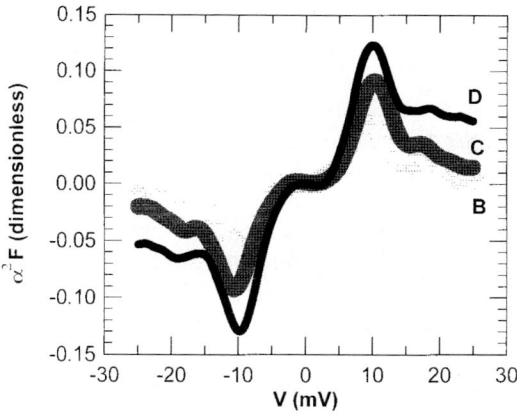

Fig. 14.7. Point contact spectroscopy curves at 4.2 K for Au nanocontacts. The *black curve* corresponds to a contact such as D in Fig. 14.6. The *gray curve* corresponds to type C and the *light gray curve* to type B. Reprinted with permission from [52]. Copyright (1997) by the American Physical Society

A comparison with the plastic deformation length λ for these necks at the largest radius (approximately 1.9 nm) corresponding to contacts with conductance 120 G_0 shows large plastic deformation lengths in contact with high crystalline order. For curve A, $\lambda = 0.2$ nm; for curve B, $\lambda = 1$ nm; for curve C, $\lambda = 2.8$ nm. In addition, for curve D λ about 6 nm. Given that λ is related to the amount of material involved in the plastic deformation in the contact D there are many layers involved in the process of plastic deformation, many more than those involved in the case of the contact A.

14.5
Suspended Chains of Single Gold Atoms

Experiments by Ohnishi et al. using a Transmission Electron Microscope [48] and Yanson et al. [57] using a MCBJ and a STM, showed that when a gold nanocontact is stretched, it may form a chain of single atoms. These structures are not only the ultimate nanowire, but they are also an ideal test bed for mesoscopic physics. Atomic chains of gold atoms have distinctive physical properties: they can sustain very large current densities of up to 8×10^{14} A m^{-2}. This fact supports a main assumption of ballistic transport, since it is possible only if the transmission through the chain is ballistic and when most of the power is dissipated in the electrodes, far away from the contact. Furthermore, given that these atomic gold structures are stable for as long as 1 h or even longer at low temperature, it is possible to test peculiar physical properties since atomic chains are close to ideal one-dimensional metallic systems. Finally, not only gold can form chains: Smit et al. [58] reported that platinum and iridium spontaneously can form chains of atoms when pulling a one atom-contact.

14.5.1
Fabrication of Chains of Atoms Using Local Probes

The evolution of conductance during the breaking of a gold nanocontact is shown in Fig. 14.8. The conductance decreases while pulling from the contact, down to a conductance value close 1 G_0. As discussed already in Sect. 14.3, a conductance of 1 G_0 in gold corresponds to a contact with a cross-section of one atom. The experiments show that a one-atom contact accommodates a maximum elastic deformation below 0.25 nm. However, it was discovered that gold nanocontacts exhibit sometimes a different behavior: this one atom-contact can be further stretched by a distance larger than 1 nm, without the conductance deviating appreciably from 1 G_0, showing up in Fig. 14.8 as a very long conductance plateau.

As was discussed in Sect. 14.4, the breaking process of metallic nanocontacts by controlled separation of the electrodes takes place in a sequence of elastic deformation and abrupt yield stages resulting in a non-continuous reduction of the minimal cross-section of the contact. It has been shown that for some metals this process takes place down to the smallest contact, a single atom contact between the electrodes [4]. Further separation of the electrodes usually results in breakage of the metallic contact and an abrupt jump into the tunneling regime. The explanation is that this stretching takes place at the position of this one atom contact. Yanson et al. [57] concluded that a single atom does not have to break, but that two atoms forming a short chain between the electrodes can replace it. As the contact is stretched even further, this chain will often break, but it does have a finite probability to be replaced by a chain containing more atoms. Once the chain starts being pulled the conductance never exceeds 1 G_0, confirming that the chain acts as one-dimensional nanowire. When the chain finally breaks, the electrodes have to travel back a return distance to reestablish metallic contact. This return distance is almost equal to the length of the

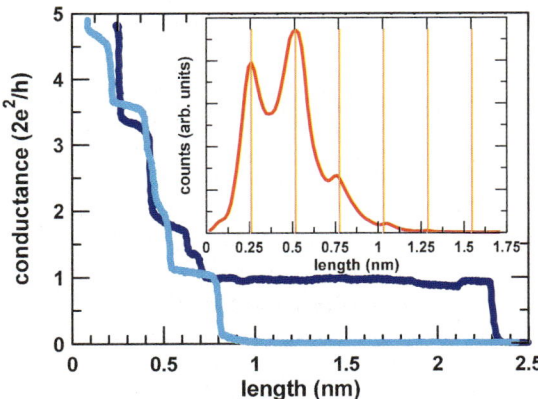

Fig. 14.8. Evolution of the conductance while pulling from the gold electrodes, extracting an atomic chain (*black curve*) at 4.2 K. The last plateau of the conductance corresponds to an atomic chain about six atoms long. *Inset*: histogram of the last plateau length made from 10,000 indentation cycles that exhibit peaks at 0.25 nm. Reprinted with permission from [79]. Copyright (2002) by the American Physical Society

last plateau itself, suggesting that after the chain breaks, its constituent atoms collapse onto the electrodes on either side. In addition to that, Yanson et al. [57] showed by the use of a STM that they can swing one of the electrodes sideways at any given position at the last plateau. They found that the lateral displacement that the structure can support increases as the length of the plateau increases, just the behavior that one would expect for a chain.

The conductance traces for successive nanocontact ruptures do not reproduce in detail, as they depend on the exact atomic positions in the contact and not every contact rupture results in the formation of a chain of atoms [57]. The definition of the length of the plateau, and thus the length of the chain, was therefore given as the distance between the points at which the conductance drops below $1.1\,G_0$ and at which conductance drops below $0.5\,G_0$. The probability of formation of such a structure can be quantified by constructing a histogram of the last plateau length. As is shown in Fig. 14.8, the histogram exhibits a series of equidistant peaks rather than a smooth distribution. The peaked structure of the histogram shows that atomic chains tend to be elongated by integer multiples of $0.25 \pm 0.2\,\mathrm{nm}$, which is close to the nearest neighbor spacing of gold atoms in the crystal. In addition, it is shown that the probability of pulling a chain of length L decreases rapidly for large L.

Despite the low probability of formation of chains, once an atomic chain is pulled, the retraction of the electrode can be stopped and the chain remains very stable at liquid helium temperature (4.2 K): some of the longest chains obtained in the experiments have been held stable for at least 1 h. This makes atomic chains suitable for investigation of one-dimensional electron transport and for studies of wear and fracture on low-coordinated metallic nanostructures.

14.5.2
Mechanical Processes During Formation of Atomic Chains

The mechanical processes involved during the formation and rupture of atomic chains of gold have been studied both experimentally and theoretically, see [3] and references therein. A STM supplemented with a force sensor (see Sect. 14.2) has been used to study the force in gold chains at low-temperature [13]. Using this probe it is feasible to study the force evolution simultaneously with the conductance while drawing out the chain.

Figure 14.9 shows a simultaneous measurement of force and conductance. The force shows a sequence of linear stages separated by sudden relaxations. The conductance on the last plateau remains just below to $1\,G_0$. There are small conductance jumps related to force relaxations, but their magnitude is much smaller than $1\,G_0$. In stages with a linearly growing tensile force the chain is stretched, while at the force jumps abrupt atomic rearrangements occur.

If an atomic chain is formed or not while pulling a nanocontact relates to the relative strength of different bonds in specific atomic configurations. The reason is that breaking a nanocontact involves breaking many individual atomic bonds. One simple model consists of a chain connected to the electrodes on each side. In an elastic stage, as the chain is drawing out the force increases. When the system is stretched so that the tensile force value is higher than a critical force the weakest of

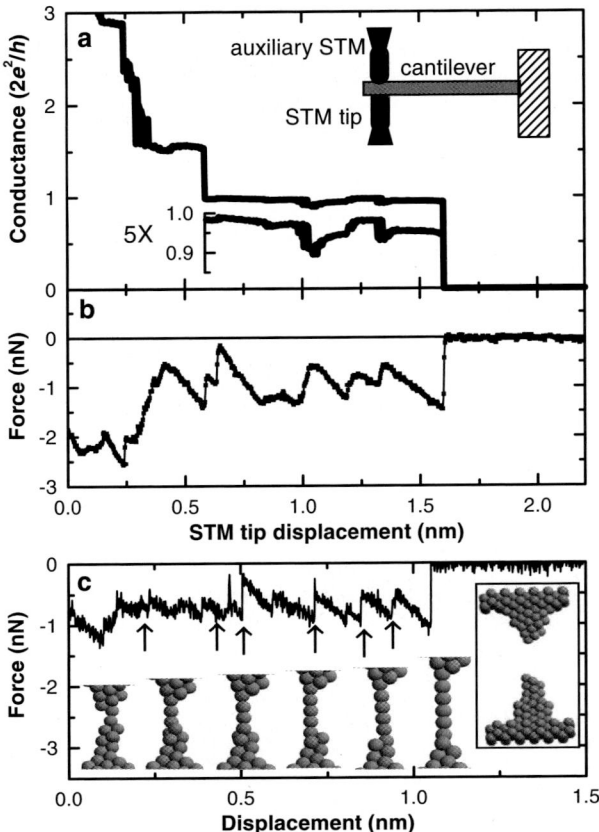

Fig. 14.9. *Top panel*: conductance (**a**) and tensile force (**b**) measured simultaneously while making and breaking a chain of gold atoms at 4.2 K. *Bottom panel*: force calculated from a MD simulation. *Arrows* indicate where a new atom pops into the chain and snapshots of the structure at these positions are shown. *Inset*: scheme of the experimental setup, a combined STM-AFM. (**c**) Calculated force during MD simulations. Reprinted with permission from [13]. Copyright (2001) by the American Physical Society

the bonds will break. If the slip of an atom from an electrode into the chain requires a smaller force than the force needed to break the chain, then the atom is added and the chain grows. This is not usually the case because pulling an atom out of a surface implies breaking more bonds than to break the chain. Nevertheless, MD simulations sustain the hypothesis that an atom could slip into the chain at lower forces [13].

In order to examine a more realistic model it is important to take in account the fact that the bond strength increases as the coordination number is reduced. Suppose a contact where the electrodes have a pyramidal shape. The atom sitting at the apex of one of these pyramids has three bonds with its neighbors placed in the pyramid and one with the chain. Given that the bond strength increases as the coordination number is reduced it may be favorable for the atom in the apex to break a bond with the underlying atoms. Moreover, in this model this atom can be incorporated into the

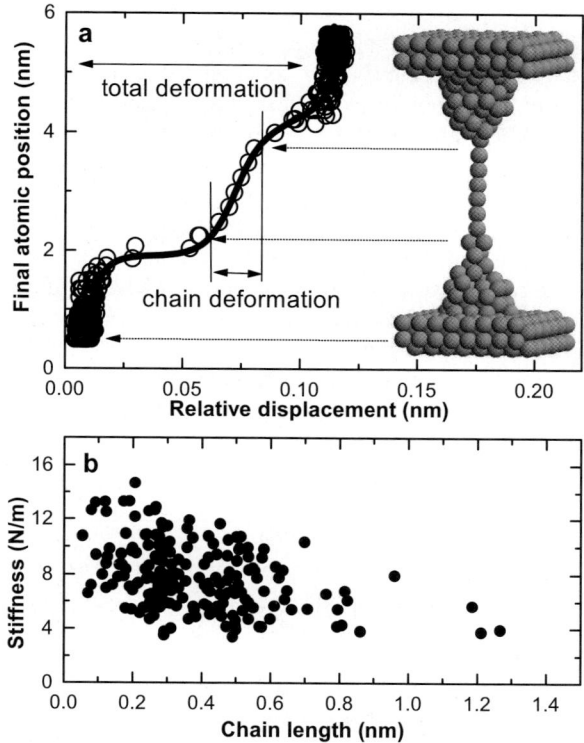

Fig. 14.10. MD simulation of the relative displacement during an elastic deformation stage. The *bottom panel* shows the measured chain stiffness just before rupture as a function of chain length. Reprinted with permission from [13]. Copyright (2001) by the American Physical Society

chain breaking only one bond. MD calculations (see [13, 59] and references in [3]) show that generally larger force jumps correspond to the incorporation of an atom into the bridging atomic chain. An experimental fact that supports this hypothesis is that mechanical relaxations take place at force values smaller than the final breaking force. A further analysis needs to take into consideration the detailed configuration of atoms at the apex of the electrode as well as the relative strength of their bonds.

MD simulations [13] show that the elastic deformation is accumulated in the electrodes, see Fig. 14.10. This is explainable because the electrodes can be deformed not only by stretching of the interatomic distance. It is also possible that atoms in the electrodes are sitting in arrangements where the breaking of bonds is the result of a more concerted motion of atoms that requires smaller forces given its longer paths. In addition, the atoms in the chain have stronger bonds due to its low coordinated situation. The sum of these two effects results in a peculiar feature of this nanostructure: thinner is actually stronger.

Some further mechanisms behind the formation of an atomic chain have been recently studied in Pt chains at low temperature [60]. Using a MCBJ supplemented with a force sensor, the authors found a correlation between the stiffness and the number of atoms in the chain. A comparison between the mean total stiffness (i.e. the electrodes plus the chain) is found to be higher if the chain breaks when pulling than if one atom is added to the chain. They concluded that longer chains have lower stiffness. This could be supported by the fact that a chain can sustain a maximum force that is independent of its length, as was reported by force measurements on Au chains. Roughly speaking, for the same elongation a larger force acts on the chain that has a larger starting stiffness, breaking it at a shorter length.

Rubio Bollinger et al. [12] have used the MCBJ technique and obtained high-resolution measurements of the contact stiffness. Figure 14.11 shows simultaneous measurement of the conductance and the force gradient while pulling a chain of gold atoms. The atomic rearrangements result in sudden drops in the stiffness that reflect bond weakening due to extreme strains close to yield. These drops are also observed during the atomic chain formation, whose conductance, in contrast remains close to 1 G_0. This mechanical behavior agrees with direct force measurements in atomic chains and is consistent with the observed longitudinal phonon frequency decrease in one-atom contacts and chains, as is shown in the next section.

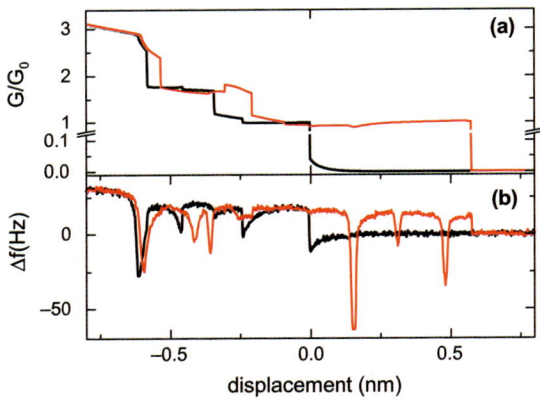

Fig. 14.11. Simultaneous measurement of conductance (**a**) and the stiffness (**b**) during elongation of a chain of atoms. Reprinted with permission from [12]. Copyright (2004) by the American Physical Society

14.5.3
Phonons in Atomic Chains

Dynamical mechanical properties of nanocontacts and chains can be probed measuring electron transport and using point-contact spectroscopy, because of finite electron-phonon scattering. Using PCS techniques, Agraït et al. have studied the evolution of these phonon modes when a chain of gold atoms is stretched [61].

In atomic chains of gold, the electron flow should be ballistic. As the mean free path of the electrons is larger than the device length, electrons are supposed to move freely: there is neither scattering nor defects to inhibit resistance-free current. In a simplified picture, the zero-bias conductance $G(0)$ of an atomic chain is the conductance quantum G_0 [62–64]. Neglecting further effects [65], this zero-bias conductance is a consequence of the coupling between the chain and the reservoirs of electrons, the electrodes. This is clear from the Landauer approach to electronic transport [66, 67], which can be applied to atomic chains. When a small bias voltage is applied between the two electrodes, there is an imbalance of population at the electrodes and a net current flows. Subsequently, there is heat dissipation from the resistance. The chain is considered to be a perfect conductor, and this dissipation is attributed to the relaxation of the electrons to the Fermi level in the electrodes and far away from the chain.

Experimentally, it has been found that the zero-bias conductance of an atomic gold chain at low temperatures is close to 1 G_0 [57]. This chain has one open channel in agreement with theoretical calculations [62–64]. In experiments at liquid-He temperatures by Agraït et al. [61, 68], the conductance was measured using the lock-in technique, and the derivative of the differential conductance dG/dV was calculated numerically. As soon as the voltage was swept over a range of 20 mV from zero, the conductance was found to depend on the voltage. Typical differential conductance curves for short and long atomic wires are shown in Fig. 14.12b, c, d. The differential conductance G features a hump at zero bias, dropping about 1% in the range (20 mV). Often the differential conductance curves look asymmetrical, showing oscillations due to elastic scattering [69, 70].

A voltage-dependent conductance is observed in ballistic point-contacts of much larger size, where it is associated to inelastic scattering of electrons with phonons and other elementary excitations taking place in the bulk [53, 71]. These processes are voltage-dependent because the electrons have to be injected with enough energy to emit an excitation. In the case of phonons, the derivative of the conductance shows peaks that correspond to peaks in the phonon density of states. The amplitude of the signal for contacts of different sizes is proportional to $G(0)^{3/2}$. This reflects the fact that only those electrons scattered in the immediate vicinity of the contact have a significant probability of coming back through the contact. The standard spectra for Au have peaks at 10 and 18 mV, corresponding to the maxima in the transverse and longitudinal phonon Density of States (DOS). In those spectra, the transverse peak is found to be stronger than the longitudinal peak.

Atomic chains exhibit markedly different features. First, the peaks in the PC spectrum vary for each chain as is shown in Fig. 14.12. In the case of a one-atom-long atomic wire, labeled S, the hump in conductance signal is narrower than in a larger wire, labeled M, supposed to be 0.3-nm longer. This distance is enough to accommodate one more atom in the chain. The peak of the spectrum is about three-times larger than that given by the semiclassical theory of PC spectroscopy [53, 71].

The wire labeled L corresponds to a long wire about seven atoms long, see Fig. 14.12a. In this long wire, Agraït et al. [61, 68] recorded the curve $G(V)$ at different elongations of the chain. The conductance curves, and subsequently the spectra, show a behavior that is considered to be a fingerprint of a one-dimensional system. The conductance curves show drops that take place quite sharply, and result in sharp peaks in the spectra.

Figure 14.12d shows that the position of the peak shifts as a function of the strain in the wire. Like the pitch of a guitar string, there is a shift as a function of the tension, but for atomic wires the frequency decreases because of the decreasing bond strength

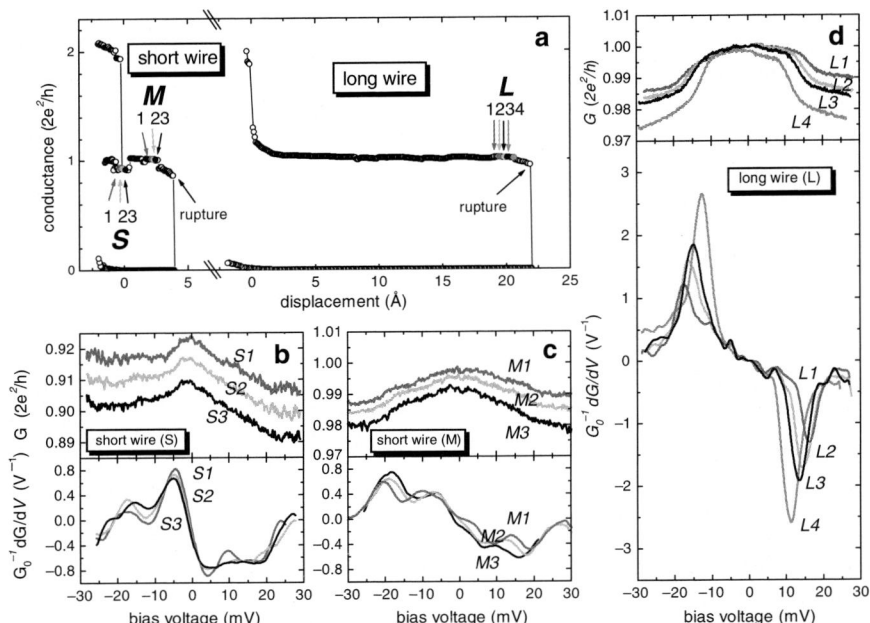

Fig. 14.12. Panel (**a**): conductance at finite bias of two gold chains of different lengths. The *short chain* is about 0.4 nm and the *long chain* is 2.2-nm long. Panels (**b**), (**c**), and (**d**): differential conductance and its derivative at points *S*, *M*, and *L*, respectively, marked by the *arrows*. The various *curves* in (**b**), (**c**), and (**d**) were acquired at intervals of 0.03, 0.03, and 0.05 nm, respectively. Note that the vertical scale these panels are identical. Reprinted with permission from [61]. Copyright (2002) by the American Physical Society

between the atoms. As the frequency decreases, the amplitude increases in an elastic stage until an atomic rearrangement takes place. The increase in the amplitude is attributed partially to the softening of the phonon modes with tension.

In a one-dimensional system at zero temperature, momentum is conserved, and, consequently, electrons can only excite longitudinal vibrations. Since momentum is conserved, electrons can only excite longitudinal vibrations of the atomic chain whose wave number is twice the Fermi wave number k_F, hence only one phonon mode will be shown in the spectra. The Fermi wavevector k_F in a linear atomic chain with interatomic distance a and one conduction electron per atom is equal to $\pi/2a$. Then, if the energy of electrons is below a threshold voltage $\hbar\omega_{2k_F}/e$, they will not interact with phonons (see Fig. 14.13). A sudden decrease in conductance marks the onset of the phonon emission process. In the experimental conductance curves the drop in conductance is somewhat rounded due to the non-zero temperature, the thermal smearing is 2 meV for 4.2 K, see [72].

Given that the chain length is finite, there is an uncertainty in the momentum of the electron that also contributes to round the drop in the conductance signal. A linear monoatomic chain of N atoms will have N longitudinal vibrational modes. For an atomic chain of length L coupled to rigid electrodes, the wavelengths of the different

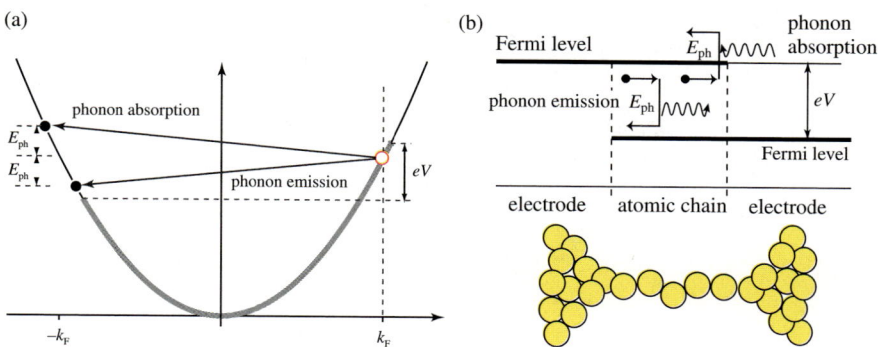

Fig. 14.13. *Left Panel*: Allowed inelastic transitions in a one-dimensional band. The only allowed transitions are from a state with wavenumber k_F to a state with wavenumber $-k_F$, that is, electrons can only interact with one phonon whose wavenumber is $2k_F$. For an atomic wire with a single conduction electron per atom, $k_F = \pi/2a$, where a is the interatomic distance. Note that the figure is not drawn to scale: phonon energies are about two orders of magnitude smaller than electron energies. *Right Panel*: Schematic representation of phonon emission and absorption processes in a one-dimensional ballistic wire. In the wire, there are two Fermi levels, which define the occupation of electronic states: for the *right-going electron* states up to the Fermi level of the *left electrode* are occupied, while for the *left-going electrons* the occupation of states is up to the Fermi level of the *right electrode*. The separation of these Fermi levels is eV, where V is the voltage difference applied to the electrodes. Electronic transitions are possible when the final electronic state is unoccupied; consequently, phonon emission is possible only for energies higher than a threshold $\hbar\omega_{2kF}$.

modes n are simply given by $\lambda_n = 2L/n$. However, there is a non-rigid coupling to the electrodes that results in broader resonances. The width of the resonances is linked to the mechanical coupling between the electrodes and the chain.

Furthermore, momentum conservation in the electron-phonon interaction will not hold strictly in a finite system. Therefore, electrons may interact with phonons at different energies and then, other vibrational modes appear in the spectra. This effect is more important in shorter chains.

The position of the peak in the spectra V_{ph} gives the frequency of the ω_{2kF} phonon, and its height A_{ph}, which is related to the conductance drop, is proportional to the probability of the phonon emission process. The magnitude of the conductance drop (about 1% for a chain of 2-nm in length) is consistent with an inelastic mean-free path of about 200 nm in an infinite wire, which is reasonable for a metal at low temperatures.

Experimentally, it is observed that stretching the wire results in an increase in the emission probability. The increase in the emission probability, see Fig. 14.14, indicates an enhancement of the electron-phonon interaction. A plot of the amplitude for many different atomic wires is shown in Fig. 14.14. The emission probability increases with the length of the wire, and the variations due to stretching are much larger than for short wires [61,68]. The background conductance as well as the zero bias conductance remains mostly unchanged. This behavior strongly suggests that

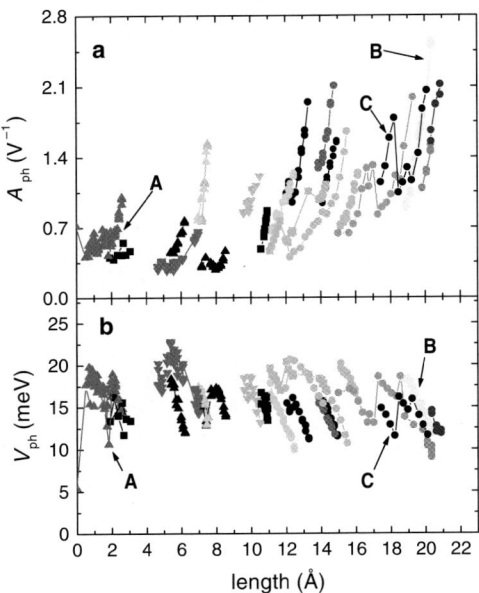

Fig. 14.14. (**a**) Magnitude A_{ph} and (**b**) position V_{ph} of the phonon peak in the PC spectrum as a function of chain length. For clarity, only 22 representative chains out of more than 100 studied are shown. Each chain is represented by a different symbol. The length of the chain is estimated from the length of the last plateau. Reprinted with permission from [61]. Copyright (2002) by the American Physical Society

the background features correspond to backscattering processes in the bulk while the peak is related to backscattering processes in the chain itself.

14.6
Metallic Adhesion in Atomic-Sized Tunneling Junctions

The study of mechanical properties of small tunneling junctions between single asperities or tips is of fundamental importance. These forces are involved in STM imaging mechanisms and tribological imaging of surfaces at the nanoscale [73, 74]. Moreover, the same situation occurs both after the rupture and before the formation of a one-atom contact. When the separation between two tips is small enough, there is a spontaneous jump to contact. Beyond a critical distance, the electrode separation becomes unstable and consequently a one-atom contact is established. This is reflected as a sudden increase of the conductance. The process described above is common in the formation of a metallic contact despite there being some exceptions [75]. For example, in tungsten contacts there is sometimes a continuous increase of current from tunneling to contact [76].

This section is focused on experiments between two gold tips, where a spontaneous jump to contact occurs. Both the force gradient and the tunneling current has been measured simultaneously using a MCBJ supplemented with a force sensor described in Sect. 14.2 [12,31]. The MCBJ technique provides freshly fractured surfaces that are only exposed to cryogenic vacuum, reducing their contamination to an absolute minimum. The force gradient of the interaction between the electrodes is proportional to the shift of the resonance frequency of the TF if excited with vanishing amplitude. This resonant frequency is tracked by implementing a phase-locked loop oscillator, and results in a high sensitivity measurement. With such a tool different contact realizations modifies the atomic configuration of the tip apex. Hence, different tips are fabricated by making a large contact and breaking it again.

In Fig. 14.15, an indentation cycle is shown. The stiffness of the contact can be obtained straightforwardly from the curve. It decreases while pulling the contact. The stiffness of the one-atom contact is $5.8\,\mathrm{N\,m^{-1}}$ in this case. When the one-atom contact is broken, there is an abrupt jump in both the current and the force gradient (jump out contact). After that, the two tips are approached from the tunneling regime. The forces acting in these regimes are attractive since the frequency shift, and consequently the force gradient, is always negative during the approach. Note that the decay length of this force is below 0.05 nm, indicating a short-range metallic interaction. This metallic adhesion is due to the overlap of the electronic wave functions. Various theoretical models [77,78] for metallic adhesion predict a decay length similar to the tunneling current decay length. For smaller distances, there is a spontaneous jump to contact and again we recover a one-atom contact.

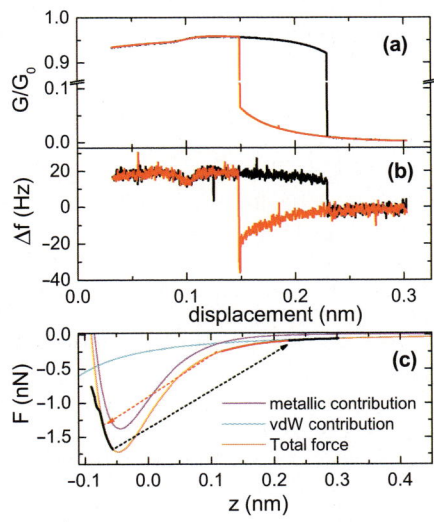

Fig. 14.15. Simultaneous (**a**) conductance and (**b**) frequency shift measurements for a one-atom junction. (**c**) Force, with van der Waals and metallic force components. Reprinted with permission from [12]. Copyright (2004) by the American Physical Society

Acknowledgments. We would like to acknowledge fruitful discussions with C. Untiedt, R.H.M. Smit, and P. Joyez. This work was partially supported by MEC (MAT2004-11982) MAT 2008-01735, CONSOLIDER CSD 2007-00010 and by Comunidad de Madrid (Spain) through program Citecnomik (S-0505/ESP/0337).

References

1. Dürig U, Züger O, Pohl DW (1990) Phys Rev Lett 65:349
2. Gimzewski JK, Möller R (1987) Phys Rev B 36:1284
3. Agraït N, Yeyati AL, van Ruitenbeek JM (2003) Phys Rep-Rev Sect Phys Lett 377:81
4. Rubio G, Agraït N, Vieira S (1996) Phys Rev Lett 76:2302
5. Landman U, Luedtke WD, Burnham NA, Colton RJ (1990) Science 248:454
6. Avouris P, Chen Z, Perebeinos V (2007) Nat Nano 2:605
7. attoAFM. Attocube systems AG
8. CryogenicSFM. Omicron Nanotechnolgy GmbH.
9. LT-SPM. NanoMagnetics Instruments
10. Binnig G, Rohrer H (1982) Helvetica Physica Acta 55:726
11. Binnig G, Quate CF, Gerber C (1986) Phys Rev Lett 56:930
12. Rubio-Bollinger G, Joyez P, Agraït N (2004) Phys Rev Lett 93:116803
13. Rubio-Bollinger G, Bahn SR, Agraït N, Jacobsen KW, Vieira S (2001) Phys Rev Lett 87:26101
14. Jarvis SP, Lantz MA, Ogiso H, Tokumoto H, Durig U (1999) Appl Phys Lett 75:3132
15. Meyer G, Amer NM (1988) Appl Phys Lett 53:1045
16. Rugar D, Mamin HJ, Erlandsson R, Stern JE, Terris BD (1988) Rev Sci Instrum 59:2337
17. Rugar D, Mamin HJ, Guethner P (1989) Appl Phys Lett 55:2588
18. Stahl U, Yuan CW, Delozanne AL, Tortonese M (1994) Appl Phys Lett 65:2878
19. Garcia R, Perez R (2002) Surf Sci Rep 47:197
20. Giessibl FJ (2003) Rev Mod Phys 75:949
21. Giessibl FJ (1998) Appl Phys Lett 73:3956
22. Rychen J, Ihn T, Studerus P, Herrmann A, Ensslin K (1999) Rev Sci Instrum 70:2765
23. Besocke K (1987) Surf Sci 181:145
24. Pohl DW (1987) Rev Sci Instrum 58:54
25. Renner C, Niedermann P, Kent AD, Fischer O (1990) Rev Sci Instrum 61:965
26. Altfeder IB, Volodin AP (1993) Rev Sci Instrum 64:3157
27. Pan SH (1993) Patent WO 9319494
28. Pan SH, Hudson EW, Davis JC (1999) Rev Sci Instrum 70:1459
29. Moreland J, Ekin JW (1985) J Appl Phys 58:3888
30. Muller CJ, Vanruitenbeek JM, Dejongh LJ (1992) Physica C 191:485
31. Valkering AMC, Mares AI, Untiedt C, Gavan KB, Oosterkamp TH, van Ruitenbeek JM (2005) Rev Sci Instrum 76
32. van Ruitenbeek JM, Alvarez A, Pineyro I, Grahmann C, Joyez P, Devoret MH, Esteve D, Urbina C (1996) Rev Sci Instrum 67:108
33. Albrecht TR, Grutter P, Horne D, Rugar D (1991) J Appl Phys 69:668
34. Giessibl FJ (2000) Appl Phys Lett 76:1470
35. Rensen WHJ, van Hulst NF, Ruiter AGT, West PE (1999) Appl Phys Lett 75:1640
36. Rychen J, Ihn T, Studerus P, Herrmann A, Ensslin K, Hug HJ, van Schendel PJA, Guntherodt HJ (2000) Rev Sci Instrum 71:1695
37. Edwards H, Taylor L, Duncan W, Melmed AJ (1997) J Appl Phys 82:980
38. Atia WA, Davis CC (1997) Appl Phys Lett 70:405
39. Karrai K, Grober RD (1995) In: Paesler MA, Moyer PJ (eds) Near-Field Optics (Proc. SPIE) 2535:69
40. Giessibl FJ, Hembacher S, Herz M, Schiller C, Mannhart J (2004) Nanotechnology 15:S79
41. Rychen J, Ihn T, Studerus P, Herrmann A, Ensslin K, Hug HJ, van Schendel PJA, Guntherodt HJ (2000) Appl Surf Sci 157:290
42. Smit RHM, Grande R, Lasanta B, Riquelme JJ, Rubio-Bollinger G, Agraït N (2007) Rev Sci Instrum 78

43. Maxwell JC (1954) A treatise on electricity and magnetism, vol. 1. Courier Dover Publications
44. Knudsen M (1934) The kinetic theory of gases, Methuen
45. Sharvin YV (1965) Zh Eksp Teor Fiz 48:984
46. Torres JA, Pascual JI, Saenz JJ (1994) Phys Rev B 49:16581
47. Scheer E, Agraït N, Cuevas JC, Yeyati AL, Ludoph B, Martin-Rodero A, Bollinger GR, van Ruitenbeek JM, Urbina C (1998) Nature 394:154
48. Ohnishi H, Kondo Y, Takayanagi K (1998) Nature 395:780
49. Rodrigues V, Ugarte D (2001) Phys Rev B 6307
50. Sadd MH (2005) Elasticity theory, applications, and numerics, Elsevier Butterworth Heinemann
51. Frenkel J (1926) Z Physik 37:572
52. Untiedt C, Rubio G, Vieira S, Agraït N (1997) Phys Rev B 56:2154
53. Duif AM, Jansen AGM, Wyder P (1989) J Phys-Condens Mat 1:3157
54. Jansen AGM, Vangelder AP, Wyder P (1980) J Phys C-Solid State Phys 13:6073
55. Jansen AGM, Mueller FM, Wyder P (1977) Phys Rev B 16:1325
56. Lyndenbell RM (1994) Science 263:1704
57. Yanson AI, Bollinger R (1998) Nature 395:783
58. Smit RHM, Untiedt C, Yanson AI, van Ruitenbeek JM (2001) Phys Rev Lett 87:266102
59. da Silva EZ, Novaes FD, da Silva AJR, Fazzio A (2004) Phys Rev B 69
60. Shiota T, Mares AI, Valkering AMC, Oosterkamp TH, van Ruitenbeek JM (2007) eprint arXiv: 0707.4555
61. Agraït N, Untiedt C, Rubio-Bollinger G, Vieira S (2002) Phys Rev Lett 88:216803
62. Brandbyge M, Kobayashi N, Tsukada M (1999) Phys Rev B 60:17064
63. Emberly EG, Kirczenow G (1999) Phys Rev B 60:6028
64. Okamoto M, Takayanagi K (1999) Phys Rev B 60:7808
65. Smit RHM, Untiedt C, Rubio-Bollinger G, Segers RC, van Ruitenbeek JM (2003) Phys Rev Lett 91:76805
66. Landauer R (1957) Ibm J Res Dev 1:223
67. Landauer R (1970) Philos Mag 21:863
68. Agraït N, Untiedt C, Rubio-Bollinger G, Vieira S (2002) Chem Phys 281:231
69. Ludoph B, Devoret MH, Esteve D, Urbina C, van Ruitenbeek JM (1999) Phys Rev Lett 82:1530
70. Untiedt C, Bollinger GR, Vieira S, Agraït N (2000) Phys Rev B 62:9962
71. Yanson IK (1974) Zhur Eksper Teoret Fiziki 66:1035
72. Khotkevich AV, Yanson IK (1995) Atlas of point contact spectra of electron-phonon interactions in metals. Kluwer Academic Publishers, Boston
73. Pfeiffer O, Nony L, Bennewitz R, Baratoff A, Meyer E (2004) Nanotechnology 15:S101
74. Stalder A, Durig U (1996) Appl Phys Lett 68:637
75. Untiedt C, Caturla MJ, Calvo MR, Palacios JJ, Segers RC, van Ruitenbeek JM (2007) Phys Rev Lett 98:206801
76. Halbritter A, Csonka S, Mihaly G, Jurdik E, Kolesnychenko OY, Shklyarevskii OI, Speller S, van Kempen H (2003) Phys Rev B 68
77. Chen CJ (1991) J Phys-Condens Mat 3:1227
78. Rose JH, Smith JR, Ferrante J (1983) Phys Rev B 28:1835
79. Untiedt C, Yanson AI, Grande R, Rubio-Bollinger G, Agraït N, Vieira S, van Ruitenbeek JM (2002) Phys Rev B 66:85418

15 Dynamic AFM in Liquids: Viscous Damping and Applications to the Study of Confined Liquids

Abdelhamid Maali · Touria Cohen-Bouhacina · Cedric Hurth · Cédric Jai · R. Boisgard · Jean-Pierre Aimé

Abstract. We present a study of dynamic atomic force microscopy (AFM) in liquid medium. In the first part, we investigate hydrodynamic viscous damping. Using an experimental analysis of the thermal noise motion of the cantilever, the dependency of the hydrodynamic damping versus the cantilever–surface separation was evaluated. In the second part, we present an improvement of the excitation holder that allows us to remove the spurious peaks not corresponding to the resonance frequency of the cantilever mechanical response. An analytical description of the motion of an acoustic-driven cantilever in liquid is given in the third part.

In the last part, we have used dynamic AFM to study a simple liquid confined between the cantilever tip and a graphite surface. The results show that the liquid undergoes ordering to layers with a characteristic size given by the molecular diameter.

Key words: Dynamic AFM, Liquid medium, Hydrodynamic damping, Liquid ordering.

Abbreviations

AFM	Atomic force microscopy
AM-AFM	Amplitude modulation atomic force microscopy
SFA	Surface force apparatus
OMCTS	Octamethylcylotetrasiloxane
HOPG	Highly oriented pyrolytic graphite

15.1 Introduction

Since the development of atomic force microscopy (AFM) [1], the topography of organic and inorganic surfaces has been imaged. High-resolution images were obtained in vacuum [2], in air and in liquid [3]. Pioneering imaging in liquid was done in the contact mode [3–6] where a constant force acts on the sample during the acquisition of an image. However, this mode can not be used to image soft isolated objects, because the lateral force can move or even damage the imaged sample. To reduce the perturbation induced by the tip on the sample, dynamic AFM was developed in liquid [7–9]. In this mode, the cantilever is oscillating during the lateral scan, which makes the contact time between tip and sample very short and thus reduces perturbation of the sample. This paper is a review of some of our work on the dynamic AFM in liquid medium. We will focus on:

The viscous hydrodynamic damping of the cantilever in water;
Improving the acoustic excitation of the cantilever-tip;
A theoretical description of the motion of an acoustic-driven cantilever in liquid;
An atomic force microscopy study of the molecular ordering of a confined liquid.

15.2
Viscous Hydrodynamic Damping of the Cantilever in a Water Medium

In a liquid environment, the performance of an AFM cantilever in dynamic mode strongly depends on its size and shape. For typical commercial cantilevers, with a length ranging from 100 to 200 μm and a width of about 20-μm, hydrodynamic damping induces forces that can reach values as high as several nN close to the surface. For biological applications, forces in the pN range need to be measured; therefore hydrodynamic forces produce a large background that can mask the more specific interaction between the tip apex and the biological sample. In tapping mode [10], both height and phase images are recorded. Since the phase image provides direct information on the damping, the phase image contrast will be drastically reduced by a large viscous damping.

Figure 15.1 shows the spectra of a rectangular cantilever (nominal dimensions of this cantilever are: length $l = 100\,\mu$m, width $W = 20\,\mu$m, and thickness $e = 0.8\,\mu$m) measured in two different mediums (air and water). In air the resonance is very narrow, however in the water medium due to the large viscosity of the water, the hydrodynamic force opposing the motion of the cantilever induces a damping that broadens the resonance frequency of the cantilever.

Roughly the added mass and the damping for a rectangular cantilever are given by:

$$\begin{cases} m_{added} = \dfrac{\pi}{4}\rho_{fluid} W^2 l \\[2mm] \gamma_{diss} = \dfrac{\sqrt{2}}{2}\pi Wl\sqrt{2\eta\rho_{flui}\,\omega} \end{cases}.$$

When the cantilever oscillates near a surface, the fluid is squeezed out of the region directly beneath the tip resulting in an additional hydrodynamic force acting on the cantilever and opposing its motion [11–15].

Figure 15.2 shows the resonance spectra of a cantilever immersed in water at different distances D from a mica surface. Two observations can be made from Fig. 15.2. First, the resonance frequency f_0 shifts to a lower value as the cantilever approaches the surface. Second, the noise spectrum peak broadens as the distance D is decreased.

The peaks of the noise spectral density signal were fitted to the amplitude response function of a simple harmonic oscillator (solid lines in Fig. 15.2):

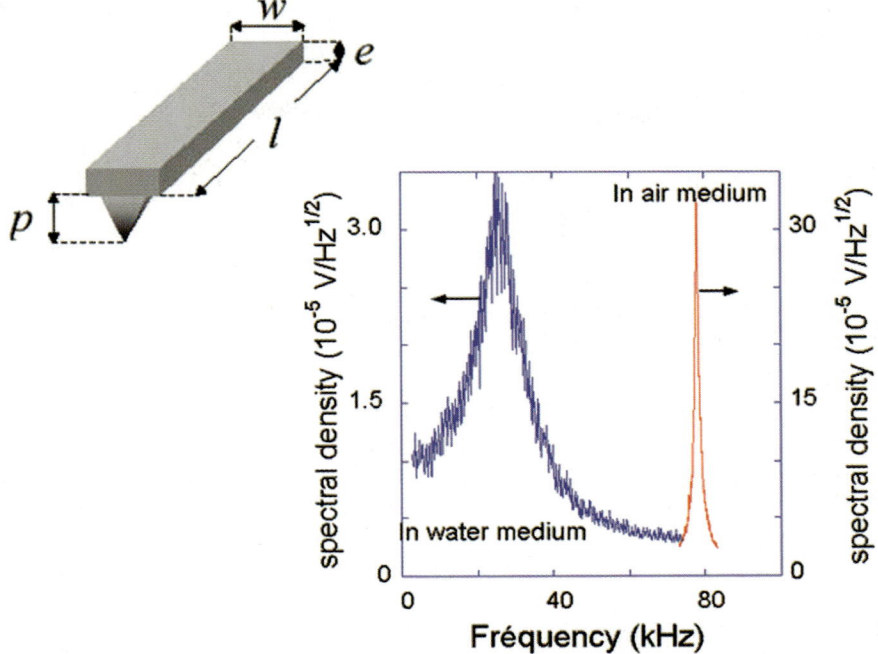

Fig. 15.1. Geometrical parameters of the rectangular cantilever (width W, length l, thickness e, and tip height p). Typical spectra measured on the same cantilever in air and water. Notice here the resonance shift to lower frequencies in water due to the added mass and the broadening of the resonance due to the large viscous damping induced by the viscosity of the water

$$A(\omega) = \frac{A_0}{\sqrt{\left(1 - \dfrac{\omega^2}{\omega_0^2}\right)^2 + \dfrac{1}{Q_0^2}\left(\dfrac{\omega}{\omega_0}\right)^2}}$$

where A_0 is the thermal noise coefficient of the fundamental vibration mode. In the present work, since we are mostly interested in the determination of the quality factor Q and the frequency $\omega_0 = 2\pi f_0$, the coefficient A_0 is set as an arbitrary fitted constant.

For each cantilever surface distance tested, values of the resonant frequency ω_0 and the quality factor Q, extracted from fits, were used to calculate the added mass and the damping coefficient of the cantilever as follows:

$$\begin{cases} m_{addedwater} = m_{cantilever}\left[\left(\dfrac{\omega_{air}}{\omega_{water}}\right)^2 - 1\right] \\[3mm] \gamma_{air} = \dfrac{m_{cantilever}\omega_{air}}{Q_{air}} \\[3mm] \gamma_{water} = \dfrac{(m_{cantilever} + m_{addedwater})\,\omega_{water}}{Q_{water}} \end{cases}$$

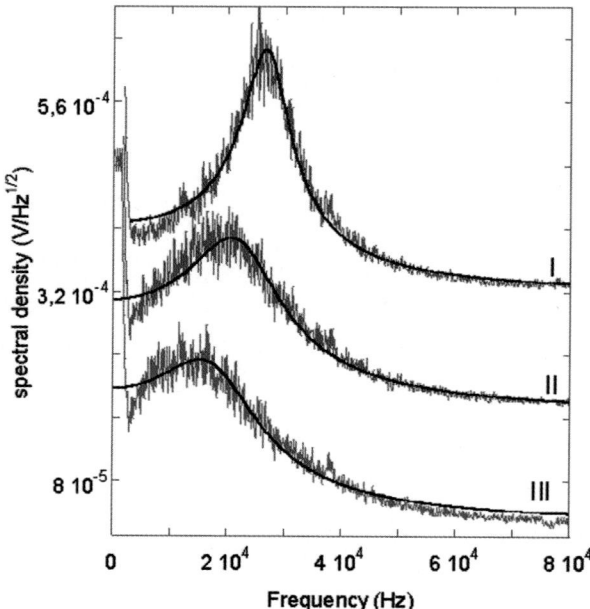

Fig. 15.2. Experimental data of noise spectra measured in water at several cantilever–surface separations D. These results were obtained with the same cantilever. The *solid curves* are Lorentzian fits. For more clarity, the spectra are shifted by an arbitrary but constant offset. For each distance D, the resonance frequency f_0 and the quality factor Q were extracted from fits of the noise spectrum density data. The obtained values of these parameters for each of the curves shown are respectively: *Curve I* ($D = 41.6\,\mu m$, $f_0 = 27.134\,kHz$, $Q = 3.45$); *Curve II* ($D = 1.2\,\mu m$, $f_0 = 23.506\,kHz$, $Q = 1.4$); *Curve III* ($D = 0.025\,\mu m$, $f_0 = 19.967\,kHz$, $Q = 1.09$)

where $m_{cantilever} = \rho_{lever}L\,w\,e$, is the mass of the cantilever and $m_{addedwater}$ is an additional mass that corresponds to the fluid dragged by the cantilever motion. In air, the added mass is not larger than 1% and was neglected for the calculation of the damping coefficient in air γ_{air}. The damping coefficient in water versus the distance d is given in Fig. 15.3.

In the case where the width of the cantilever is much smaller than its length ($W \ll l$) and much larger than the distance d between the cantilever and the surface ($W \gg d$), Vinogradova et al. used a lubrication approximation to obtain an analytical expression for the dependence of the hydrodynamic force F_S on the distance [12]
$F_S = -\gamma_s . V$.

γ_S is the damping due to the squeezed fluid and V is the velocity of the cantilever.

$$\gamma_S = \frac{\eta . L.}{2} \left(\frac{w}{D}\right)^3 g^*$$

with

$$g^* = \beta \left(1 - \frac{3\beta}{2} + 3\beta^2 - 3\beta^3 \ln\left(1 + \frac{1}{\beta}\right)\right)$$

Fig. 15.3. Damping coefficient in water of the cantilever as a function of the cantilever–surface separation distances

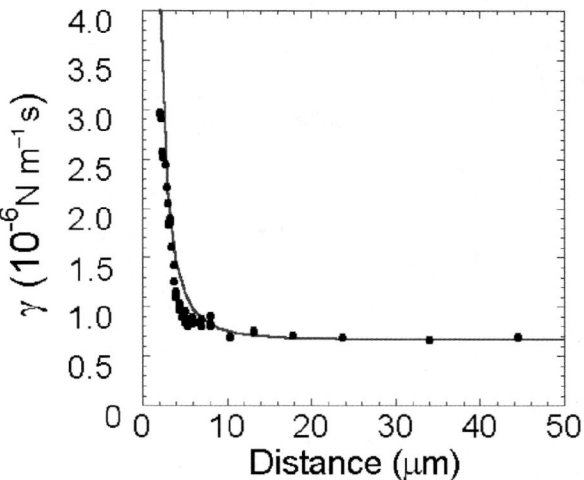

where $\beta = D/(L \sin \alpha)$. η is the viscosity of the medium, and α is the tilt angle of the cantilever. The force is neither concentrated at a point nor distributed uniformly on the cantilever. The total damping coefficient γ is defined by: $\gamma = \gamma_s + \gamma_\infty$, γ_∞ is the bulk damping [16] (for the cantilever used in this experiment: $\gamma_\infty = 6.6 \times 10^{-7}\,\mathrm{N\,m^{-1}}$ s). The calculated total damping using the above equation (Vinogradova expression) is presented in Fig. 15.3 as a continuous line. The calculated damping values give a good quantitative agreement with the experimental measurements.

15.3
Improving the Acoustic Excitation of the Cantilever-Tip

In order to properly put the cantilever of an atomic force microscope working in the dynamic mode into motion, a variety of excitation techniques have been used. However, the excitation of the cantilever is more challenging in liquid media because of the hydrodynamic drag acting on the cantilever as we have seen in the previous section. In viscous liquid media, the cantilever can essentially be driven by thermal noise [16, 17], by a piezoelectric-actuator [7–9], or magnetically by either attaching a magnetic particle to the cantilever [18] or coating the cantilever [19, 20]. Up to now, magnetic excitation is the only procedure that allows proper excitation of the cantilever in a liquid environment at its resonance frequencies. However, it requires specifically prepared cantilevers that are more expensive and more time-consuming to prepare. Moreover, the coating process may contaminate the cantilever tip resulting in an increase of the tip radius which makes the resolution very poor. The acoustic excitation supplied by a piezo-actuator is the most widespread for dynamic AFM in air and vacuum. In most liquid AFM experiments, the piezoelectric actuator used to excite the cantilever is placed in a cell far away from the cantilever. This method results in a multitude of spurious peaks related to the liquid cell eigenfrequencies

Fig. 15.4. Comparison of the acoustic excitation spectrum obtained using the commercial liquid cell (*continuous line*) and the thermal noise spectrum (*dotted line*)

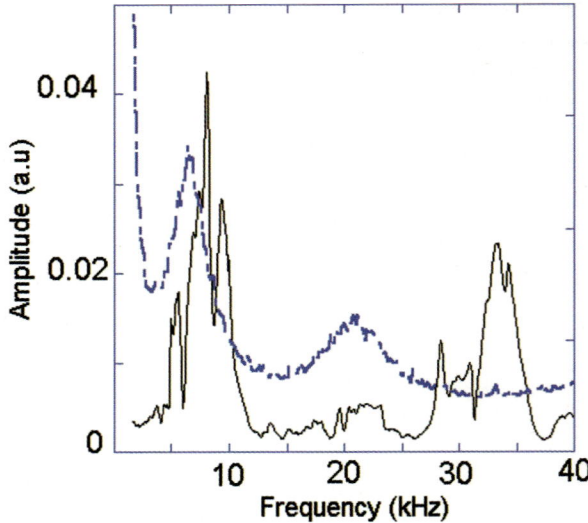

[9, 20, 21]. Consequently, it is difficult to clearly observe the resonance frequencies of the cantilever [9, 20, 21] and the usual approximation of the damped harmonic oscillator is wrong and quantitative interpretation of the measurements becomes very intricate (Fig. 15.4).

We recently carried out a simple modification of an existing commercial cantilever holder (part number MMMC, Veeco Instruments) used to perform Tapping-Mode AFM in ambient air or in vacuum. The piezoelectric actuator in this holder is very close to the cantilever. The holder is carved above the piezo-element [21] in order to insert a piece of a microscope cover glass (1-mm thickness, 12 × 12 mm). The liquid drop (200 µL) is then confined between the glass slide and the sample surface (Fig. 15.5). The electric wires and the piezo actuator are insulated from the conductive aqueous solutions by an evaporated Teflon film. To reduce the backscattered laser light from the glass slide which may induce an additional noise, we coated the glass with an antireflective layer for a wavelength of 640 nm. The excitation spectrum obtained using this apparatus is shown in Fig. 15.6.

Fig. 15.5. Section of the modified cantilever holder. The holder is carved above the piezo-element in order to insert a piece of a microscope cover glass (1-mm thickness, 12 × 12 mm)

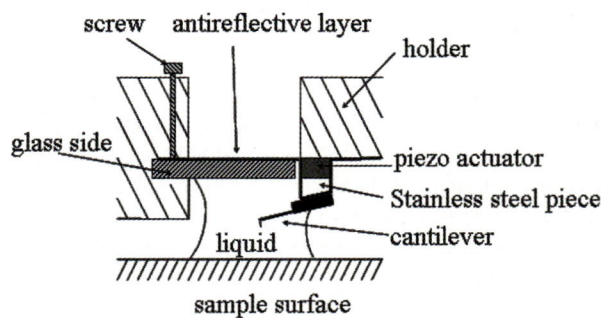

Fig. 15.6. The acoustic excitation using the modified commercial piezo actuator (*continuous line*) of a 450-μm-long rectangular cantilever (Nanosensors TL-450) in pure deionized water (Milli-Q^{TM}). With the modification described here, the second and the third oscillation modes of the cantilever are clearly observed; the spurious vibration modes of the liquid cell are removed

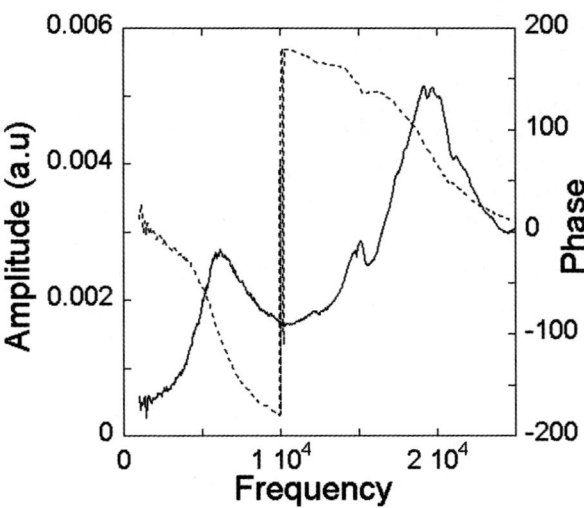

15.4
Theoretical Description of the Motion of an Acoustically Driven Cantilever in Liquid

The amplitude modulation atomic force microscopy AM-AFM is one of the most commonly used modes [22–26]. It is usually referred to as tapping-mode AFM. The cantilever is put in oscillation by an external signal and the amplitude and phase of the cantilever are measured as it interacts with a sample. The most widespread detection method of AFM tip motion is optical beam deflection [27, 28]. A light beam is focused to an optical spot on the cantilever, and the angular deflection of the reflected beam is measured by a position-sensitive photodetector (quadrant photodiode). This method is therefore sensitive to the deflection of the cantilever but not to its displacement.

For cantilevers having high quality factors as is the case in vacuum and air, the displacement of the cantilever base is negligible compared to the displacement of the tip when the cantilever is excited close to its resonance. Several authors have derived expressions that give the interaction damping as a function of the measured amplitude and phase of the cantilever [29–32].

Herein, we show that for cantilevers having low quality factors the cantilever oscillation amplitude is comparable to the base displacement. Thus, the whole motion of the tip is the sum of the base displacement and the deflection oscillation amplitude of the cantilever [33]. Since the measured value in most AFM is only the deflection motion whereas forces acting on the tip depend on the whole motion, one may expect additional terms in the effective driving force.

We used a point mass model to describe the steady-state response of the cantilever to an acoustic excitation. We show that the effective driving force that acts on the deflection motion depends also on the damping [34].

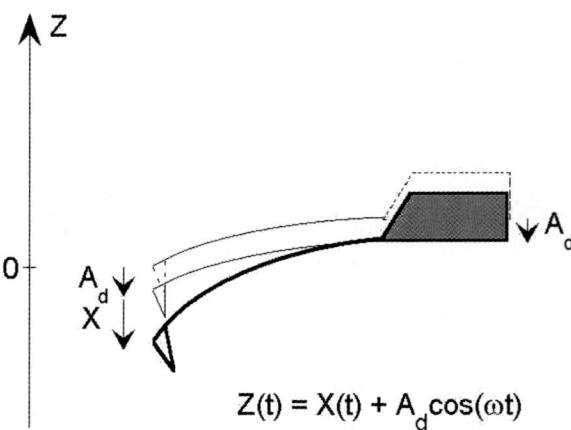

Fig. 15.7. Schematic diagram of the cantilever configuration. The instantaneous tip position is the sum of the measured deflection X and the base displacement: $Z = X + A_d \cos(\omega t)$. For a low quality factor the base displacement A_d is comparable to the deflection X

$$Z(t) = X(t) + A_d \cos(\omega t)$$

Figure 15.7 shows the model for a cantilever in tapping-mode atomic force microscopy. The base of the cantilever beam is excited by the piezoelectric actuator with a constant displacement amplitude A_d and a constant frequency ω so that the instantaneous displacement of the base is $A_d \cos(\omega t)$.

The instantaneous tip position is the sum of the measured deflection X and the base displacement:

$$Z = X + A_d \cos(\omega t) \tag{15.1}$$

Without any interaction, the equation of tip motion is:

$$m^* \ddot{Z} = -\gamma_0 \dot{Z} - k_l X \tag{15.2}$$

where m^* is the effective mass of the cantilever, k_l is the cantilever force constant, and γ_0 is the bulk damping coefficient.

When the tip interacts with a sample, we have to add to the second term of the above equation the interaction force. For a cantilever oscillating with a small amplitude compared to the range of the interaction length, the instantaneous force can be linearized and has two contributions; one is a conservative term ($- k_{\text{int}} \cdot z$) and the other one is a dissipative term ($- \gamma_{\text{int}} \cdot \dot{z}$) where k_{int} and γ_{int} are the interaction stiffness and damping coefficient, respectively.

Taking into account the interaction force and rearranging Eq. (15.2) one gets:

$$m^* \ddot{X} + \gamma_{tot} \dot{X} + (k_l + k_{\text{int}})X = A_d \sqrt{(m^* \omega^2 - k_{\text{int}})^2 + (\gamma_{tot}\omega)^2} \cos(\omega t - \phi) \tag{15.3}$$

where $\gamma_{tot} = \gamma_0 + \gamma_{\text{int}}$

and

$$\sin(\phi) = \frac{\gamma_{tot}\omega}{\sqrt{(m^* \omega^2 - k_{\text{int}})^2 + (\gamma_{tot}\omega)^2}} \tag{15.4a}$$

$$\cos(\phi) = \frac{m^*\omega^2 - k_{\text{int}}}{\sqrt{(m^*\omega^2 - k_{\text{int}})^2 + (\gamma_{tot}\omega)^2}} \tag{15.4b}$$

Equation (15.3) describes the instantaneous tip deflection measured by the AFM quadrant photodiodes. A similar equation has been recently derived by Yagasaki [35] using a continuous model to describe the cantilever beam deflection.

It can be noticed that for a given frequency, the effective driving force of the tip deflection is not constant but depends on the damping. During the interaction, the force increases as the damping increases. Additionally, the driving force is not in phase with the base displacement.

Particularly, if the cantilever is far away from the interaction region and is excited at its resonance frequency ω_0, the driving amplitude is no longer $A_d\omega_0^2 m^*$ but $A_d\omega_0^2 m^* \sqrt{1 + {}^1\!/_{Q_0}{}^2}$ and the dephasing is $\tan\phi = \frac{1}{Q_0}$.

Where $Q_0 = \frac{m^*\omega_0}{\gamma_0}$ is the quality factor of the cantilever without any interaction, for instance far away from the surface. Let us look for a study solution of the above equation in the form $X = A\cos(\omega t + \varphi)$ where A is the amplitude of the tip deflection and φ is the phase of the tip motion compared to the base displacement. Equation (15.3) becomes:

$$(-m^*\omega^2 + k_l + k_{\text{int}})A\cos(\omega t + \varphi) - \gamma_{tot}\omega A\sin(\omega t + \varphi)$$

$$= A_d\sqrt{(m^*\omega^2 - k_{\text{int}})^2 + (\gamma_{tot}\omega)^2}\cos(\omega t - \phi) \tag{15.5}$$

Multiplying both sides of the above equation by $\sin(\omega t + \varphi)$ and by $\cos(\omega t + \varphi)$ and integrating over an oscillation period, we obtain the time-independent analytical relations:

$$-\gamma_{tot}\omega A = A_d\sqrt{(m^*\omega^2 - k_{\text{int}})^2 + (\gamma_{tot}\omega)^2}\sin(\phi + \varphi) \tag{15.6a}$$

$$(-m^*\omega^2 + k_l + k_{\text{int}})A = A_d\sqrt{(m^*\omega^2 - k_{\text{int}})^2 + (\gamma_{tot}\omega)^2}\cos(\phi + \varphi) \tag{15.6b}$$

From this expression, we calculate the expressions of the deflection amplitude and phase:

$$A = \frac{A_d\sqrt{\left(\omega^2 - \frac{k_{\text{int}}}{m^*}\right)^2 + \left(\frac{\gamma_{tot}\omega\omega_0}{\gamma_0 Q_0}\right)^2}}{\sqrt{\left(\frac{k_{\text{int}}}{m^*} + \omega_0^2 - \omega^2\right)^2 + \left(\frac{\gamma_{tot}\omega\omega_0}{\gamma_0 Q_0}\right)^2}}, \tag{15.7}$$

with $k_l = m^*\omega_0^2$

$$\varphi = -\arccos\left(\frac{m^*\omega^2 - k_{\text{int}}}{\sqrt{(m^*\omega^2 - k_{\text{int}})^2 + (\gamma_{tot}\omega)^2}}\right)$$

$$+ \arccos\left(\frac{-m^*\omega^2 + k_l + k_{\text{int}}}{\sqrt{(m^*\omega^2 - k_l - k_{\text{int}})^2 + (\gamma_{tot}\omega)^2}}\right) \tag{15.8}$$

The above expressions give the amplitude and the phase of the cantilever deflection measured by the lock-in amplifier taking into account the tip–sample interaction. In particular, when the tip is far away from the interaction region, the above expressions become:

$$A = \frac{A_d \omega \sqrt{(\omega^2) + \left(\dfrac{\omega_0}{Q_0}\right)^2}}{\sqrt{(\omega_0^2 - \omega^2)^2 + \left(\dfrac{\omega \omega_0}{Q_0}\right)^2}} \tag{15.9a}$$

and

$$\varphi = -\arccos\left(\frac{\omega^2}{\sqrt{\omega^4 + \left(\dfrac{\omega \omega_0}{Q_0}\right)^2}}\right) + \arccos\left(\frac{\omega_0^2 - \omega^2}{\sqrt{(\omega^2 - \omega_0^2)^2 + \left(\dfrac{\omega \omega_0}{Q_0}\right)^2}}\right) \tag{15.9b}$$

At the resonance frequency:

$$A_0 = A(\omega = \omega_0) = A_d \sqrt{1 + Q_0^2}, \tag{15.10a}$$

$$\varphi = -\frac{\pi}{2} - \arccos\left(\frac{Q_0}{\sqrt{1 + Q_0^2}}\right) \tag{15.10b}$$

For cantilevers having a high quality factor ($Q_0 \approx 500$), A_0 is about 500-times larger than A_d and the phase at resonance is $\varphi = -\dfrac{\pi}{2}$ ($-90°$). However, for cantilevers with a low quality factor ($Q_0 \leq 3$), A_0 is comparable to the base displacement A_d ($A_0 = 3.16A_d$). Moreover, at the resonance, the phase of the deflection is not $-\dfrac{\pi}{2}(-90°)$ but $-\dfrac{\pi}{2} - 0.32$ ($-108°$). This is because of the contribution of the damping to the effective forcing acting on the tip motion as we expect from Eq. (15.3).

Let us now calculate the damping and the stiffness as the tip interacts with the sample. From Eqs. (15.6b), (15.4a), and (15.4b), one obtains:

$$m^* \omega^2 - k_{int} = \frac{k_l A + \omega A_d \gamma_{tot} \sin(\varphi)}{A + A_d \cos(\varphi)} \tag{15.11}$$

By inserting this expression into Eq. (15.6a), one obtains:

$$\gamma_{tot} = \frac{-A_d A k_l \sin(\varphi)}{\omega(A^2 + 2AA_d \cos(\varphi) + A_d^2)} \tag{15.12}$$

Equation (15.12) can be calculated directly by equating the external energy supplied to the cantilever by the actuator and the energy dissipated during the motion over one cycle.

After rearranging the above equation, one obtains:

$$\frac{\gamma_{tot}}{\gamma_0} = -\frac{\omega_0}{\omega}\frac{A_0}{A}\sin(\varphi)\left(\frac{Q_0}{\sqrt{1+Q_0^2}+2\frac{A_0}{A}\cos(\varphi)+\frac{A_0^2}{A^2\sqrt{1+Q_0^2}}}\right) \qquad (15.13)$$

A_0 is the deflection amplitude at resonance far away from the interaction region. Let us emphasize that for very high quality factors ($Q_0 \to \infty$) we recover the usual expression for the damping coefficient: $\frac{\gamma_{tot}}{\gamma_0} = -\frac{\omega_0}{\omega}\frac{A_0}{A}\sin(\varphi)$.

By inserting Eq. (15.12) into Eq. (15.11), one gets the expression of the interaction stiffness:

$$\frac{k_{int}}{k_l} = \left(\frac{\omega}{\omega_0}\right)^2 - 1 + \frac{A_0}{A}\left(\frac{\cos(\varphi)+\dfrac{A_0}{A\sqrt{1+Q_0^2}}}{\sqrt{1+Q_0^2}+2\dfrac{A_0\cos(\varphi)}{A}+\dfrac{A_0^2}{A^2\sqrt{1+Q_0^2}}}\right)$$

$$(15.14)$$

As a particular case for ($Q_0 \to \infty$) $\dfrac{k_{int}}{k_l} = \dfrac{A_0\cos(\varphi)}{AQ_0} + \left(\dfrac{\omega}{\omega_0}\right)^2 - 1$, which is the usual equation used for amplitude modulation AFM for a cantilever with a high quality factor.

In practice, there is a phase lag between the displacement of the cantilever base and the excitation voltage coming from the lock-in amplifier. To get rid of this lag phase, one can measure the quality factor of the cantilever far away from the interaction region and then set the phase of the cantilever according to Eq. (15.9b) at large tip–sample separations. For example, if the working frequency is ($\omega = \omega_0$) the phase data has to be shifted in order to get a phase at large separations equal to:

$$\varphi(d = \infty) = -\frac{\pi}{2} - \arccos\left(\frac{Q_0}{\sqrt{1+Q_0^2}}\right).$$

In this section we have presented an analytical description that enables determination of the motion of an acoustic-driven atomic force microscope cantilever. We showed that for low quality factors the displacement of the excitation base has to be taken into account. We derived equations that give the amplitude and phase of the deflection motion of the cantilever and we give the expression of the interaction damping and stiffness.

15.5
Atomic Force Microscopy Study of the Molecular Ordering of a Confined Liquid

The physical properties of materials at the nanometer scale can be completely different from those of the bulk [36]. A good example of this is provided by a liquid at a solid interface in which the liquid undergoes some ordering at scales of the order of the molecular size due to interactions with the solid wall. Ordering of liquids at interfaces is a phenomenon of fundamental importance and has interested several fields of research such as tribology [37], nanofluidics, and biology [38]. It has been extensively studied using the Surface Force Apparatus (SFA) [38–45] and Atomic Force Microscopy [46–54].

Several pioneering works [38–50] reported that the so-called solvation force (force acting on the surfaces confining a fluid) has an oscillatory profile versus the distance separating the surfaces. Moreover, the viscosity measured by shearing the confined liquid laterally was reported to increase strongly when decreasing the thickness of the confined liquid [40, 41, 43–45].

Recently, our group demonstrated that a simple liquid like Octamethylcyclotetrasiloxane (OMCTS) confined between the tip of an atomic force microscope and a graphite surface (Fig. 15.8) undergoes structuring and presents dissipation with an oscillatory profile [54]. The measurements were obtained by vibrating a cantilever near resonance at very small amplitude (much smaller than the molecular size) and simultaneously recording the amplitude and the phase of the oscillating cantilever (Fig. 15.9). Our experimental data show that the damping coefficient and the viscosity present oscillations as the gap between the cantilever tip and the surface is diminished (Fig. 15.10). Such an experimental result is different from what has been reported earlier using the surface force apparatus or an AFM where only a continuous increase of the damping and the viscosity are observed.

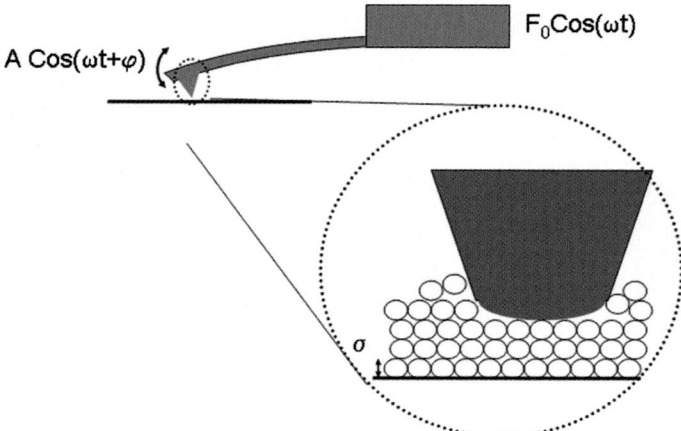

Fig. 15.8. Schematic representation of the experiment performed. The cantilever was driven at a fixed forcing and the phase and the amplitude of the cantilever are measured simultaneously as the tip approaches the surface

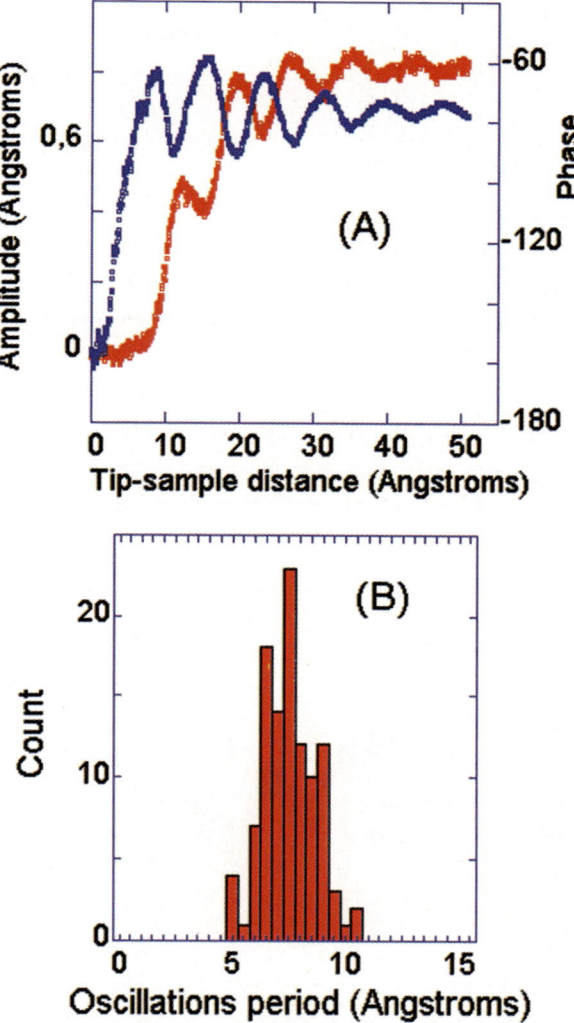

Fig. 15.9. (**A**) The amplitude and the phase of the cantilever are measured simultaneously as the tip approaches the surface. The forcing was adjusted to have the cantilever vibrating at an amplitude of $A_0 = 0.8$ Å far away from the surface. The amplitude is shown in *red* and the phase appears in *blue*. Other values of amplitude were used and it was found that the modulation period does not depend on the vibration amplitude. (**B**) Histogram of the measured oscillation period in the amplitude signal. The average periodicity measured over 20 approach and retraction cycles is 0.78 Å, which is consistent with value previously reported using SFA experiments [27]

The experiments were performed in tapping mode and close to the cantilever resonance frequency in order to increase the sensitivity. The liquid confined between the tip and a freshly cleaved surface of HOPG (Highly Oriented Pyrolytic Graphite) is Octamethylcyclotetrasiloxane. The experiments were carried out in ambient air at room temperature using the excitation setup described in the previous section.

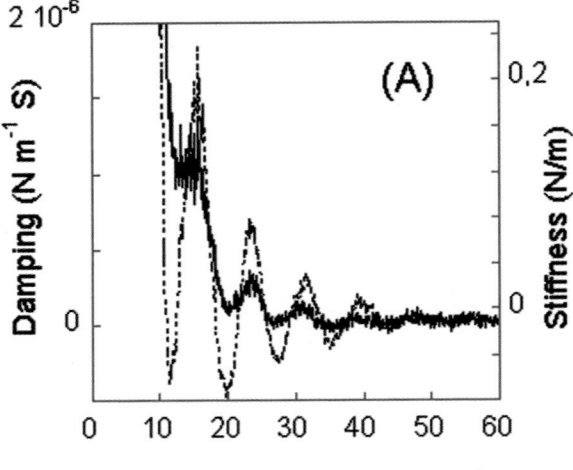

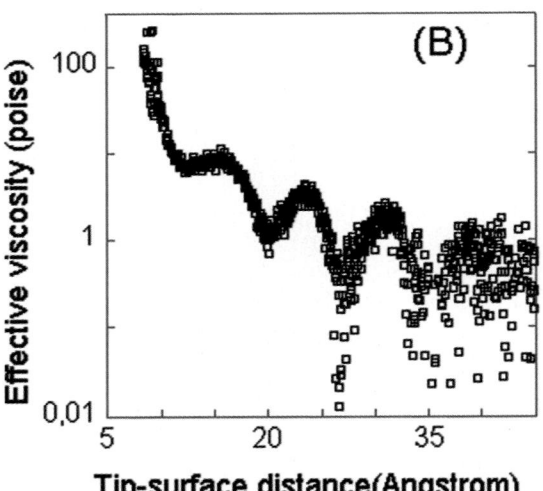

Fig. 15.10. (A) Damping versus the distance between the cantilever-tip and the surface. For distances greater than 50 Å, the interaction damping is zero. As the tip approaches the surface, the damping shows two features: a periodic variation and an increase. The damping modulation period is equal to the molecular diameter. Notice that the damping is in phase with the stiffness curve. When the tip–surface distance is equal to a multiple of the molecular diameter (distances corresponding to the maxima of the stiffness), the damping is higher and for distances corresponding to a multiple and a half of the diameter the damping is at a minimum. **(B)** The effective viscosity is extracted from the damping data using $\eta_{eff} = \dfrac{\gamma_{int} D}{6\pi R^2}$. The viscosity is not only increasing as reported earlier in SFA experiments but is also modulated. The viscosity modulation length is equal to the molecular diameter

Note here that in contrast to SFA, experiments where one measures the shear stress by imposing a shear strain and derives the viscosity by assuming Couette flow geometry, the viscosity in this AFM experiment is measured as the tip approaches the surface perpendicularly.

After reducing the thermal drift, it should be possible to measure the shear viscosity of the confined liquid by laterally moving the lower surface. These results can then be compared to those obtained in SFA experiments and to those shown above. Such studies will allow us to examine the possible anisotropy of the confined liquid viscosity which should be important for the case of nonspherical molecules.

Another interesting aspect is the possible link between the wetting properties of the fluid on the surface and the ordering of the molecules at the solid–liquid interface. The temporal response of the liquid ordering to an external perturbation is also interesting to investigate. Issues such as the role of orientational degrees of freedom may turn out to have a role in the dissipation near interfaces.

15.6
Conclusion

In summary, we have presented some experimental work on dynamic AFM in liquid. We have evaluated the hydrodynamic drag force acting on the cantilever close to the surface and we have briefly presented the improvements of the holder that excites the vibration. The analytical description that enables determination of the motion of an acoustic driven atomic force microscope cantilever in liquid was also described. For low quality factors the effective driving force that acts on the deflection motion depends on the damping. We derived equations that accurately give the amplitude and phase of the cantilever deflection and we give also the expressions of the damping and stiffness of the interaction.

In the last part we used dynamic AFM to study a simple liquid confined between the cantilever tip and a graphite surface. The results show that the liquid undergoes layer structuring and presents oscillatory dissipation as the thickness of the confined liquid is diminished.

References

1. Binnig G, Quate CF, Gerber Ch (1986) Phys Rev Lett 56:930
2. Giessibl JF (1995) Science 267:68
3. Ohnesorge F, Binnig G (1993) Science 260:1451
4. Schabert FA, Henn C, Engel A (1995) Science 268:92
5. Hoh JH, Sosinsky GE, Revel JP, Hansma PK (1993) Biophys J 65:149
6. Müller DJ, Schabert FA, Büldt G, Engel A (1995) Biophys J 68:1681
7. Putman CAJ, Vanderwerf KO, De Grooth BG, Vanhulst NF, Greve J (1994) Appl Phys Lett 64:2454
8. Hansma PK, Cleveland JP, Radmacher M, Walters DA, Hillner PE, Bezanilla M, Fritz M, Vie D, Hansma HG, Prater CB, Massie J, Fukunaga L, Gurley J, Elling V (1994) Appl Phys Lett 64:1738
9. Shâffer TE, Cleveland JP, Ohnesorge F, walters DA, Hansma PK (1996) J Appl Phys 80:3622

10. Magonov SN, Elings V, Whangbo MH (1997) Surf Sci 375:L385
11. Vinagradova OI, Butt HJ, Yakubov GE, Feuillebois F (2001) Rev Sci Instrum 72:2330
12. Vinogradova OI, Yakubov GE (2003) Langmuir 19:1227
13. Rankl C, Pastushenko V, Kienberger F, Stroh CM, Hinterdorfer P (2004) Ultramicroscopy 100:301
14. Naik T, Longmire EK, Mantell SC (2003) Sensor Actuat A 102:240
15. Benmouna F, Johannsmann D (2002) Eur Phys J E 9:435
16. Maali A, Hurth C, Boisgard R, Jai C, Cohen-Bouhacina T, Aimé JP (2005) J Appl Phys 97:074907
17. Chon JWM, Mulvaney P, Sader JE (2000) J Appl Phys 87:526
18. Florin EL, Radmacher M, Fleck B, Gaub HE (1994) Rev Sci Instr 65:639
19. Han W, Lindsay SM, Jing T (1996) Appl Phys Lett 69:4111
20. Revenko I, Proksch R (2000) J Appl Phys 87:526
21. Maali A, Hurth C, Cohen-Bouhacina T, Couturier G, Aimé JP (2006) Appl Phys Lett 88:163504
22. Zhong Q, Imniss D, Kjoller K, Elings VB (1993) Surf Sci 290:L688
23. Tamayo J, García R (1996) Langmuir 12:4430
24. Sasaki K, Koike Y, Azehara H, Horaki H, Fujihira M (1998) Appl Phys A: Mater Sci Process 66:1275
25. Bar G, Thomann Y, Whangbo M-H (1998) Langmuir 14:1219
26. Garcia R, Pérez R (2002) Surf Sci Rep 47:197
27. Meyer G, Amer NM (1988) Appl Phys Lett 53:1045
28. Alexander S, Hellemans L, Marti O, Schneir J, Elings V, Hansma PK, Longmire M, Gurley J (1989) J Appl Phys 65:164
29. Tamayo J, García R (1998) Appl Phys Lett 73:2926
30. Cleveland JP, Anczykowski B, Schmid E, Elings V (1998) Appl Phys Lett 72:2613
31. Aime JP, Boisgard R, Nony L, Couturier G (2001) J Chem Phys 114:4945
32. San Paulo A, García R (2001) Phys Rev B 64:193411
33. Lantz M, Liu YZ, Cui XD, Tokumoto H, Lindsay SM (1999) Surf Interface Anal 27:354
34. Jai C, Cohen-Bouhacina T, Maali A (2007) Appl Phys Lett 90:113512
35. Yagasaki K (2004) Phys Rev B 70:245419
36. Bhushan B (2007) Springer handbook of nanotechnology, 2nd edn, Springer-Verlag, Heidelberg, Germany
37. Persson BNJ (2000) Sliding friction: physical principles and applications, Springer, Heidelberg
38. Israelachvili J, Winnerström H (1996) Nature 379:219
39. Horn RG, Israelachvili J (1981) J Chem Phys 75:1400
40. Gee ML, McGuiggan PM, Israelachvili JN (1990) J Chem Phys 93:1895
41. Granick S (1991) Science 253:1374
42. Israelachvili J (1992) Intermolecular and surfaces forces, Academic, London
43. Klein J, Kumacheva E (1998) J Chem Phys 108:6996
44. Kumacheva E, Klein J (1995) Science 269:816
45. Demirel AL, Granick S (1996) Phys Rev Lett 77:2261
46. O'Shea SJ, Welland WE, Pethica JB (1994) Chem Phys Lett 223:336
47. Han W, Lindsay SM (1998) Appl Phys Lett 72:1656
48. Lim R, Li SFY, O'Shea SJ (2002) Langmuir 18:6116
49. Lim R, O'Schea SJ (2002) Phys Rev Lett 88:246101
50. Antognozzi M, Humphris ADL, Milles MJ (2001) Appl Phys Lett 78:300
51. Jeffery S et al (2004) Phys Rev B 70:054114
52. Sader JE, Jarvis SP (2004) Appl Phys Lett 84:1801
53. Uchihashi T et al (2004) Appl Phys Lett 85:3575
54. Maali A, Cohen-Bouhacina T, Couturier G, Aimé JP (2006) Phys Rev Lett 96:086105

16 Microtensile Tests Using In Situ Atomic Force Microscopy

Udo Lang · Jurg Dual

Abstract. In recent years a new field in the micromechanical characterization of materials has emerged. Researchers started to integrate atomic force microscopes (AFM) into microtensile tests. This allowed to investigate surface deformation of layers with thicknesses in the range of micrometers. In the first part of this article experiments on organic samples are presented followed by developments on anorganic specimens. In the second part of the paper latest developments at the Center of Mechanics of ETH Zurich are presented. The setup allows to monitor crack growth with micrometer resolution. At the same time forces can be measured in the millinewton range. Specimens are made from photodefinable polyimide. The stress-crack- length diagrams of two experiments are presented which enables to identify different stages of crack growth and therefore of fracture behaviour. Finally, possible extensions of the setup employing digital image correlation (DIC) are envisioned by analyzing the displacement field around the crack tip.

Key words: Microtensile testing, AFM, In situ, Force sensing, Digital image correlation

Abbreviations

SEM	Scanning electron microscopy
AFM	Atomic force microscopy
PET	Polyethylene terephtalate
LDPE	Low density polyethylene
HDPE	High density polyethylene
ME	Metal evaporated
MP	Metal particle
LVDT	Linear variable displacement transducer
DIC	Digital image correlation
DLC	Diamond-like carbon
PI	Polyimide
CT	Compact tension
HEPP	Hard elastic poly-propylene
TPV	Thermoplastic vulanizates
PB	Poly(1-butene)

16.1
Introduction

Crack initiation and growth are key issues when it comes to the mechanical reliability of microelectronic devices and microelectromechanical systems (MEMS). Especially in organic electronics where flexible substrates will play a major role these issues will become of utmost importance. It is therefore necessary to develop methods which in situ allow the experimental investigation of surface deformation and fracture processes in thin layers at a micro and nanometer scale. While scanning electron microscopy (SEM) might be used it is also associated with some major experimental drawbacks. First of all if polymers are investigated they usually have to be coated with a metal layer due to their commonly non-conductive nature. Additionally they might be damaged by the electron beam of the microscope or the vacuum might cause outgasing of solvents or evaporation of water and thus change material properties. Furthermore, for all kinds of materials a considerable amount of experimental effort is necessary to build a tensile testing machine that fits into the chamber. Therefore, a very promising alternative to SEM is based on the use of an atomic force microscope (AFM) to observe in situ surface deformation processes during straining of a specimen. First steps towards this goal were shown in the 1990s in [1–4] but none of these approaches truly was a microtensile test with sample thicknesses in the range of micrometers. To the authors' knowledge, this was shown for the first time by Hild et al. in [5].

16.2
Literature Review

16.2.1
Organic Samples

The fundamental idea of Hild et al. was to integrate an AFM into a microtensile test setup and to scan in situ the sample surface after each straining step. In their studies a 25-μm thick sample of hard elastic poly-propylene (HEPP) was first clamped and then stepwise stretched with a micrometer screw. Simultaneously the force acting on the specimen was measured by a force sensor and the surface was scanned after each stretching step by an AFM in contact mode. Their setup allowed them to differentiate between crystalline and non-crystalline regions and to analyze deformation induced structural changes such as cracks and voids. Their tensile setup was then borrowed by Oderkerk et al. in [6] who investigated in tapping mode 20-μm thick films of thermoplastic vulanizates (TPVs). The AFM scans revealed a heterogenous morphology of rubber parts dispersed in a nylon matrix. The scans also showed how upon stretching of the sample the deformation was localized in small nylon ligaments. Nishino et al. presented a very similar setup in [7] (see Fig. 16.1). The main idea of Nishino's setup was that a sample is strained with a dead force and then an AFM scans the surface of the specimen in contact mode. Because of fabrication fluctuations there were irregular small patterns on the surface.

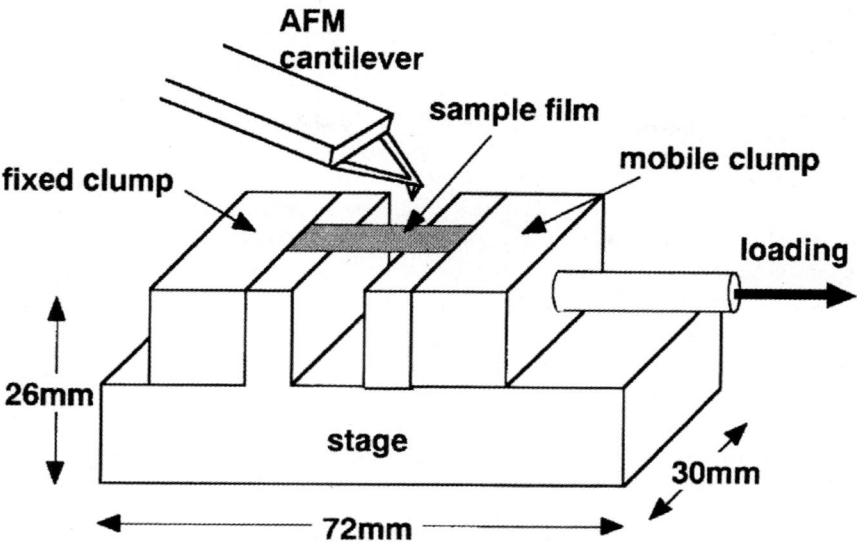

Fig. 16.1. Setup presented by Nishino et al. in [7]. A sample film is strained while an AFM is used to determine the surface displacements (reused with permission from Takashi Nishino [7]. Copyright 2000, American Institute of Physics)

Upon tensile loading the relative distances between these structures changed. This effect was used for the determination of strain. The mobile clamp (see Fig. 16.1) was connected to a load cell which then permitted the measurement of tensile forces and therefore stress. Nishino et al. used their setup to determine the mechanical properties of thin films of polyethylene terephtalate (PET) with a thickness of $100\,\mu$m. A similar setup but with a different intent was proposed by Opdahl and Somorjai in [8]. They strained low density and high density polyethylene (LDPE and HDPE) with a thickness of 1 mm and then applied an out-of-plane force by pushing the surface with an AFM in contact mode. This allowed them to evaluate structural stiffness and could therefore be used to characterize surface morphological changes due to tensile loading. Bhushan et al. also used a very similar setup (see Fig. 16.2) in [9, 10] to observe for the first time crack growth in thin films by in situ atomic force microscopy. This became possible by a left-right lead screw and two sliders which kept the control area at approximately the same place during testing. The compounds they investigated are normally used in magnetic tapes and were made from a approx. $6\,\mu$m thick PET film with metallic magnetic and non-magnetic layers deposited on it. The total thicknesses of these metal layers were approx. $0.2\,\mu$m for metal evaporated (ME) tapes and of approx. $1.5\,\mu$m for metal particle (MP) tapes. The experiments showed that under tensile loads cracks in the metal layers can be identified with submicron resolution (see Fig. 16.3). On the basis of these AFM scans in tapping mode the authors could develop a model to predict spacing between cracks in the metal layers. Using the same setup in [11], Bhushan et al. were able to determine Poisson's ratios of PET (6 and $14\,\mu$m thick) and polyethylene naphtalate (PEN) films ($6\,\mu$m thick). The global strain was determined from the number of steps through which the stepper motor was rotated. By tracking surface defects and analyzing their

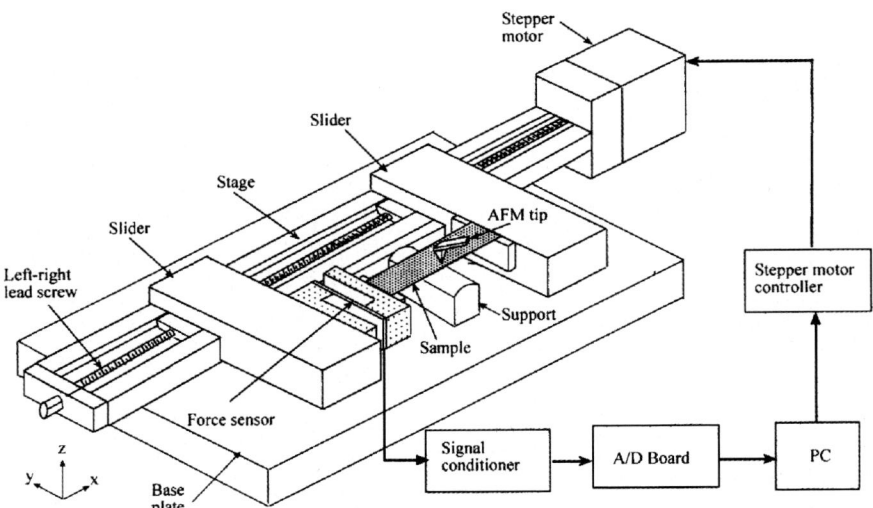

Fig. 16.2. Setup presented by Bhushan et al. in [9]. They introduced a left-right lead screw and two sliders in their setup. This kept the control area at the same place during pulling and enabled crack growth monitoring (reused with permission from Bharat Bhushan [9]. Copyright 2001, Materials Research Society)

relative positions Poisson's ratio could be determined. Roggemann and Williams concentrated in [12] on the development of image analysis algorithms to map surface displacements. As a test material they used an epoxy resin blended with 10 wt% of a polybutadiene. They could show that their algorithm was able to detect displacements in the vicinity of a crack. Li et al. used a commercial digital image correlation DIC software in [13] to determine local deformation and cracking in magnetic ME tapes made from 6 μm thick PEN or 4. 4 μm thick PET tapes, each coated with magnetic metal films with a total thickness of about 100 nm. The samples were strained using a custom built microtensile stage while the AFM scans were obtained in tapping mode after stretching. Using the same approach in [14], Li et al. determined strain fields in PEN and PET films. Furthermore, particles on the surface were used to investigate surface deformation mechanisms. Increased efforts in the mechanical design of the setup were presented by Bamberg et al. in [15]. The setup presented there included both, a complete LabVIEW controlled micro tensile testing machine and an integral AFM in tapping mode to investigate the surface of the tensile specimens during testing. As a test material they used the triblock copolymer polystyrene-b-isoprene-b-styrene (SIS). Height and phase image then revealed the deformation mechanisms under tensile load. Recently, a new study has been published by Thomas et al. in [16]. They investigated 100 μm thick films of poly(1-butene) (PB). Optical micrographs showed how the material was composed of spherulites with diameters of about 100–200 μm. Both clamps of the tensile test setup were stepwise actuated and after a relaxation time of 10 min, AFM scans in tapping mode were taken. As both clamps were actuated simultaneously the region of interest remained at the same location and the progress of deformation could be recorded. Scans with high magnification showed how crazes developed inside the spherulites under tensile loading.

MP Tape

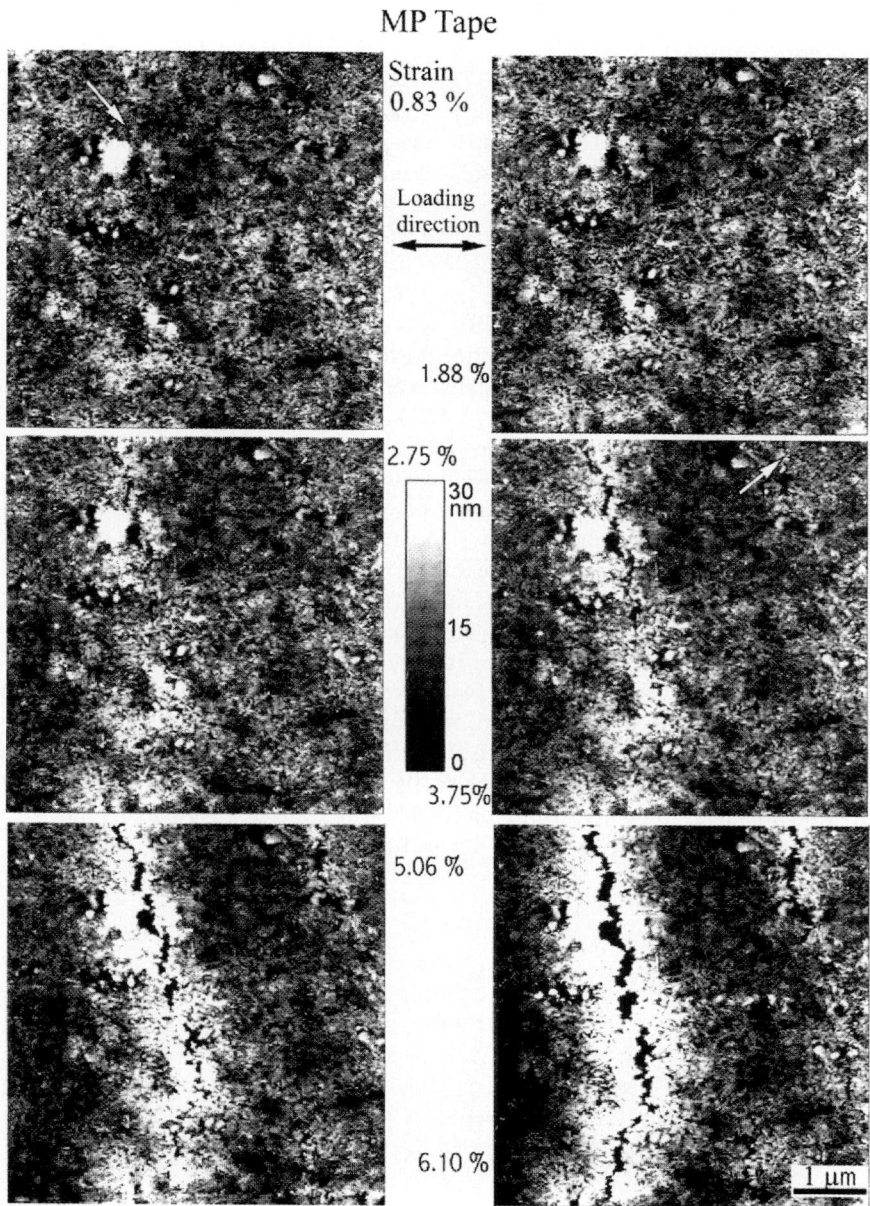

Fig. 16.3. Observation of crack growth in ME tape with submicron resolution as presented by Bhushan's group in [9] (reused with permission from Bharat Bhushan, [9]. Copyright 2001, Materials Research Society)

16.2.2
Anorganic Samples

Tensile testing of surface micromachined polysilicon samples using an AFM in non-contact mode for strain field determination was published by Chasiotis and Knauss in [17]. The test cross section of the samples was $2 \times 50\,\mu m$ in size. The authors also developed a new gripping process for samples that were bent due to residual stresses from fabrication. The idea was that the samples had large paddles at their ends. This enabled an electrostatically controlled UV adhesive gripping of the very thin polysilicon samples for the application of the tensile loads. Figure 16.4 shows this gripping process. The tensile loading during testing was done by an inchworm motion system. Chasiotis and Knauss additionally applied for the first time digital image correlation (DIC) to AFM scans of tensile tested samples and could thus derive the displacement field on the surface of the specimens. An AFM scan as well as the corresponding displacement field obtained from DIC are shown in Fig. 16.5. The displacement field and the force values from a load cell were then used to determine the Young's modulus of polysilicon. In a subsequent study based on the same method (presented in [18]) Chasiotis et al. developed a new sample geometry with a circular hole in the sample which permitted the evaluation of the mechanical properties of polysilicon. The advantage of this method was that both elastic constants (Young's modulus and Poisson's ratio) could be derived from only small sample areas and the displacement had only to be known in one direction. In [19] the Chasiotis' group investigated the minimum grain number necessary for isotropic behaviour of polysilicon. Their setup was then applied to fracture mechanics investigations of polysilicon in [20, 21]. By applying a tensile load to a precracked polysilicon specimen and by monitoring the displacement field in the vicinity of the crack, the fracture toughness of polysilicon could be obtained. In another study presented in [22], Young's modulus, Poisson's ratio and tensile strength of diamond-like carbon (DLC) could be derived.

(a)   (b)

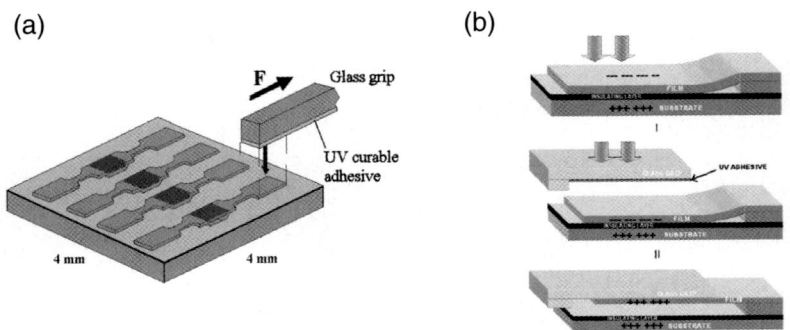

Fig. 16.4. (a) The surface micromachined polysilicon specimens with $2\,\mu m$ thickness on a silicon chip, (b) procedure for gripping polysilicon microsamples as presented by Chasiotis and Knauss in [17]. In step I a film buckled by residual stress is made to lie flat on the surface by electrostatic forces. In step II a glass grip with UV curable glue approaches the sample. By repelling electrostatic forces the sample can move freely again in step III (reused with permission from Ioannis Chasiotis [17, 19]. Copyright 2002 and 2007, Springer Boston)

Fig. 16.5. Surface scan (*left*) and corresponding displacement field (*right*) for polysilicon sample strained in vertical direction (from [17]). The displacement field was calculated with a commercial DIC package (reused with permission from Ioannis Chasiotis [17]. Copyright 2002, Springer Boston)

Isono et al. investigated $15\,\mu m$ thick nickel (Ni) and $20\,\mu m$ thick single crystal silicon (SCS) samples in [23]. The micromachined specimens had line patterns

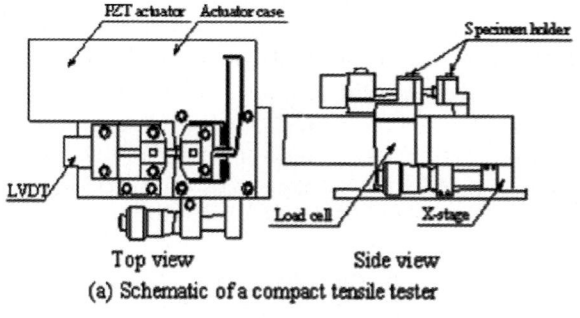

(a) Schematic of a compact tensile tester

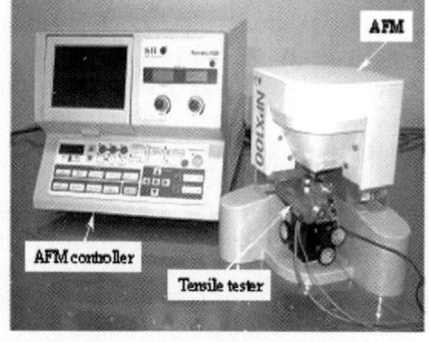

(b) A compact tensile tester built in an AFM

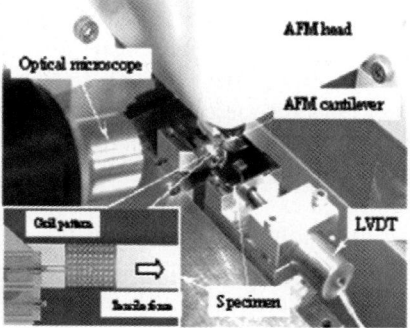

(c) During tensile testing

Fig. 16.6. Schematic of the setup presented by Isono et al. in [23, 24] (reused with permission from Yoshitada Isono [24]. Copyright 2006, IEEE)

made from silicon dioxide for the SCS specimens and made from photoresist for the Ni specimens on their surface. They were then clamped in a specimen holder and subsequently loaded (see Fig. 16.6 for their setup).

The change of spacing between the pattern elements during pulling allowed them to determine strain in the sample. As the setup also included a linear variable displacement transducer (LVDT) the differences between local strain in the test section determined by AFM and the global strain measured by the LVDT could be shown. Finally, in combination with the values obtained from the load cell also the mechanical properties such as Young's modulus and fracture strength could be determined. In [24] Isono et al. tested with the same setup diamond-like carbon films with submicron thickness on $19\,\mu m$ thick SCS. It was thus possible to determine Young's modulus, fracture strength, and Poisson's ratio of DLC.

16.2.3
Summary

Table 16.1 summarizes the state of the art of microtensile testing using in situ AFM for determining surface deformation fields.

16.3
Recent Developments at the Center of Mechanics of ETH Zurich

16.3.1
Setup

At the Center of Mechanics of ETH Zurich a microtensile test setup with integrated in situ AFM measurements has recently been developed (see Fig. 16.7). In contrast to all approaches shown so far where no or only external load cells were used, in the research presented here the force measurement is integrated on wafer level into the microfabricated samples as presented by Haque and Saif in [25]. This allows to build samples with thicknesses down to $1\,\mu m$ and to measure forces in the range of millinewtons. The force sensing is based on the deformation observation of a double fixed silicon beam by using an optical microscope. By applying textbook beam mechanics the force acting on the specimen and causing the observed deformation can be determined. In our particular case where in situ AFM measurements are to be done the force sensing beam is hidden underneath the AFM. Therefore an extension rod which can be considered a rigid body is connected to the force sensing beam. Because of this fact the deformation of the silicon force sensing beam is transferred directly to this extension rod. The deflection of the beam can then be determined by measuring the distance between the extension beam and the rigid silicon frame. The tensile samples in this research are made from the photodefinable polyimide PI2723 from HD Microsystems (Wilmington, DE, USA). Polyimide was chosen because it is widely used as a passivation layer to protect microelectronic devices from moisture and corrosion. Additionally polyimides are often used as stress buffer layers during packaging of dies [26]. Furthermore, polyimides are promising candidates for

Table 16.1. Summary of state-of-the-art microtensile tests with in-situ AFM measurements. If there was no specific data available in a publication it is correspondingly indicated by a "—"

Authors	Material	AFM mode	Actuation	Force sensor	Strain analysis
Hild et al. [5]	20 μm thick HEPP film	contact	screw	–	distance between clamps
Oderkerk et al. [6]	20 μm thick TPV	tapping	screw	–	distance between clamps
Nishino et al. [7]	100 μm thick PET film	contact	dead load	load cell	distance between surface defects
Opdahl et al. [8]	1 mm thick LDPE or HDPE	contact	–	–	–
Bhushan et al. [9–11]	ME and MP tapes made of PEN and PET with thicknesses 6–14 μm	tapping	stepper motor	beam type	stepper motor increment
Roggemann and Williams	1 mm thick blend of epoxy resin and polybutadiene	non-contact	–	load cell	globally: dial gauge; locally: DIC
Li et al. [13, 14]	PEN and PET ME tapes, 6 and 4.4 μm thick	tapping	–	–	Commercial DIC package
Bamberg et al. [15]	300 μm thick triblock copolymer	tapping	stepper motor	load cell	encoder of stepper motor
Thomas et al. [16]	100 μm thick PB	tapping	stepper motor	no force measurement	distance between clamps
Chasiotis et al. [17–21]	2 μm thick polysilicon	non-contact	inchworm	load cell	DIC
Chasiotis et al. [22]	≈ 1.5 μm thick DLC	non-contact	inchworm	load cell	DIC
Isono et al. [23, 24]	15 μm thick Ni, 20 μm thick SCS and 19 μm thick Si coated with DLC	–	PZT actuator	load cell	locally: deformation of surface pattern; globally: LVDT

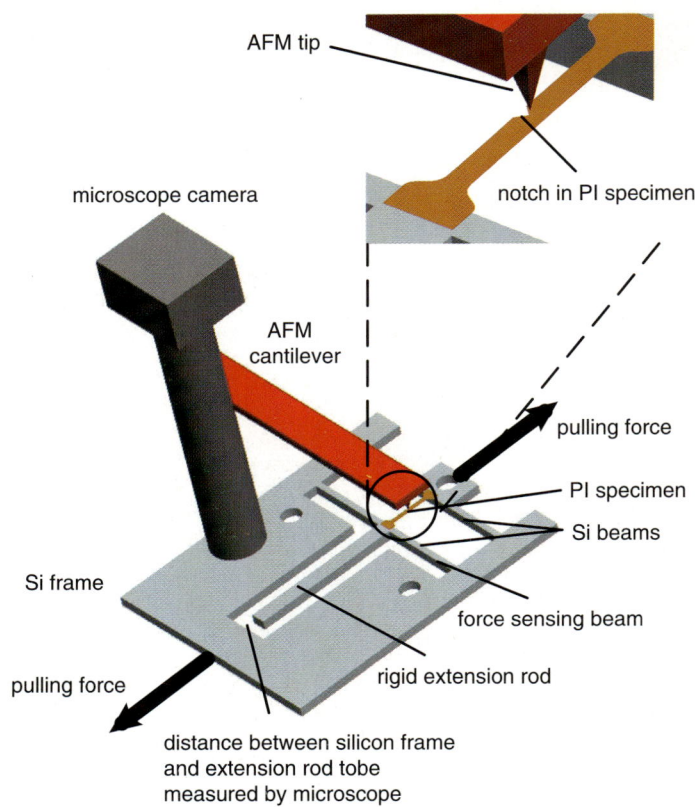

Fig. 16.7. Principle of the setup: the specimen is strained between two silicon beams while crack growth is monitored in situ by an AFM. A photolithographically structured notch results in a locally well defined crack initiation. The deformation of a force sensing beam is determined by an optical microscope. The rigid extension rod is necessary because the actual force sensing beam is hidden underneath the AFM. The three pin holes in the Si frame are used for force transmission by small pins from external piezo actuators to the specimen (from [34])

flexible substrates in organic electronics [27, 28] and in the MEMS community as a membrane material [29–31]. The specimen itself consists of a freestanding $3\,\mu m$ thick, $185\,\mu m$ wide and 1-mm long tensile probe and a silicon frame with a thickness of $525\,\mu m$. The geometry of the PI tensile probe is based on an international norm on tensile tests of polymers [32] with an additional single notch with a depth of $48\,\mu m$ on one side. The notch is necessary for a locally well-defined crack growth. This enables a simplified monitoring of crack behaviour under load. The fabrication of the specimens is mainly based on the Bosch deep dry etching process and is explained in [33] in detail. An overview of the process is given in Fig. 16.8. An easyScan AFM (Nanosurf AG, Liestal, Switzerland) is used in the experiments. Its main characteristics are a maximum scan range of $100\,\mu m$ in x- and y- directions and of $20\,\mu m$ in the z- direction. Scans are taken in contact mode with a typical load of 11 nN and automatic z-offset adjustment. The cantilevers are of type Contr-16 (Nanosensors AG, Neuchâtel, Switzerland) and have a typical tip diameter of less

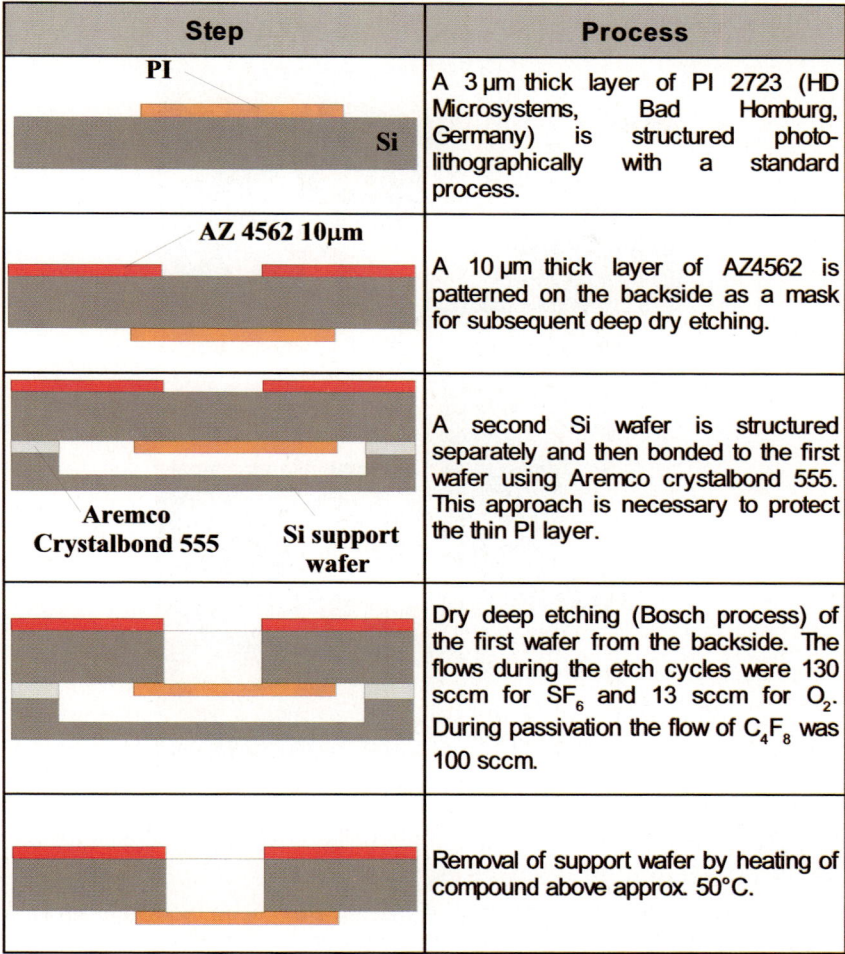

Step	Process
PI **Si**	A 3 μm thick layer of PI 2723 (HD Microsystems, Bad Homburg, Germany) is structured photo-lithographically with a standard process.
AZ 4562 10μm	A 10 μm thick layer of AZ4562 is patterned on the backside as a mask for subsequent deep dry etching.
Aremco Crystalbond 555 **Si support wafer**	A second Si wafer is structured separately and then bonded to the first wafer using Aremco crystalbond 555. This approach is necessary to protect the thin PI layer.
	Dry deep etching (Bosch process) of the first wafer from the backside. The flows during the etch cycles were 130 sccm for SF_6 and 13 sccm for O_2. During passivation the flow of C_4F_8 was 100 sccm.
	Removal of support wafer by heating of compound above approx. 50°C.

Fig. 16.8. Process flow for the fabrication of thin freestanding polyimide tensile specimens [33] (from [34])

than 10 nm and a tip height of about 10–15 μm. The AFM scans were analyzed by carefully determining the distance from the bottom of the notch to the tip of the crack with the analysis program Motic Images Plus 2.0. The actual deflection of the extension rod during experiments is monitored by a high magnification zoom lens (1–12x) (Navitar Inc., Rochester, USA) with a 2 megapixel CCD camera (Moticam 2000, Motic Deutschland GmbH, Wetzlar, Germany) attached to it (see Fig. 16.7). The deflection is then determined by measuring the distance between the rod and the frame using again Motic Images Plus 2.0. For every AFM scan therefore also a picture of the extension rod is taken. The actual experiment is thus conducted step-wise: the specimen is strained for approximately 5 μm then an AFM scan and a picture of the extension rod are taken. Then the specimen is strained for an additional 5 μm and again scans and pictures are taken and so forth until rupture of the sample. Figure 16.9 shows the actual setup. For the experiments the samples are

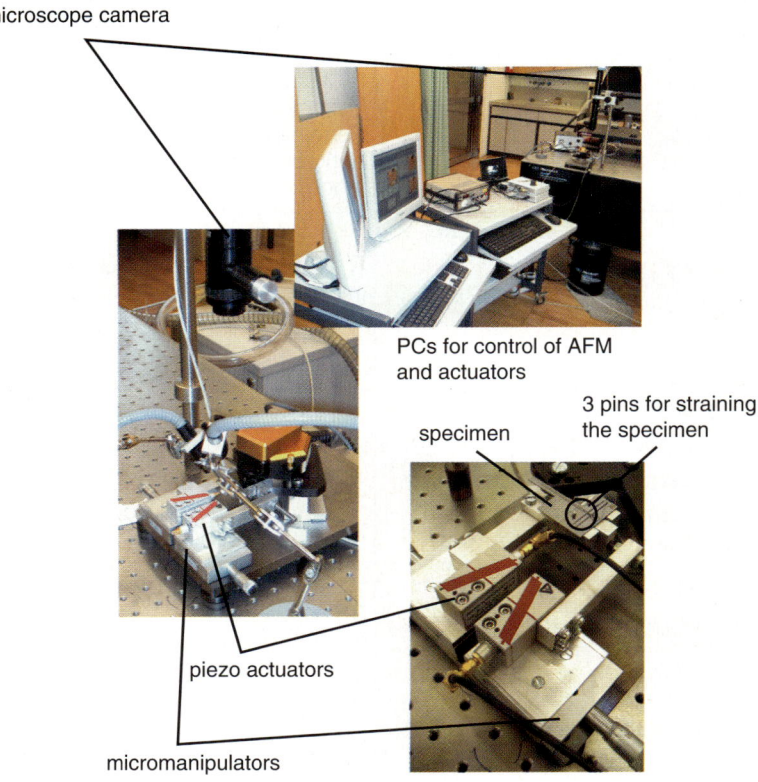

Fig. 16.9. Experimental setup. Two micromanipulators and two piezo actuators are used. They point pairwise into opposite directions which results in larger maximum displacements (from [34])

put on the pins which transfer the motion generated by the micromanipulators and piezo actuators (model P-280, 100 μm range, PI GmbH, Karlsruhe, Germany) to the specimens. The micromanipulators are necessary to bring the pins in contact with the specimen to make sure that the whole range of the piezo actuators can be used for precisely straining the specimen. During manual actuation it might happen that the two micromanipulators are not actuated exactly the same way and thus the crack might be moved away from the central position of the AFM scan. This can then be accounted for in the AFM scan software by readjusting the scan area.

16.3.2
Results

Two experiments were conducted at 23 °C and 25% relative humidity. Figure 16.10 shows the nominal stress within the sample far enough away from the notch (i.e. the normal stress at a width of 185 μm) versus the crack length starting from the bottom of the notch. The results show the same behaviour for both of the tests which means that they are not random but are connected to inherent material properties i.e. fracture

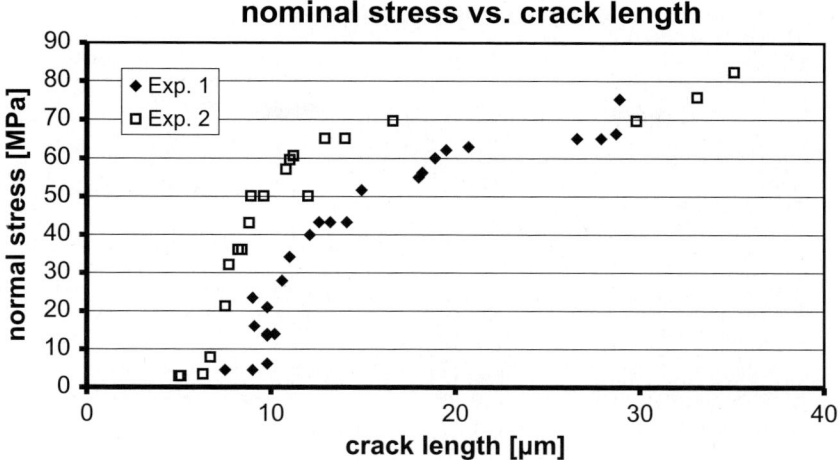

Fig. 16.10. Nominal stress (i.e. normal stress at a sample width of 185 μm) vs. crack length diagrams for the two conducted experiments. Both samples showed basically the same behaviour (from [34])

toughness. Evidently there are different stages during crack growth. Figure 16.11 shows in 2D and 3D view how the AFM scans can be used to analyze these different stages. These phases during crack growth might be explained as follows:

- Initial crack length
  For both of the samples there is an initial crack length even without external forces applied to the setup. In order to explain this phenomenon one has to take into consideration that during fabrication the compound of the silicon substrate and the polyimide layer is cooled down to room temperature after baking at 350 °C. Because of the large mismatch in the coefficient of thermal expansion between the two materials high thermal tensile stresses develop in the polyimide layer. After releasing the sample from the substrate during dry etching these thermal stresses are partially relieved as the sample is free to contract, partially close the thermal crack and deform the silicon beams. If during the experiment the samples are strained the initial crack is opened again without applying any additional force. This explains why there is an apparent crack "growth" between the first two data points without any increase in external force and therefore nominal stresses.
- Stationary phase
  The crack length remains stable while the external force is increased. The elastic energy stored in the material is not yet large enough to activate crack growth.
- Stable crack growth
  The crack starts to grow and two very important observations can be made. Firstly, during crack opening no crazes can be found and secondly material becomes visible at the bottom of the crack. This can also be seen in the three-dimensional scans (Fig. 16.11). At this point the assumption is that a corner crack (for a geometrical description of a corner crack see e.g. [35]) growing vertically and laterally from the top middle of the notch into the specimen could be seen.

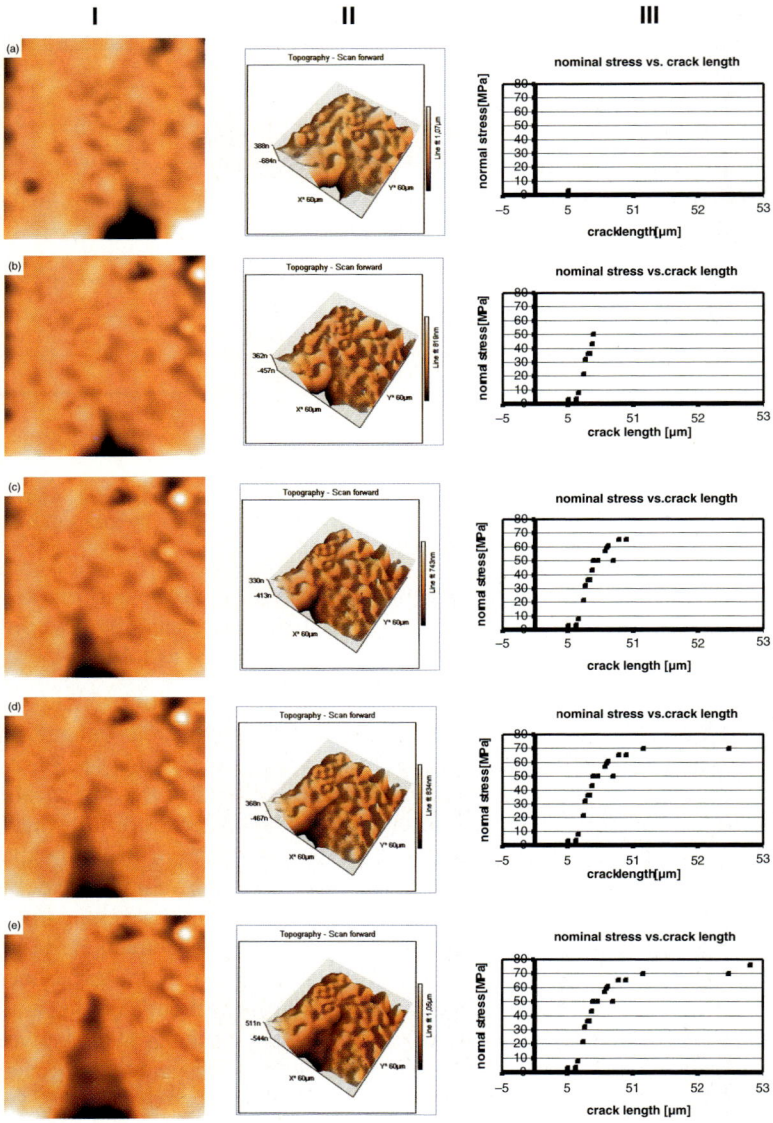

Fig. 16.11. Typical situations during crack propagation. *Column I* shows 2D scans with side lengths of 60 μm, *column II* are the corresponding 3D views and *column III* shows the stress-crack length diagrams. The AFM scans always refer to the corresponding last data point of the diagrams. The stages of crack growth are: (**a**) initial crack length, (**b**) stationary state i.e. no crack propagation at increasing stress, (**c**) stable crack growth, (**d**) unstable crack growth i.e. without increase of nominal stress the crack grows for a considerable distance and (**e**) before failure the crack growth is stopped again and a final increase of stress is necessary to initiate rupture. Note the appearance of material at the *bottom* of the crack for the last data points. This is an indication for a crack growing vertically and laterally from the corner of the middle of the top of the notch into the specimen (from [34])

- Unstable crack growth
  The crack propagates vastly into the material without almost any additional load.
- Pop-in
  The unstable crack growth is stopped shortly before final failure of the sample. Such a behaviour is normally called pop-in for macroscopic samples. This term will also be used here. In order to be able to stop a propagating crack, plastic deformations need to take place to dissipate energy. Only from the AFM scans such plastic processes could not be analyzed as they occur right before catastrophic failure of the samples and are therefore not visible on the scans. For an analysis of these processes SEM pictures of the fracture zone of the ruptured samples have to be made after testing. This will also be a focus of future research.

16.3.3
Outlook

While in these experiments the crack length was determined, the application of digital image correlation software could greatly enhance the possibilities of the setup. Figure 16.12 shows the results obtained by the DIC software package VEDDAC (Chemnitzer Werkstoffmechanik GmbH, Chemnitz, Germany). Not only could the crack tip opening displacement be observed but also the displacement field ahead

Fig. 16.12. Displacement field (*white*) at the crack tip obtained with the DIC software VEDDAC. It shows the state right before final rupture (from [34])

of the tip could in principle be analyzed. Both pieces of information would be of great use in the application of fracture mechanics theories to the results and have partly already been reported in the literature for a resin polymer compact tension (CT) specimen [36]. Further improvements of the experiments seem to be possible by changing the layout of the PI specimens by applying rules set by Feddersen [37]. Instead of the single notched specimen as used in this study, middle tension specimens with a center notch and defined ratio of initial notch length to specimen width would be used. All of these measures would allow for a simplified quantitative determination of fracture toughness.

16.4
Conclusions

An overview of the new and expanding field of microtensile testing with in situ AFM was given. Several research groups have entered this field and are investigating both organic and anorganic materials. Especially for polymeric materials this approach is very promising as it omits disadvantages usually associated with electron microscopy. It was shown with a newly developed setup at ETH Zurich that the AFM could also be used to observe crack propagation with micrometer resolution in polyimide microspecimens. By applying beam mechanics to the deformation of a double fixed silicon beam connected to the specimens forces in the range of millinewtons could be measured. By combining the two data sets of stresses and AFM scans the determination of stress-crack-length diagrams was possible. Different stages during crack growth could therefore be identified. Future experiments could be done with different geometries and by using DIC software and might therefore allow a reliable quantitative determination of fracture toughness. Other possible tracks of research include crack investigations at higher magnifications to take advantage of the nanomechanical potential of atomic force microscopy. This could eventually help explore fracture processes at a molecular level.

Acknowledgments. The authors would like to express their gratitude to Prof. Bharat Bhushan from the Ohio State University, Prof. Taher Saif from the University of Illinois at Urbana-Champaign, Dr. Hans-Jakob Schindler from Mat-Tec AG (Winterthur, Switzerland), Dr. Pieter van Schendel from Nanosurf AG (Liestal, Switzerland), Dipl.-Ing. Bettina Seiler and Dr. Michael Dost from CWM GmbH (Chemnitz, Germany), and finally Dr. Nicola Naujoks from the Nanotechnology Group of ETH Zurich for very helpful discussions. Prof. Ralph Spolenak and Prof. Andreas Stemmer both from ETH Zurich are gratefully acknowledged for proofreading this chapter. Special thanks go to Tobias Süss and Nina Wojtas for their experimental efforts and their ideas during their thesis on this subject.

References

1. Small MK, Coupeau C, Grilh J (1995) Atomic-force microscopy of in-situ deformed LiF. Scripta Metall Mater 32(10):1573–1578
2. Goken M, Vehoff H, Neumann P (1996) Atomic force microscopy investigations of loaded crack tips in nial. J Vac Sci Technol B 14(2):1157–1161

3. Tong W, Hector LG, Weiland H, Wieserman LF (1997) In-situ surface characterization of a binary aluminum alloy during tensile deformation. Scripta Mater 36(11):1339–1344
4. Bhushan B (1999) Wear and mechanical characterisation on micro- to picoscales using AFM. Int Mater Rev 44(3):105–117
5. Hild S, Gutmannsbauer W, Luth R, Fuhrmann J, Guntherodt HJ (1996) A nanoscopic view of structure and deformation of hard elastic polypropylene with scanning force microscopy. J Polym Sci Pt B-Polym Phys 34(12):1953–1959
6. Oderkerk J, de Schaetzen G, Goderis B, Hellemans L, Groeninckx G (2002) Micromechanical deformation and recovery processes of nylon-6 rubber thermoplastic vulcanizates as studied by atomic force microscopy and transmission electron microscopy. Macromolecules 35(17):6623–6629
7. Nishino T, Nozawa A, Kotera M, Nakamae K (2000) In situ observation of surface deformation of polymer films by atomic force microscopy. Rev Sci Instrum 71(5):2094–2096
8. Opdahl A, Somorjai GA (2001) Stretched polymer surfaces: Atomic force microscopy measurement of the surface deformation and surface elastic properties of stretched polyethylene. J Polym Sci Pt B-Polym Phys 39(19):2263–2274
9. Bobji MS, Bhushan B (2001) In situ microscopic surface characterization studies of polymeric thin films during tensile deformation using atomic force microscopy. J Mater Res 16(3):844–855
10. Tambe NS, Bhushan B (2004) In situ study of nano-cracking in multilayered magnetic tapes under monotonic and fatigue loading using an AFM. Ultramicroscopy 100(3–4):359–373
11. Bhushan B, Mokashi PS, Ma T (2003) A technique to measure Poisson's ratio of ultrathin polymeric films using atomic force microscopy. Rev Sci Instrum 74(2):1043–1047
12. Roggemann MC, Williams JG (2002) Use of an atomic force microscope to measure surface deformations in polymeric systems. J Adhes Sci Technol 16(7):905–920
13. Li XD, Xu WJ, Sutton MA, Mello M (2006) Nanoscale deformation and cracking studies of advanced metal evaporated magnetic tapes using atomic force microscopy and digital image correlation techniques. Mater Sci Technol 22(7):835–844
14. Li XD, Xu WJ, Sutton MA, Mello M (2007) In situ nanoscale in-plane deformation studies of ultrathin polymeric films during tensile deformation using atomic force microscopy and digital image correlation techniques. IEEE Trans Nanotechnol 6(1):4–12
15. Bamberg E, Grippo CP, Wanakamol P, Slocum AH, Boyce MC, Thomas EL (2006) A tensile test device for in situ atomic force microscope mechanical testing. Precis Eng-J Int Soc Precis Eng Nanotechnol 30(1):71–84
16. Thomas C, Ferreiro V, Coulon G, Seguela R (2007) In situ AFM investigation of crazing in polybutene spherulites under tensile drawing. Polymer 48(20):6041–6048
17. Chasiotis I, Knauss WG (2002) A new microtensile tester for the study of MEMS materials with the aid of atomic force microscopy. Exp Mech 42(1):51–57
18. Cho SW, Cardenas-Garcia JF, Chasiotis I (2005) Measurement of nanodisplacements and elastic properties of MEMS via the microscopic hole method. Sensor Actuat A-Phys 120(1):163–171
19. Cho SW, Chasiotis I (2007) Elastic properties and representative volume element of polycrystalline silicon for MEMS. Exp Mech 47(1):37–49
20. Chasiotis I, Cho SW, Jonnalagadda K (2006) Fracture toughness and subcritical crack growth in polycrystalline silicon. J Appl Mech-Trans ASME 73(5):714–722
21. Cho SW, Jonnalagadda K, Chasiotis I (2007) Mode I and mixed mode fracture of polysilicon for MEMS. Fatigue Fract Eng Mater Struct 30(1):21–31
22. Cho S, Chasiotis I, Friedmann TA, Sullivan JP (2005) Young's modulus, Poisson's ratio and failure properties of tetrahedral amorphous diamond-like carbon for MEMS devices. J Micromech Microeng 15(4):728–735

23. Lee Y, Tada J, Isono Y (2005) Mechanical characterization of single crystal silicon and UV-LIGA nickel thin films using tensile tester operated in AFM. Fatigue Fract Eng Mater Struct 28(8):675–686
24. Isono Y, Namazu T, Terayama N (2006) Development of AFM tensile test technique for evaluating mechanical properties of sub-micron thick DLC films. J Microelectromech Syst 15(1):169–180
25. Haque MA, Saif MTA (2002) In-situ tensile testing of nano-scale specimens in SEM and TEM. Exp Mech 42(1):123–128
26. LLC HD MicroSystems. Pyralin PI2720 Processing Guidelines. 1998.
27. Kajii H, Taneda T, Ohmori Y (2003) Organic light-emitting diode fabricated on a polymer substrate for optical links. Thin Solid Films 438:334–338
28. Lee JG, Seol YG, Lee NE (2006) Polymer thin film transistor with electroplated source and drain electrodes on a flexible substrate. Thin Solid Films 515(2):805–809
29. Tung S, Witherspoon SR, Roe LA, Silano A, Maynard DP, Ferraro N (2001) A MEMS-based flexible sensor and actuator system for space inflatable structures. Smart Mater Struct 10(6):1230–1239
30. Aslam M, Gregory C, Hatfield JV (2004) Polyimide membrane for micro-heated gas sensor array. Sens Actuat B-Chem 103(1–2):153–157
31. Kuoni A, Holzherr R, Boillat M, de Rooij NF (2003) Polyimide membrane with ZnO piezoelectric thin film pressure transducers as a differential pressure liquid flow sensor. J Micromech Microeng 13(4):S103–S107
32. ISO527-3. Plastics-determination of tensile properties-part 3. Technical report, 1995.
33. Lang U, Reichen M, Dual J (2006) Fabrication of a tensile test for polymer micromechanics. Microelectron Eng 83(4–9):1182–1184
34. Lang U, Dual J Observing crack propagation in polyimide microtensile specimens by in situ atomic force microscopy, 2008. Submitted to Exp Mech
35. Anderson TL (1995) Fracture mechanics: Fundamentals and applications, CRC Press, Boca Raton, p 630.
36. Keller J, Vogel D, Schubert A, Michel B (2004) Displacement and strain field measurements from SPM images. In: Bhushan B, Fuchs H, Hosaka S (eds) Applied scanning probe methods, volume I of Nanoscience and technology, Springer pp 253–276
37. Feddersen CE (1971) Evaluation and prediction of residual strength of center cracked tension panels. In: Rosenfield MS (ed) Damage tolerance in aircraft structures, vol ASTM STP 486. pp 50–86

17 Scanning Tunneling Microscopy of the Si(111)-7×7 Surface and Adsorbed Ge Nanostructures

Haiming Guo · Yeliang Wang · Hongjun Gao

Abstract. In this chapter, we review advances towards a comprehensive understanding of the Si(111)-7×7 surface and adsorbed Ge nanostructures by scanning tunneling microscopy. We first present our recent results on the highest resolution STM imaging on Si(111)-7×7 surfaces, in which all the six rest atoms and 12 adatoms are resolved simultaneously with unprecedented high-contrast. In the second part, STM was used as an atomic manipulation tool to fabricate groove structures with atomically straight edges and uniform lateral size by extracting atoms one by one from the Si(111)-7×7 surfaces. The critical current under various voltages for fabricating such grooves is measured, and a modification mechanism is discussed. Lastly, the behaviors of various Ge nanostructures created on the Si(111)-7×7surface, ranging from the initial adsorption sites of single Ge atoms, to the evolution and aggregation patterns of Ge nanoclusters, and then to the formation of 2-D extended Ge islands are comprehensively investigated by STM experimental studies and theoretical calculations.

Key words: Scanning tunneling microscopy, Si(111)-7×7 surface, Atomic manipulation, Germanium nanostructures

Abbreviations

0D, 1D, 2D, 3D	zero, one, two or three dimensions (or dimensional)
AES	Auger electron spectroscopy
AFM	Atomic force microscopy
CITS	Current imaging tunneling spectroscopy
DAS	Dimer-adatom-stacking fault model
DFT	Density functional theory
dI/dV	Differential current-voltage
DOS	Density of states
E_F	Fermi energy
FHUC	Faulted half unit cell
GGA	Generalized gradient approximation
LDOS	Local density of states
LEED	Low-energy electron diffraction
LT	Low temperature
I–V	Current (I) – voltage(V)
MBE	Molecular beam epitaxy
ML	Monolayer

RT   Room temperature
STM   Scanning tunneling microscopy
STS   Scanning tunneling spectroscopy
UHUC   Unfaulted half unit cell
UHV   Ultra high vacuum
VASP   Vienna ab initio software package
XSW   X-ray standing-wave

17.1
Introduction

When the lattice of a bulk crystal is terminated at the surface, it will lead to the destruction of periodicity and a loss of symmetry, production of dangling bonds and increasing surface energy. To reduce the surface energy, the surface atoms usually need to rearrange themselves and the surface layers form a new structure, this is called surface reconstruction. Since the surface layers are in close contact with the bulk, the periodicity changes are likely to be, if not the same, a simple multiple, sub-multiple or rational fraction of the lateral bulk periodicity (a, b). This provides convenient classification means. A variety of reconstruction structures have been found and investigated on the metal and semiconductor surfaces.

Among these structures, the 7×7 reconstruction of Si(111) is probably the most well known and complicated example of a reconstructed surface. Just as it is said that God created the bulk and Satan added the surface, actually we couldn't even believe nature could choose such a complex and fascinating arrangement as we try to learn this structure for the first time. It results in a large surface unit mesh and extends several layers deep from the surface. The atomic arrangement of the reconstruction surface can be described as the dimer-adatom-stacking fault (DAS) model of Si(111)-7×7 structure, as schematically shown in Fig. 17.1 [1]. This model consists of nine dimer bonds, 12 adatoms, and a stacking fault between the second and third layer atoms in each unit cell. Additional features are corner holes and six unsaturated dangling bonds of six exposed "rest atoms" in the layer below the adatoms. The 12 adatoms and six rest atoms are evenly distributed in the faulted half unit cell (FHUC) and unfaulted half unit cell (UHUC).

The Si(111)-7×7 has been the most important surface in surface science due to the importance of silicon-based electronics for many decades. It shows metallic surface electronic states. The large unit cell size ($2.7 \times 2.7 \, nm^2$) makes it an ideal template for the growth of well-ordered nanostructures. This reconstructed surface has also provided a platform for the testing of the unprecedented resolution of STM as a novel powerful apparatus since the 1980s [2]. To date, it has been extensively used and studied in various research fields ranging from surface science and material science to nanotechnology.

In this chapter, we will first present our recent achievements in exploring the "ultimate" atomic resolution on this surface by scanning tunneling microscopy, in which all the rest atoms and adatoms of the Si(111)-7×7 surface can be observed simultaneously with unprecedented high-contrast. In the second part, STM was

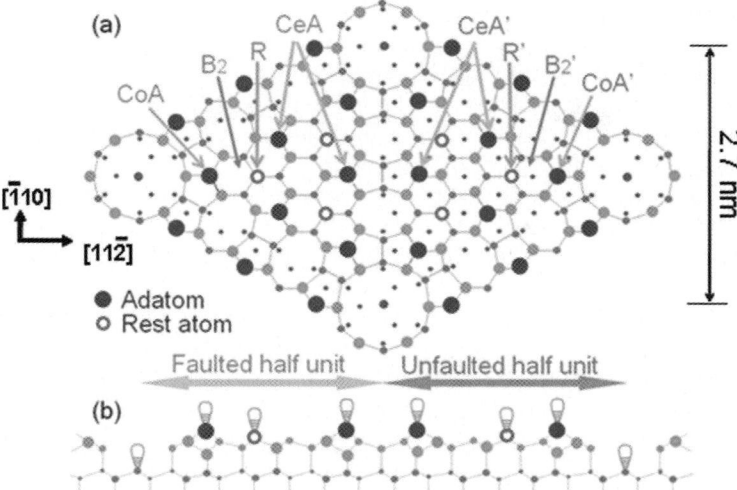

Fig. 17.1. Schematic diagram for Si(111)-7×7 'DAS' model [1]. (**a**) *Top view*: Atoms on (111) layers at decreasing heights indicated by *dots* of decreasing sizes. (**b**) *Side view*: Dangling bonds located at the topmost of all adatoms, rest atoms, and *holes*. The sites of corner-adatom, center-adatom, and rest-atom in FHUC (labeled CoA, CeA, and R, respectively), and corresponding sites in UHUC (labeled CoA', CeA' and R', respectively), are identified by *arrows*. Positions B_2 in FHUC, and B_2' in UHUC are also denoted by *arrows*

used as an atomic manipulation tool to fabricate groove structures with atomically straight edges and uniform lateral size by extracting atoms one by one from the Si(111)-7×7 surface. Lastly, the behaviors of various Ge nanostructures created on the Si(111)-7×7 surface, ranging from the initial adsorption sites of single Ge atoms, to the evolution and aggregation patterns of Ge nanoclusters, and then to the formation of 2-D extended Ge islands are investigated comprehensively and systematically with STM.

17.2
STM Imaging on Si(111)-7×7: Resolving the Rest Atoms

17.2.1
The Familiarity: Si(111)-7×7 Structure and STM

The history of STM and the Si(111)-7×7 surface is ever indivisible. STM is one typical application of the concept of electron tunneling. In 1982, Binnig, Rohrer, and Gerber invented the first STM and observed tunneling through a controlled vacuum gap. One year later, the Si(111)-7×7 reconstructions were imaged in real space for the first time by this new instrument in 1983 and then a new field was born [3]. Over the years, STM has been proven to be one of the powerful modern research techniques that allow investigation of the morphology and local properties of the solid body surface with high spatial resolution. On the other hand, the discovery and investigation of the 7×7 surface was a milestone in the study of surface crystallography of semiconductors, and its DAS structure was finally determined in 1985 by Takayanagi et al. after a detailed analysis of transmission electron diffraction data in combination with low-energy electron diffraction (LEED) results and real-space STM images [1]. Today, Si(111)-7×7 is the standard surface for verification of the operation of STM in UHV (ultra high vacuum).

Since then, with this powerful tool and the later family of scanning probe microscopes, the structure of the Si(111)-7×7 surface has been extensively investigated [4–9]. The STM demonstrations on the atomic topography of a clean Si(111)-7×7 surface commonly show the topmost adatoms, as shown in Fig. 17.2. In the empty state STM image, i.e., with a positive sample bias voltage, the 12 adatoms in one unit cell have the same brightness, while the occupied state image reveals that the adatoms on the faulted half are brighter than that on the unfaulted one. Many ab initio calculations for the electronic structure on the 7×7 unit have been carried out. The surface is electronically metallic, with the corner atoms on the faulted half having the highest DOS near the Fermi energy. Furthermore, the state of dangling bonds of the adatoms is about 0.4 eV below E_F, while the states of dangling bonds of rest atoms is about 0.8 eV below E_F [10]. The tunneling current in STM of the Si(111)-7×7 surface originates from these dangling bonds. Voltage-dependent imaging showed that the protrusions depended on the polarity and did not represent the positions of the atoms but rather the maxima of the local density of states (LDOS). STM is very sensitive to states closest to the sample Fermi energy (E_F), so the mapping of rest atoms whose dangling bond states are far from E_F is difficult to carry out.

On the mapping of rest atoms of the Si(111)-7×7 surface, some saddle points at the position expected for the rest atoms were reported by Avouris et al. [4] and Nishikawa et al. [5] using STM. Recently, some special techniques were used to obtain images of the Si(111)-7×7 surface with atomic-scale resolution, such as, Lantz et al. [6] using scanning force microscopy and Giessibl et al. [7,8] using atomic force microscopy. Sutter et al. [9] have mapped selectively the rest atoms at a price

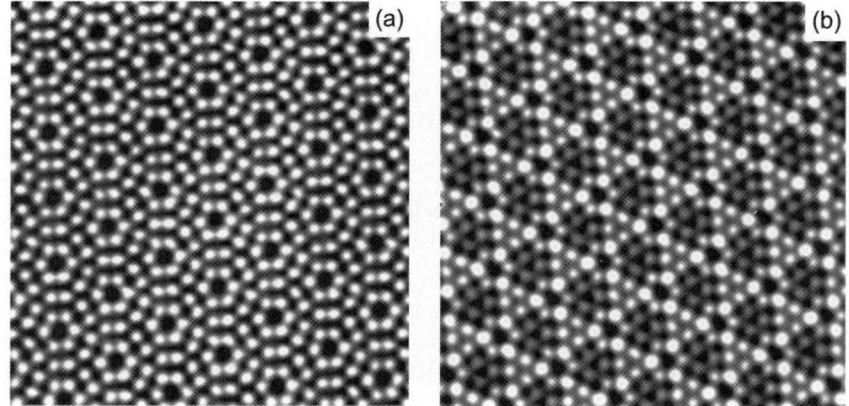

Fig. 17.2. Typical atomic resolution STM images of a Si(111)-7×7 surface. The *bright spots* mark only the "dangling bonds" of adatoms. (**a**) Empty states. (**b**) Occupied states. The adatoms on the faulted half are brighter than that on the unfaulted half

of suppressing the adatom spots with a monocrystalline semiconductor tip since its energy gap can suppress the tunneling from the adatoms at certain sample bias. Up to now, the adatoms and the rest atoms of the Si(111)-7×7 surface have not been clearly distinguished simultaneously in an STM image. This inability has led to the perception that the measured tunneling current for semiconductor materials comes from mostly states near the Fermi level instead of the states further away, due to the exponential dependence of the tunneling probability on the energy level position [11].

We revisit this surface by using UHV-STM. The resultant images clearly show not only the 12 adatoms but also the six rest atoms per (7×7) unit cell of the Si(111) surface with high contrast. A careful preparation of the STM tips (reducing the radius of the apex) may be the key to our success, as our first-principles calculations reveal a geometric hindrance effect of the apex for such complex surfaces.

17.2.2
The Simultaneous Imaging of the Rest Atoms and Adatoms

The experiments were performed using an ultrahigh-vacuum STM system with a base pressure of $\sim 5 \times 10^{-11}$ mbar. N-doped Si(111) substrate ($\rho \sim 0.03 \; \Omega \cdot$ cm, thickness ~ 0.5 mm) was degassed at about 650 °C in the UHV chamber for several hours and subsequently flashed at 1,200 °C for 20 s, rapidly lowering the temperature to about 900 °C, and then slowly decreasing the temperature at a pace of 1–2 °C/s to room temperature (RT). Finally, an atomically flat and clean (7×7) reconstruction was obtained. Sharp STM tips were made of a polycrystalline tungsten wire etched electrochemically in NaOH.

Figure 17.3a shows STM images with high-contrast between adatoms and rest atoms. They demonstrate simultaneously the adatoms and the rest atoms, that is, 18 topographic maxima per (7×7) unit cell. By zooming in, a high-resolution image with a scanning area of 8×8 nm^2 presents more clearly all of the adatoms and rest atoms, as shown in Fig. 17.3b. In the UHUC side, the rest atoms appear to have almost the same brightness as the central adatoms, whereas in the FHUC, the rest atoms appear to have considerably less brightness than the central adatoms. The line profile in Fig. 17.3c shows the positions and height differences of the six distinct types of atoms (labeled 1–6) along the solid line in Fig. 17.3b. The ranking of height of these atoms is: 1 is the highest, then 3 and 6 follow, with 2, 4, and 5 being the lowest. The rest atom (site 2) in the FHUC side is at the same level as the rest atom (site 5) in the UHUC side, and they are both even at the same level as the central adatom (site 4) in the UHUC half side. The high-contrast between rest atoms and adatoms is even better than the previous results obtained by using scanning force microscopies [6–8]. To our best knowledge, it is the first time that all the rest atoms and adatoms on a Si(111)-7×7 surface have been simultaneously demonstrated with high-contrast in an STM topographic image. The emergence of rest atoms will be further rationalized in the following by theoretic simulations.

There is a defect seen in Fig. 17.3b where one corner adatom is missing, although it shows no influence on its adjacent rest atom, and the rest atom is still visible and stays in its normal position without any lateral distortion. So the absence of a local adatom does not affect the geometric structures of its surrounding atoms in a (7×7)

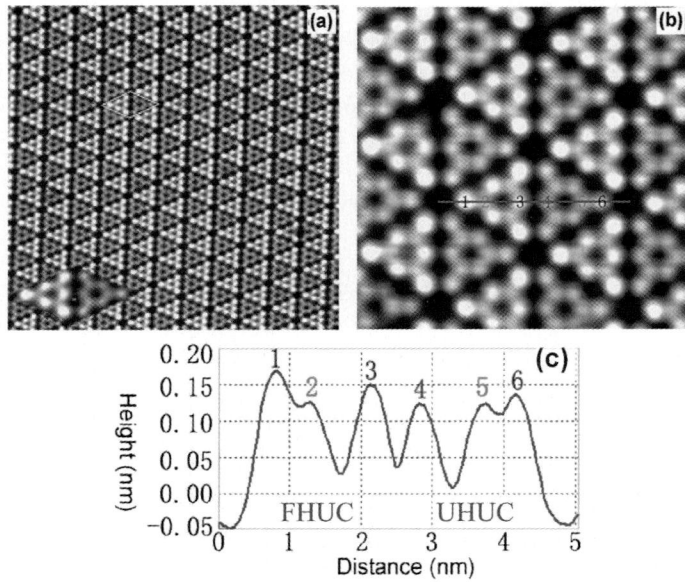

Fig. 17.3. Filled states STM images of the Si(111)-7×7 surface, revealing 12 adatoms and six rest atoms per unit cell. (**a**) The image extends over an area of 30 × 30 nm^2. The amplificatory (7×7) unit cell was indicated in the inset, and a (7×7) unit cell is depicted with a rhombus; (**b**) amplified image with scanning area 8 × 8 nm^2. Both of the images are recorded by sample bias voltage of −1.5 V and a tunneling current of 0.3 nA. (**c**) The *line* profile taken along the *line* in (**b**), labels "1", "2", "3" denote the corner adatom, rest atom, center adatom in the faulted half unit (FHUC), and labels "4", "5", "6" denote the center adatom, rest atom, corner adatom in the unfaulted half unit (UHUC), respectively

unit cell. This result coincides with the recent reports about the local structures of adatom vacancies in Si(111)-7×7 surfaces [12]. There Chen et al. conducted STM dI/dV mappings on adatom vacancies and found that the adatom vacancies showed different local electronic structures than the normal adatom but no effect to the geometric or electronic structures of the nearby rest atom.

17.2.3
Voltage-Dependent Imaging of Rest Atoms

A series of STM images obtained at different sample bias voltage illustrate that the emergence of rest atoms is dependent on the sample bias voltage, as shown in Fig. 17.4. At lower bias voltage of −0.5 V and −0.6 V, the images (see Fig. 17.4a and b) only show 12 adatoms in each (7×7) unit cell. This suggests the electronic states of adatoms are closer to the Fermi level than that of the rest atoms. The absence of the rest atoms suggests the electronic states of the rest atoms are outside the range of the bias when the value of sample bias is kept very low. By increasing the value of bias voltage, the rest atom spots may be visible at the sample bias voltage of less than −0.7 V, as shown in Fig. 17.4c–f. This clearly reveals that the dangling bond states

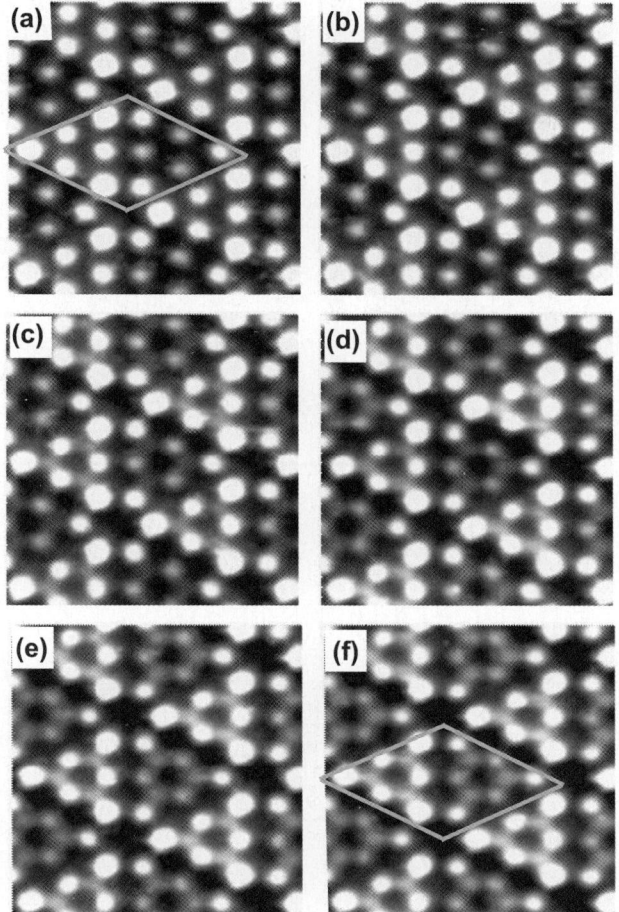

Fig. 17.4. STM images of the Si (111)-7×7 surface with different sample bias voltages: (**a**) −0.5 V; (**b**) −0.6 V; (**c**) −0.7 V; (**d**) −0.8 V; (**e**) −0.9 V; (**f**) −1.0 V, respectively. The rest atoms appear at the sample voltages less than −0.7 V. All images are taken at a tunneling current of 0.4 nA in a scanning area of $5 \times 5 \, nm^2$

of the rest atoms are located at about 0.7 eV below the Fermi energy (E_F), which is in excellent agreement with the experimental results measured by the method of current imaging tunneling spectroscopy (CITS). In 1989, Hamers et al. measured the electronic banding structure of the Si(111)-7×7 surface by using CITS. This provided information on the dangling bond states of adatoms (about 0.35 eV below the E_F) and rest atoms (about 0.8 eV below the E_F) [10].

The STM observations presented here are in sharp contrast to previous STM studies, which in most cases showed images similar to Fig. 17.5a with 12 protrusions in each (7×7) unit, irrespective of the bias voltages (somewhere between −2 V and 2 V). A common explanation [11] for the absence of the rest atom spots in the images relies on the fact that the tunneling probability depends on the thickness of

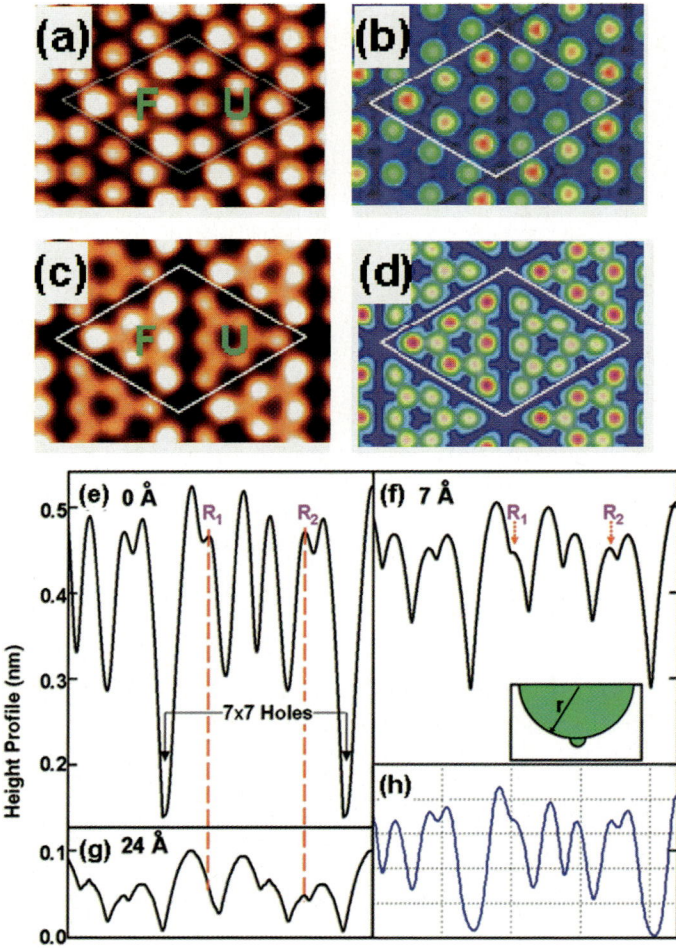

Fig. 17.5. (**a**), (**c**) The experimental STM images with bias voltage of −0.57 and −1.5 V, and tunneling currents of 0.3 and 0.41 nA, respectively. F and U depict the FHUC and UHUC, respectively. (**b**), (**d**) Calculated STM images for Si(111)-7×7 with bias voltages of −0.57 and −1.5 V, respectively. The *red* peaks are about 2Å above the *dark blue* borderlines. (**e**), (**f**), and (**g**) are the calculated height profiles along the diagonal of the (7×7) unit cell with a tip apex radius $r = 0.0, 7.0$, and 24.0 Å, respectively. (**h**) The experimental profile. *Inset* schematically shows an STM tip with an adsorbed cluster beneath the apex

the tunneling barrier. Because the tunneling current is inversely proportional to the exponential of the thickness, the lower electronic state located in the valence band corresponds to the smaller tunneling current. The fact that the rest atoms are invisible but the adatoms are visible may be because the former have significantly lower energies than the latter. This argument, however, contradicts the theoretical prediction that the dangling bond states of the rest atoms extended into the vacuum region like the adatoms [13]. Also, because the rest atoms are about 4.6Å away from the nearest adatoms, if one has an infinitely sharp tip positioned right above the rest atom, there is no reason to believe that the adatoms have the effect of screening the rest-atom tunneling. If the tunneling current from the rest atom were indeed weak, one can move the tip closer to the surface in a constant current STM mode. Thus, this common explanation is probably questionable.

Another possible explanation concerns tip contamination, that is, a few silicon atoms might be accidentally picked up by the tungsten tip during the scan, resulting in a semiconductor tip instead of the original metallic tip. Indeed, recently it has been shown that an InAs semiconductor tip [9] could be used to enhance rest-atom visibility by utilizing the second gap above the fundamental gap (both lie in the Brillouin zone center) of the InAs material to suppress the tunneling current from the high-lying adatom states. However, a previous study [14] also showed that the local electronic structure of a typical metal–semiconductor interface remains metallic until several monolayers into the semiconductor. Thus, this is unlikely in the present case with Si atom adsorption unless the thickness of the contaminant layer exceeds the effective screening length of Si.

17.2.4
First-Principles Calculations

It is impractical for us to experimentally determine what might have happened to the few tips that worked so remarkably well. Instead, we look for a plausible explanation from theory calculations. Our collaborators carried out the calculation by using first-principles density function theory (DFT) [15], as implemented in the VASP (Vienna ab initio software package) codes [16]. The Vanderbilt ultrasoft pseudopotential [17] was used with a cutoff energy equal to 170 eV and one special k-point in the Brillouin zone sum. The surface unit cell contains a slab of six Si layers (without counting the Si adatoms) and a vacuum layer equivalent to six Si layers. The front surface contains the (7×7) reconstruction in the Takayanagi model, whereas the back surface is passivated by hydrogen. Except for the very bottom layer, all the Si atoms are fully relaxed to minimize the system total energy.

Apparently, the actual tip morphology is also complex, possibly with additional atoms adsorbed at the end of the apex, as shown in the inset in Fig. 17.5f. Because only the lower semispherical part of the tip can be in close proximity with the surface, here the tip is replaced by a sphere of radius r. To further simplify the calculations, only the line-scans along the diagonal of the (7×7) unit cell are considered in our simulations.

Figure 17.5a shows the STM image of the Si(111)-7×7 surface at a sample bias of -0.57 V. The appearance shows a significant contrast between the FHUC and UHUC of the (7×7) unit. At this low sample bias, the electronic states of the rest atoms are outside the range of the bias. Thus, the STM topography here reveals only the 12 topmost adatoms. The adatoms in the FHUC appear noticeably brighter than those in the UHUC. In each half, the adatoms at the corners appear also slightly brighter than those near the center. These qualitative features are in good agreement with the calculated real-space charge distribution at this particular bias (Fig. 17.5b). Figure 17.5c shows the STM image at a sample bias

−1.5 V. Images of similar quality can be repeatedly reproduced over large area up to 30 × 30 nm (Fig. 17.3a). We can clearly see both the adatoms and the rest atoms. On the UHUC, the rest atoms appear to have almost the same brightness as the central adatoms, whereas on the FUHC, the rest atoms appear to have considerably less brightness than the central adatoms. These observations are again in excellent agreement with the calculated real-space charge distribution at the experimental bias in Fig. 17.5d.

Figure 17.5e shows the calculated line-scan at −1.5 V with an infinitely sharp tip, i.e., $r = 0$, as has been done before in most STM image simulations [18]. A sharp tip is also assumed in calculating the images in Fig. 17.5b and d. Now, we trace this $r = 0$ curve with a disk of radius r, which is a two-dimensional representation of the three-dimensional sphere, to explore geometric hindrance. It is assumed that at each tip position, tunneling takes place at only one spot on the disk. This is reasonable in most cases because tunneling probability diminishes exponentially with distance. However, there are a few exceptions where the disk is nearly or equally distanced from the $r = 0$ curve, i.e., at or near the local symmetry points. For simplicity, however, such a tunneling-current doubled effect is ignored in our simulation.

Our results show that for small disk radius mimicking adsorbed clusters, the line-scan is essentially the same as in Fig. 17.5e. Figure 17.5f shows the simulated result for $r = 7$ Å. At this radius, while none of the main surface topological features have been lost, the overall shape of the line-scan has been significantly modified, noticeably the depth of the profile, and the size of the atoms being noticeably larger than those in Fig. 17.5e. Figure 17.5g shows the simulated result for $r = 24$ Å. At this radius, the rest atom on the FHUC has completely vanished. Even for the UHUC, the contrast between the much-more visible rest atom spots and the adatom spots has been greatly reduced. Thus, it is clear that the attainable size of the tip apex is the crucial factor in imaging the true charge distribution on the (7×7) surfaces. Figure 17.5h shows the corresponding line-scan determined by our experiment. Despite the simplicity of the model, the calculated result for $r = 7$ Å in Fig. 17.5f is in quantitative agreement with experimental observation.

It is now understood that STM probes the real-space charge distribution near the E_F in a rather delicate way that may or may not reveal the unperturbed real-space charge distribution of the surfaces. Here, for the Si(111)-7×7 surface, we show that the calculated and experimental voltage-dependent charge distributions of the Si(111)-7×7 surface, which reveals simultaneously both the 12 adatoms and six rest atoms in each (7×7) unit cell [19]. The emergence of rest atoms is dependent on the bias voltage and the rest atom spots can be visible at the sample bias voltage of less than −0.7 V. The first-principles electronic structure calculations also show a strong dependence of the charge distribution on the bias voltage: 12 spots at −0.57 V for the 12 adatoms (see Fig. 17.5b), whereas 18 spots at −1.5 V for the 12 adatoms plus six rest atoms (see Fig. 17.5d). Our results suggest that a geometric hindrance due to the finite size of the tip apex could be the reason. This finding should invoke significant research interest in the design and fabrication of the STM tip and its applications in exploring more detailed information about surface reconstructions and nanostructures for physical properties of nanosystems.

17.3
Atomic Manipulation on Si(111)-7×7 Surfaces with STM

17.3.1
Introduction

STM has been a unique tool for manipulating atoms and molecules on surfaces since its invention. There is an increasing effort to fabricate nanometer-scale structures by STM and great progress has been made in this field. For manipulating the single atom or molecule, Eigler et al. can reposition Xenon atom physisorbed on Ni(110) surfaces one by one to define the single atom pattern at low temperatures (4 K) [20]. Both the Avouris and the Aono group have realized the invertible transfer of a single Si atom by applying a voltage pulse between the W tip and Si(111) substrate [21,22].

Modification of semiconductor surfaces is of particular interest for the fabrication of nanometer quantum devices. An important aspect is the fabrication of ordered atomic-scale structure of any desired size and shape. In the pioneering investigations, by using the popular voltage pulse method, a regular structure with lateral dimension as small as the single atom on the MoS_2 surface was formed by Hosoki et al. [23], and ordered structures on a Si(001) surface with atomically straight edges and one dimer width were fabricated by Salling et al. [24], respectively. At the same time, rough-edge grooves with a linewidth of 40 Å on Si(111) surfaces were created using tip scanning with an increasing bias voltage at constant current by Kobayashi et al. [25].

In this section, we described a series of experiments in which the grooves with atomically straight edges and with lateral features as small as a 7×7 unit cell wide have been created on Si(111) surfaces under the conditions of high tunneling current and low bias voltage [26]. The detailed investigation of the modification process shows that the grooves are formed by removing atoms one by one from the Si(111)-7×7 surfaces. The modification mechanism is proposed based on the experimental data.

17.3.2
Fabricating Groove Nanostructures on Si(111)-7×7 Surfaces

The Si(111)-7×7 substrates were fabricated with the standard process as discussed in Sect. 17.2.2. The tips were electrochemically etched tungsten tips. All of the STM images presented here were obtained in a constant current mode with a positive sample bias voltage of 2.0 V and a tunneling current of 1.0 nA using an Omicron UHV-STM system. In order to modify the sample surfaces, first the tunneling current was increased from 1.0 to 30–50 nA while the gap bias was kept constant. With the constant current mode, the tip would move toward the sample surfaces under the action of the feedback loop. Then we performed line scanning of the tip along a certain direction repeatedly at 400 nm/s for a few seconds. The same area was imaged to detect modification. Figure 17.6a shows a groove created with a few nanometers width. When the direction of modification is parallel to the basic vector of the Si(111)-7×7 surface, a groove with an atomically straight edge and the width of a

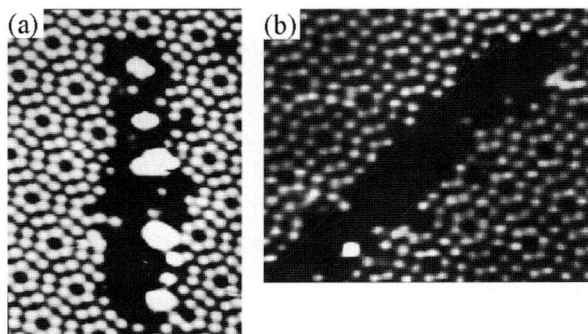

Fig. 17.6. (a) STM image of the groove formed by extracting atoms from the Si(111)-7×7 surface. (b) The groove structure has atomically straight edges and a width of one 7×7 unit cell size if the direction of modification is along the basic vector of the cell

unit cell size can be created, as shown in Fig. 17.6b. The groove is formed by extraction of individual 7×7 unit cells from the surface, so its width is just the same as a 7×7 cell (i.e., 23.3 Å).

An experimental procedure was also employed to understand the modification process. Assuming the time to form a regular structure is t_0, we divided it into n equal parts, modified the sample with action time of t_0/n; and then moved the tip slightly away and modified the sample again with an action time of $2\,t_0/n$, etc. We repeated the process until the action time was t_0, then imaged the sample in order to observe what had happened during the modification process. Figure 17.7 is an example of such a procedure with $n = 3$. Lines A, B, and C are formed with action times of $t_0/3$, $2\,t_0/3$, and t_0, respectively. From the STM image we can find that line A is composed of single-atom hole defects, and they begin to join together in line B and finally a regular groove structure is formed in line C. The process shows that the groove is fabricated by extraction of atoms one by one from the Si(111) surfaces.

The inherent structure properties of Si(111) play an important role in the grooves formed by such a modification process. It is known that a 7×7 structure is the most stable structure at room temperature. Consequently when a unit cell is broken by extracting several adatoms from it, the remaining adatoms on this unit cell are easier to remove. As a result, when the action time is appropriate, just these unit cells under

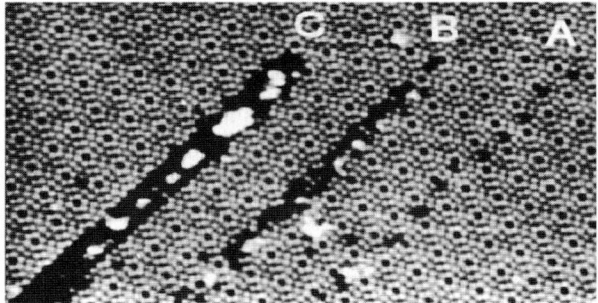

Fig. 17.7. STM image of the modification process of the groove structure. *Lines* A, B, and C are formed with action times of $t_0/3$, $2\,t_0/3$, and t_0, respectively. Here t_0 is the action time needed to fabricate a regular groove. The process shows that the groove is fabricated by extraction of atoms one by one from the Si(111) surfaces

the apex atom of the tip were completely removed and formed such regular structures; this requires that the adatoms must be removed from the surface one by one. Whereas, if the tip removed a cluster every time, then it is possible that two adjacent unit cells in the lateral dimension were broken simultaneously, and finally both were removed partially or wholly. But this situation doesn't occur in our experiments. Salling et al. have reported that the atomic scale structures fabricated by STM on Si(001) surfaces are related to its dimer structure [24]. Here the structure construction on the Si(111) surface also benefits from the inherent properties of a 7×7 unit cell in spite of its large size.

By the method mentioned above, some patterns have been defined to demonstrate the powerful ability for fabrication of ordered nanometer-scale structures. As an example, Fig. 17.8 shows a nanostructure of "100" on an Si(111) surface. The central island maintains the 7×7 reconstruction after modification. The structures showed very high stability during our experimental process.

Fig. 17.8. A groove nanostructure of "100" pattern fabricated by extracting atoms from the Si(111) surface. The structure has atomically straight edges and the central island maintains the 7×7 reconstruction

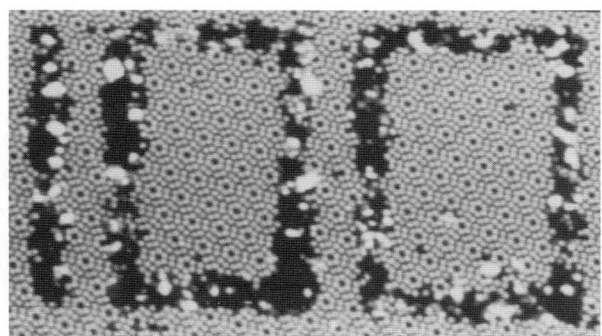

17.3.3
The Modification Mechanism

In order to investigate the modification mechanism on Si(111)-7×7 surfaces, the critical (threshold) current I_c for creating regular groove structures under various bias voltage V_b is measured. The $I_c - V_b$ curve is shown in Fig. 17.9. When $V_b < 0.5$ V, I_c increases with V_b; when $V_b > 0.8$ V, I_c decreases greatly with increase of V_b. This suggests that the tip–sample separation S stays approximately constant when $V_b < 0.5$ V, while S increases with V_b when $V_b > 0.8$ V [27]. Measurements indicate that under normal tunneling conditions, the tip–sample separation S is ∼5 Å. When $V_b < 0.5$ V, the conditions of critical current correspond to S (2.5 Å), so the electrostatic field intensity is about 0.04–0.2 V/Å (V_b from 0.1 to 0.5 V). However, under these conditions, chemical bonds have been formed between the apex atom of the tip and the sample atoms [28]. Therefore, the chemical interaction dominates the tip–sample interaction. Because the binding energy of Si to the W surface is much higher than that of the Si adatom in the 7×7 structures [29], it can extract Si adatoms from the sample surfaces. When $V_b > 0.8$ V, the chemical interaction decreases exponentially with increasing of S, so the electrostatic interaction becomes

Fig. 17.9. Curve showing the threshold current (I_c) for creating regular groove structures vs. bias voltage (V_b). I_c increases with V_b when $V_b < 0.5\,\text{V}$ and decreases greatly after $V_b > 0.8\,\text{V}$

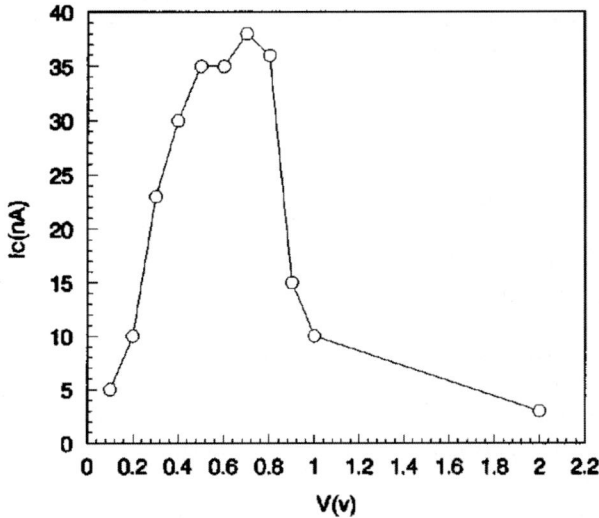

dominant and the exaction of Si atoms is field induced [29, 30]. The critical field is about 0.4 V/Å, which agrees with Kobayashi's results [25]. During the V_b range of 0.5–0.8 V, the two kinds of interactions are of equal importance approximately. Another possible interaction is that induced by current because the current density at the tunneling gap is very high. However, under critical conditions of various bias voltages, a fixed current threshold is not required. This suggests that the interaction induced by current is not a dominant one.

After a groove structure is created, if the bias voltage V_b is turned down to 0.5 V with unchanged tunneling current I_c (40–50 nA), then a ridge structure can be formed by depositing Si atoms onto the Si(111)-7×7 surface after repeated STM line scans. Figure 17.10 shows a groove (line A) and a ridge (line B) created at the side. We have demonstrated that during the redeposition process atoms can be deposited onto the substrate in the form of Si clusters and there are indeed Si atoms extracted. When V_b is decreased to 0.5 V, the tip–sample distance S becomes less than 3 Å and the

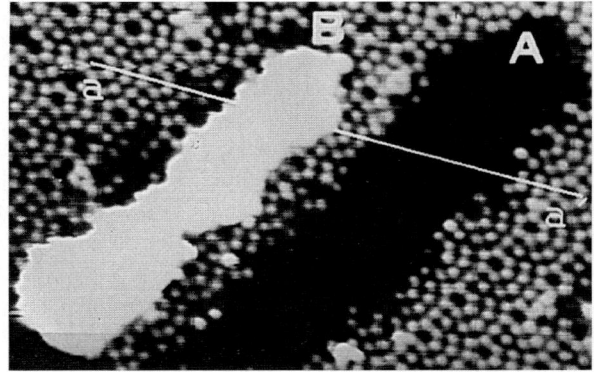

Fig. 17.10. STM image of a groove (*line* A) and a ridge structure on its side (*line* B) created by extracting Si atoms from the Si(111)-7×7 surface and redepositing Si atoms back on the substrate

electric field is less than 0.2 V/Å. Strong chemical interaction may form between the Si clusters on the tip and the substrate and become dominant in this process [31].

17.4
Ge Nanostructure Growth on Si(111)-7×7 Surfaces

17.4.1
Introduction

Low-dimensional structures can provide interesting physical and chemical properties due to their size and shape. The growth of nanostructures with reduced dimensions has been extensively studied, driven by the intrinsic interest in structures as well as the potential technological applications in quantum devices [32]. Recent studies demonstrated the feasibilities and possibilities of growing self-organized nanostructures on periodic solid surfaces. The Si(111)-7×7 surface offers a unique template for the self-assembly growth of diverse nanostructures because of the large number of distinct bonding sites. Recently, "Magic" islands and nanoclusters of semiconductor or metal have been grown on this surface [33–37]. Ordered arrays of two-dimensional nanodots/nanoclusters, including Al, Ga, In, Tl, Si, Ge, Sn, Pb, Na, Cu, Au, Ag, were successfully fabricated on the Si(111)-7×7 surface [38–48]. These self-organized structures are expected to have a smaller size and stronger confinement potentials compared to the lithographically defined clusters [49].

The adsorption of Germanium on the Si(111)-7×7 surface has been extensively studied in recent years [50–63], due to Ge-based nanostructures having potential applications in microelectronics and optoelectronics. Indeed, the Ge/Si system has the advantage of being naturally compatible with Si technology. In addition, being currently incorporated in Si structures, Ge can be used to fabricate strained Si layers with enhanced mobility. Therefore, there are renewed activities in Ge-based nanostructures grown on Si in expectation of functional devices with unique electronic and optoelectronic properties [64].

The microscopic understanding of the bonding nature of the adsorbed Ge atoms is an essential issue for the controlled fabrication of desired nanostructures, since the initial adsorption nature may affect the growth behaviors of Ge-based quantum dots and films. In spite of numerous investigations, a unified picture for the bonding structures of Ge atoms on Si(111)-7×7 surfaces has not been established. Meanwhile, the formation and transformation process of various Ge nanostructures during the initial growth stages on the Si(111)-7×7 surface is far from well understood. In particular, the lack of understanding for the formation mechanisms of Ge nanostructures impedes the further control of the growth process. Here, we provide an STM investigation on various Ge nanostructures on Si(111)-7×7 surfaces with different size and geometry, ranging from individual Ge atoms (adsorption sites), to Ge nanoclusters (evolution and aggregation patterns), and then to 2D extended Ge islands (phase and components). Especially, we provide insight into the structural characterizations as well as the transformation process and possible mechanisms of the observed Ge nanostructures in association with first-principles calculations.

17.4.2
Experimental Aspects

The preparation of the Si(111)-7×7 surface was conducted as described in
Sect. 17.2.2. Then Germanium (99.9999% purity) was deposited onto the as-
prepared Si(111)-7×7 surface by resistive evaporation. The substrate during
deposition was kept at room temperature (RT, 25 °C), or above room temperature
(ranging from 100 °C to 300 °C) to facilitate the formation of ordered structures
since at RT Ge atoms do not have enough mobility to span the dimer wall after arriving on the surface [65–67]. The system pressure was better than 5×10^{-10} mbar.
A typical deposition rate of 0.01 monolayer (ML)/min was routinely achieved. One
monolayer is defined as the atomic density of the unreconstructed Si(111) surface
($1 \, \text{ML} = 7.83 \times 10^{14}$ atoms/cm^2). Each sample was cooled down to room temperature, and then transferred to the STM chamber for measurements. All the images
were acquired in a constant-current mode with an electrochemically etched tungsten tip.

17.4.3
Initial Adsorption of Ge Atoms on the Si(111)-7×7 Surface

For the adsorption sites of Ge atoms on Si(111)-7×7 surfaces, X-ray standing-wave
(XSW) studies of submonolayer Ge-deposited at 300 °C carried out by Patel et al. in
1985 suggested that Ge atoms might occupy substitutional-like sites on the Si(111)
plane [68]. However, it was not possible to determine the precise Ge sites and the
bonding structures in their studies. Also based on XSW measurements, Dev et al. in
1986 proposed that at low coverages (< 0.5 ML) Ge atoms would prefer to occupy
the on-top sites and to bond directly to the Si adatoms and rest-atoms which were
just below the adsorbed Ge atoms [69]. Reflection electron microscopy and transmission electron diffraction investigations on Ge/Si(111)-7×7 prepared at 640 °C
by Kajiyama et al. in 1989 found evidence that Ge atoms randomly substituted any
Si atoms in the top three layers [70]. Core-level photoemission spectroscopy measurements by Carlisle et al. in 1994 provided indirect observations that there was
some preference for Ge to replace the Si adatoms for the annealed Ge/Si(111)-7×7
samples [71]. More recent measurements using near-edge X-ray absorption spectroscopy and STM did not provide conclusive descriptions of Ge bonding sites on
the Si(111)-7×7 surface [72, 73].

   Some theoretical calculations have also been reported on Ge bonding sites on
the Si(111)-7×7 surface, however, the calculations provided limited information
and showed contradictory results. Early work was semi-empirical X_α and extended
Hückel calculations with limited predictive capabilities, which provided support for
the notion that Ge atoms bond directly to rest atoms or Si adatoms [74–76]. In
contrast to the semi-empirical calculations, on the other hand, using first-principles
density functional calculations Cho and Kaxiras in 1998 reported a limited explo-
ration of bonding possibilities and found that the most stable adsorption position
for Ge on Si(111) is the high-coordination bridge B$_2$ site (see Fig. 17.1 for perti-
nent terminology of the Si(111)-7×7 surface), which was a bonding site that had not

been proposed on the basis of experimental data [77]. They introduced the so-called basins of attraction, which contain stable adsorbate positions as high-coordination sites rather than surface dangling bond sites. Their calculations showed that the rest atoms or intrinsic Si adatom sites (dangling bond T_1 sites) of the substrate were the high-energy sites, and the low-energy sites were the B_2-type sites for Si and Ge adsorption.

17.4.3.1
Ge Atom Substitution for Si Adatoms at Elevated Temperature

In the first part, we report STM observations and first-principles calculations for the initial adsorption of Ge atoms on Si(111)-7×7 surfaces at elevated temperature. Figure 17.11 shows an STM image of the Si(111)-7×7 surface with Ge coverages of 0.02 ML, 0.08 ML, and 0.10 ML deposited at 420 K, respectively. These images show that the surface lattice retains the original (7×7) reconstruction with the dimers and the adatoms. The faulted half unit cell and the unfaulted half unit cell of the (7×7)

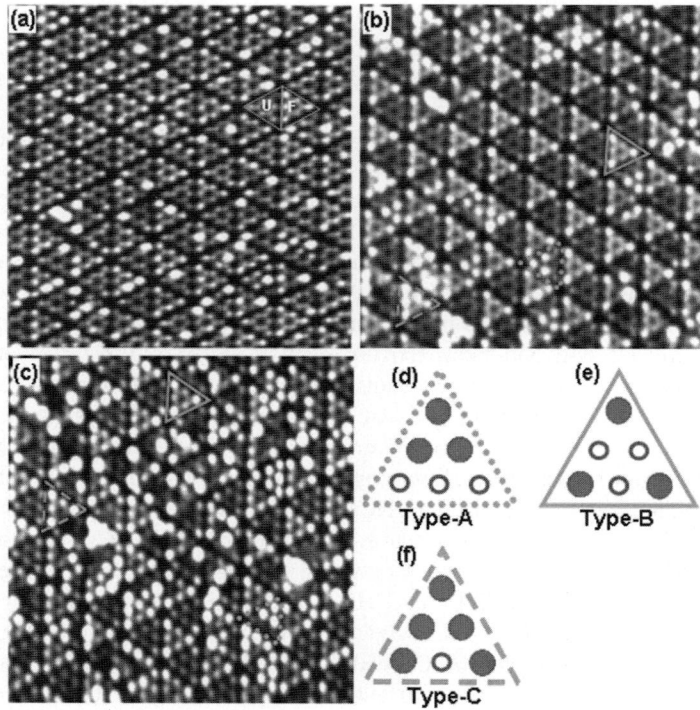

Fig. 17.11. Filled-state STM images of the Si(111)-7×7 surface with Ge coverages of (**a**) 0.02 ML; (**b**) 0.08 ML; and (**c**) 0.10 ML. A (7×7) unit cell is marked by two triangles in (**a**), respectively. Sample bias: −2. 2 V in (**a**), and −1. 5 V in (**b**) and (**c**); Tunneling current: 0.5 nA in (**a**), and 0.2 nA in (**b**) and (**c**). The scanning area is 20 × 20 nm. Three different configurations of Ge protrusion distributions are denoted in (**b**) and (**c**) by the *solid-line triangle, dot-line triangle,* and *dashed-line triangle,* respectively. The schematics for the three typical Ge patterns, type-A, type-B, and type-C, are shown in (**d**), (**e**), and (**f**), respectively

reconstruction, labeled as F and U in Fig. 17.11, respectively, are distinguished due to the different contrasts. The deposited Ge atoms appear as bright protrusions. Three significant features are presented in the STM images. First, the deposited Ge atoms are clearly resolved as individual atoms on the surface. Second, the adsorbed Ge atoms reside on the sites that were occupied by the Si adatoms on Si(111)-7×7. Finally, more Ge atoms occupy the corner Si adatom sites in the FHUC than the other Si adatom sites. No Ge atoms are found at either the rest-atom or the high-coordination surface sites.

The profile lines through the bright dots show that the height difference between the Ge atoms and the original Si adatoms is about 0.2 Å in the STM images. This data clearly shows that the Si adatom does not stay in its original position (on a clean Si(111)-7×7 surface, the Si adatom occupies a so-called T_4 site just above a second-layer Si atom) [78, 79]. In the case of Ge bonding directly atop Si adatoms, the increased height due to the Si–Ge bond length of about 2.36 Å should be reflected in the STM image. Therefore, the addition mechanism is suggested to be a questionable explanation. Moreover, the number of dangling bonds will increase to three if a Ge atom adds on the top of one Si adatom, and this isn't considered to have a suitable total energy.

The topographic height undulations of adatom sites in STM images caused by Ge–Si exchange on Si(111)-5 × 5-Ge reconstructions have been proposed by Becker et al. [80] and Fukuda et al. [81], and were also investigated by Rosei et al. [82] with current imaging tunneling spectroscopy. Here, we suggest that Ge–Si exchange can also occur during the initial adsorption stage of Ge/Si epitaxy growth due to the structural similarity of Ge and Si. The feature can be confirmed by our experiments and the following ab initio calculations.

Three dominant types of Ge protrusion patterns on the Si(111)-7×7 surface are found with a slight increasing of Ge coverage, as shown in Fig. 17.11b and c. The schematics of these three types of patterns, named type-A, type-B, and type-C, are given in Fig. 17.11d, e, and f, respectively. Type-A illustrates three Ge atoms locating at one corner adatom site and two adjacent center adatom sites in a HUC. Type-B indicates the configuration with three Ge atoms occupying corner adatom sites in a HUC. Type-C refers to the adsorption structure with five Ge atoms residing on the sites of three corner adatoms and two center adatoms in a HUC. Type-B and C patterns distribute preferentially in the FHUCs. The site distribution of the bright protrusions at the corner and center adatom sites in both the FHUCs and the UHUCs is illustrated in Fig. 17.12. At the Ge coverage of 0.02 ML, the site preference ratio is about 5.6:4.4 for the FHUC to the UHUC, and 6.1:3.9 for the corner to the center adatom sites, respectively. When the Ge coverage increases to 0.08 ML, the site preference ratios are about 9:1 for the FHUC to the UHUC, and 4:1 for the corner to the center adatom sites. The site distribution for the coverage of 0.10 ML is similar to that for the coverage of 0.08 ML. The overall conclusion is that after an initial random occupation of Si adatom sites, corner adatom sites in the FHUC are preferred and gradually type-B patterns become dominant. Type-A and Type-C patterns are more discernible at slightly higher coverages, and finally, small islands begin to appear.

Our collaborators performed first-principles density-functional calculations using the pseudo-potential method and a plane-wave basis set [15]. The Si(111) sur-

Fig. 17.12. Site distributions of Ge at various adatom positions at coverages of 0.02 ML, 0.08 ML, and 0.10 ML

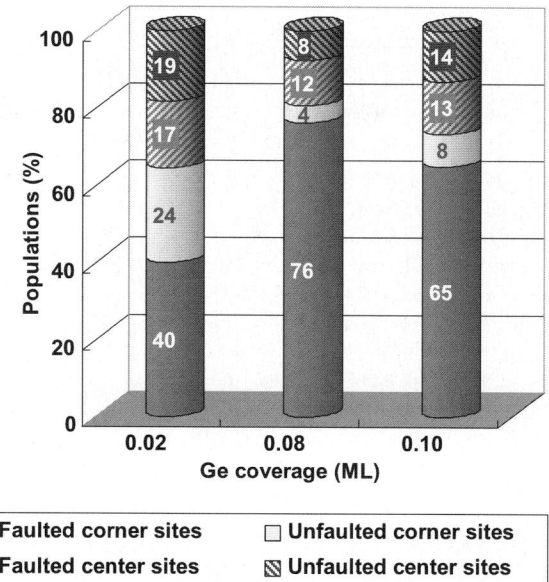

face was modeled by repeated slabs with four layers of Si atoms (each layer contained 16 Si atoms, corresponding to a 4 × 4 surface unit cell) and four Si adatoms, separated by a vacuum region of 12 Å. Two of the four rest-atoms were saturated by hydrogen, so that the ratio of the number of the adatoms to that of the rest-atoms is the same as for the 7×7 surface. Except for the Si atoms in the bottom layer, which were fixed and saturated by H atoms, all the atoms were relaxed until the forces on them were less than 0.05 eV/Å. The exchange-correlation effects were treated with the generalized gradient-corrected exchange-correlation functions (GGA) given by Perdew and Wang [83]. The Vanderbilt ultrasoft pseudopotentials are adopted. A plane-wave energy cutoff of 14.7 Ry and the point for reciprocal space sampling were used for all the calculations.

All the possible configurations with a Ge atom near an adatom or/and a rest atom were calculated. Two lowest energy configurations, shown in Fig. 17.13, were found to have essentially the same total energy (the difference in total energy is smaller than 0.02 eV). The first one consists of Ge at a B_2 site (Fig. 17.13a), as identified earlier by Cho and Kaxiras [84]. In the second configuration (Fig. 17.13b), the adsorbed Ge atom substitutes for a Si adatom and the substituted Si adatom occupies a nearby B_2 site. We refer to the Ge position in the second configuration as S_4 (substitutional site with four nearest-neighboring silicon atoms). The total energies of the configurations with Ge bonded at the on-top positions of adatoms and rest-atoms are significantly higher (2.3 eV and 1.6 eV, respectively) than the B_2 and S_4 configurations. So we can clearly rule out the possibility of such configurations, which were suggested previously on the basis of semiempirical calculations [74–75]. For both lowest energy configurations (B_2 and S_4), the atom (Si or Ge) at a bridge site may diffuse within a basin (to occupy any of the six B_2 sites near the rest atom) and across basins (to occupy the B_2 sites near different rest-atoms). The diffusion barriers within a basin

Fig. 17.13. Schematic *top* view of the calculated lowest energy configurations of a Ge atom on the Si(111) surface. (**a**) Ge at a B_2 site and the nearby Si adatom at a position away from its original site. (**b**) Ge at the substitutional S_4 site and the Si adatom at a B_2 site. (**c**) Ge atoms substituted for some of the Si adatoms and no atoms are bonded at any of the B_2 sites. The bond lengths are shown in the images in Å

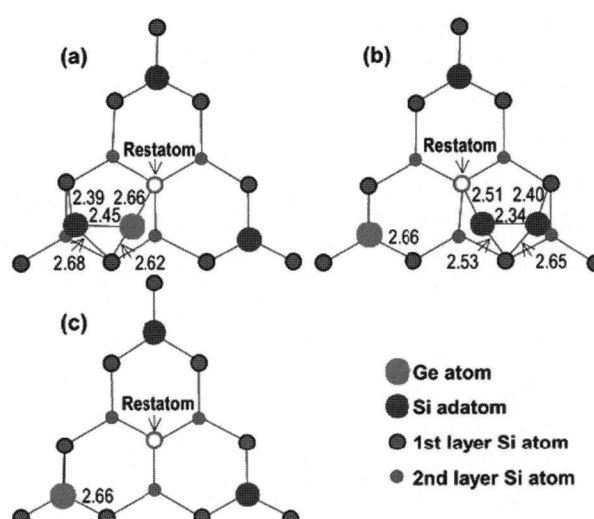

and across basins are about 0.5 eV (0.6 eV) and 1.0 eV (1.0 eV) for the Ge (Si) atoms, respectively, which is in agreement with previous first-principles calculations [76–85]. Therefore, Ge atoms in the S_4 configurations are thermodynamically more stable than in the B_2 configurations. In particular, after the atoms initially bonded at the B_2 sites migrate to step edges and/or to form islands, the surface exhibits a stable Ge-S_4 configuration, in which Ge atoms substitute for some of the Si adatoms and no atoms are bonded at any of the B_2 sites (Fig. 17.13c), as shown by our STM observations. The Ge-S_4 configuration is demonstrated again by the recent results from Xiao's and Tomitori's groups [55, 66]. In their STM measurements they also confirmed that Ge replaced Si adatoms on the Si(111)-7×7 surface at elevated temperature.

It is well known that the backbonds of the Si adatoms of the Si(111)-7×7 surface are under considerable strain. We therefore suggest that the adsorbed Ge atoms are able to break the backbonds and further replace the Si adatoms at elevated temperatures. Previous studies have established that the corner adatoms in the FHUCs are under more strain than the other adatoms, which implies that backbonds of the corner adatoms in the FHUCs are broken easier than those of the other adatoms [84, 85]. When Ge atoms are deposited on the surface, the chance for the Ge atoms to occupy the B_2 sites near a center adatom is larger than that near a corner adatom (the center adatom has two nearby rest atoms while the corner adatom has only one). Thus, the Ge-S_4 bonding structure tends to be preferentially formed at the corner adatom sites and in the FHUCs of the Si(111)-7×7 surface [86].

Finally, the relaxed Ge-S_4 configuration obtained from our calculations shows that the Ge atom resides at a position higher by 0.24 Å than the original Si adatom that has been replaced by Ge, which is in good agreement with our STM data.

17.4.3.2
Initial Adsorption Sites of Ge Atoms at Room Temperature Deposition

When the single Ge atom deposits onto the Si(111)-7×7 surface at room temperature, it will show totally different adsorption behaviors. Zhao et al. have

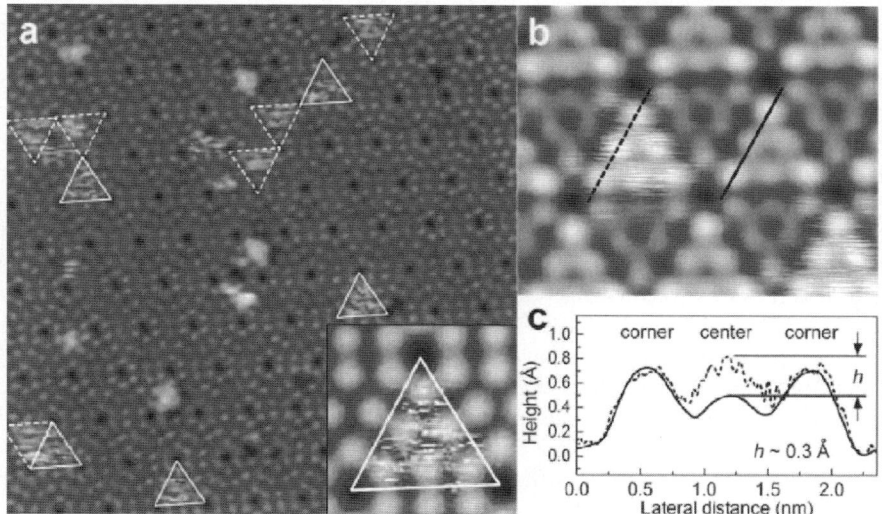

Fig. 17.14. (**a**) RT empty state STM image (bias $+1.5$ V) of a 28×28 nm^2 area showing single Ge atoms (fuzzy image) and small Ge clusters adsorbed on the Si(111)-7×7 surface. The *inset* is a zoom-in empty state STM image of a single Ge atom adsorbed in a FHUC. The fuzzy noise indicates the frequent hopping of a single Ge atom within the FHUC. The scanning speed was 75 nm/s. (**b**) A filled-state image (bias -1.5 V) of single hopping Ge atoms adsorbed in FHUCs at a scanning speed of 25 nm/s. (**c**) Line profiles along the *solid* and the *dashed lines* in (**b**) (from [66])

performed detailed experimental investigations with scanning tunneling microscopy [66]. Figure 17.14a shows a RT STM image of low Ge coverage on a Si(111)-7×7 surface. They observed many fuzzy triangular patterns in both FHUC and UHUC of the surface, and these noisy patterns are likely to originate from a single Ge atom hopping frequently from site to site within the half-unit cell, similar to the activity of noble metal atoms [87–90]. No evidence of Ge atom substitution for the original Si adatoms was found. The inset of Fig. 17.14a shows an empty state STM image of a single Ge atom adsorbed in a FHUC, where the Si adatoms underneath the fuzzy noise can be clearly identified. Figure 17.14b shows the filled-state STM image of single Ge atoms diffusing in FHUC obtained at a relatively slow scanning speed. Under this slower scanning speed, it is expected that the more favorable adsorption sites will appear to be brighter than those less favorable sites due to different average residence times of Ge at different sites. In Fig. 17.14c, the line profiles along the solid and dashed lines across the corner-center-corner Si adatoms (Fig. 17.14b) in two neighboring FHUCs, one with a Ge atom and the other without a Ge atom, are shown. This remarkable height difference (about 0.3 Å) between the center Si adatoms suggests that the adsorbed Ge atoms prefer to stay at the adsorption sites near a center Si adatom than at those near a corner Si adatom during diffusion.

A time-dependent tunneling current study was also applied to study quantitative information about the hopping rates and residence times of Ge atoms at the adsorption sites [90]. Zhao et al. obtained the decay time constant of $\sim 2.9 \pm 0.2$ ms and thus the hopping rate out of the adsorption site near the corner Si adatom is about

340 Hz. This hopping rate of Ge within the half unit cell is much slower than that of noble metal atoms (e.g., slower than Cu by a factor of ~10 [90]) but comparable with that of Si at room temperature, implying a similar interaction for Ge and Si with the Si(111)-7×7 surface. These observations are quite similar to what has been observed in the diffusion of single Si atoms on Si(111)-7×7 at RT by Sato et al. [91] using atom-tracking scanning tunneling microscopy.

In order to identify the adsorption sites of Ge atoms by RT deposition, a low temperature study is required. Figure 17.15 shows filled and empty-state images of the same area for Ge atoms adsorbed on a Si(111)-7×7 surface at 78 K. Two typical states were observed for single Ge atoms: one is the state of stable adsorption near a corner Si adatom and the other is the state of hopping within a half-unit cell. The probabilities of a Ge atom in these two states are found to be approximately equal. In each state, the events that occur in FHUC and UHUC are also about the same. With first-principles calculations using the VASP code [92,93] within the framework of density functional theory (DFT), they found that the lowest energy sites are the high-coordination G' sites near the center Si adatoms, and the second lowest energy sites are the G sites near the corner Si adatoms, where the G' (G) site is laterally close to the B_2' (B_2) bridge sites but has a lower symmetry.

When the imaging temperature is reduced to 78 K, Ge atoms are found to reside at the G sites near the corner Si adatoms or to still hop among the G' sites near the center Si adatoms. They suggested that although the G' site is calculated to be the lowest energy adsorption site, the barrier between two adjacent G' sites may be low to allow the Ge atom to hop between them, resulting in nonstationary adsorption. In contrast, the stability of the G adsorption site in the STM images is perhaps kinetically originated due to a high diffusion barrier to its neighboring equivalent G site or to a neighboring G' site. Different hopping images were also observed at positive and negative bias voltages (Fig. 17.15a and b), which can be attributed to an electric-field effect. [66]

Ansari et al. [55] also studied temperature dependence of initial adsorption behaviors of Ge atoms on the Si(111)-7×7 surface, they reached the same conclu-

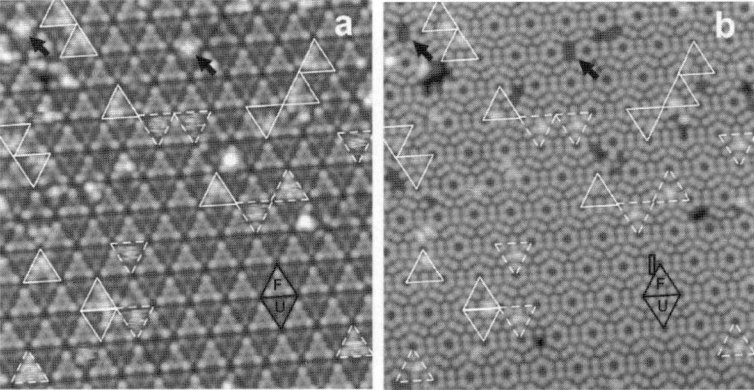

Fig. 17.15. STM images of Ge atoms on the Si(111)-7×7 surface deposited at RT but imaged at 78 K, showing two adsorption states: the stable adsorption state near a corner Si adatom (*solid triangles*) and the hopping state (*dashed triangles*) in a half-unit cell. (**a**) Filled-state image at a bias voltage of −1.5 V; (**b**) empty-state image at a bias voltage of +1.5 V (from [66])

sion that Ge atoms preferably occupy B_2 sites (G' (G) sites can be regarded as fine adjustments to the exact adsorption sites) at RT and replace Si adatoms (S_4 sites) at 100 °C, respectively. Both the adsorption sites have local minimum energies with high occupation probability. The thermal energy is required to break the backbonds and further replace a Si adatom with a deposited Ge atom at elevated temperatures.

17.4.4
Temperature Dependence of Formation and Arrangement of Ge Clusters

It is of course interesting to study formation and possible configurations of Ge nanoclusters with increasing coverage in subsequent Ge deposition. Indeed, Ge cluster growth on the Si(111)-7×7 surface has been aggressively studied by various groups. In 1991, Kohler et al. reported that irregular Ge clusters initially grow in the half unit cells of the Si(111)-7×7 surface at room temperature [67]. In our group's previous works, Yan et al. found that the periodic structure of the Si(111)-7×7 reconstruction can serve as a template to regularly arrange Ge nanoclusters, and a regular hexagonal arrangement can form with submonolayer coverages of Ge deposited at room temperature under carefully controlled conditions [62]. Recently Ansari et al. present again direct evidence of the self-organized hexagonal structures and reconfirmed these results [94]. In this section, we will expatiate on the formation and configuration of Ge clusters, and also their temperature-dependence feature.

17.4.4.1
Room Temperature Deposition

Figure 17.16 shows the STM images of the Si(111)-7×7 surface, on which 0.10 and 0.30 ML of Ge was deposited at room temperature. The irregular Ge clusters are almost all located in the triangular half-cells of the Si(111)-7 × 7 surface, where the area with the greatest dangling bond density adsorbs Ge atoms preferentially. The images also show equal probabilities of Ge clusters occupying the faulted and the unfaulted halves. It is estimated that at room temperature Ge atoms do not have enough mobility to span the dimer wall after arriving on the surface [65–67]. Cho and Kaxiras [77, 84] theoretically studied the stable sites for adsorption and diffusion of various adatoms on a Si(111)-7×7 surface using density functional theory and total energy calculations. They interestingly pointed out that Ge atoms are trapped in the basins of attractive potential around rest atom sites, which are high coordination sites. With increasing Ge coverage, there is no change in the size of the Ge clusters and only their density increases.

Further increase of Ge coverage to 0.5 ML almost covered the surface with the clusters of a half unit cell size while the corner holes and dimer rows were not covered, resulting in formation of a hexagonal mesh, as shown in Fig. 17.17a. This image directly evidences the formation of the self-organized hexagonal arrangement of Ge clusters with the periodicity of 7×7 structure, completely different from the forming of the 5 × 5 reconstruction, which was critically remarked in [95].

When the sample with a 0.5-ML Ge coverage was annealed at 400 °C for 10 min, more well-ordered cluster arrangements were obtained, as shown in Fig. 17.17b. The dimer rows on the lower and upper terraces are parallel to each other and the surface

Fig. 17.16. The topographic STM images of the Si(111)-7×7 surface on which (**a**) 0.10 ML and (**b**) 0.30 ML Ge were deposited at room temperature. Scanning area: 40 × 40 nm². $U_{bias} = -1.5\,V$, $I_t = 0.20\,nA$. The irregular Ge clusters are almost all located in the triangular half-cells of the Si(111)-7×7 surface

Fig. 17.17. STM images showing (**a**) the hexagonal arrangement of Ge clusters with a half unit cell size for 0.5 ML Ge deposition at RT, and (**b**) after annealing at 400 °C for 10 min. The *dark areas* and *dark lines* correspond to *corner holes* and *dimer rows* of the 7×7 reconstruction of the layer underneath, respectively. Tunneling current: 0.10 nA. (**a**) Sample bias: −2.0 V, 40 × 40 nm²; (**b**) sample bias: +2.0 V, 80 × 80 nm² (from [94])

is completely covered with the clusters arranged in a hexagonal pattern. The shape and size distribution of Ge clusters become more uniform compared with that in Fig. 17.17a. This implies that excess Ge atoms are no longer strongly bound within the half unit cells and so easily migrate towards a smaller cluster to form a stable circular cluster at 400 °C [94].

17.4.4.2
High-Temperature Deposition

However, if we increase the substrate temperature during Ge deposition, due to the Ge–Si substitution mechanism, evolution of Ge clusters occurs in the following steps from our STM observations. At very low Ge coverages (less than 0.1 ML), according to Sect. 17.4.3, most Ge atoms form correlated patterns replacing three or five out of the six Si adatoms in the FHUC of the Si(111)-7×7 surface (type A, B, and C), as shown in Fig. 17.11d–f) [86]. At Ge coverages of ~ 0.1–0.12 ML, a triangular lattice of three single Ge atoms forms, in which Ge atoms replace the Si adatoms at the corners of each FHUC (type-B), with only very few replacing the Si adatoms in UHUCs (Fig. 17.18a).

When the Ge coverage is larger than ~ 0.12 ML, Ge atoms begin to nucleate and form isolated nanoclusters, mostly at the center of FHUCs, with typical diameters of 1.0 nm (Fig. 17.18.), just fitting within FHUCs. The clusters are possibly formed by adsorption of extra Ge atoms to the same potential well through the weak Ge–Ge interaction. At larger local coverages, pairs of Ge nanoclusters appear in adjacent FHUCs and UHUCs. The two clusters are distinct, clearly separated by ~ 6 Å. In Fig. 17.19, a sequence of STM images at varying Ge coverages from ~ 0.15 to 0.50 ML show the continuing evolution of the Ge clusters from isolated ones into hexagonal patterns. The distribution tendency suggests the evolution of hierarchical cluster patterns from dispersing clusters to close cluster rings. Most of the clusters discretely emerge on the substrate at the low coverage of 0.10–0.15 ML with an initial preference in the faulted halves (see Fig. 17.19a). And then open cluster rings containing three, four, and five clusters, nucleated on the Si(111)-7×7 surface at a Ge coverage of ~ 0.2 ML form (Fig. 17.19b). Afterwards closed Ge hexagonal rings consisting of six clusters begin to form at a Ge coverage of ~ 0.3 ML (Fig. 17.19c). When the Ge coverage approaches 0.4 ML, most of the HUCs of both FHUC and UHUC are occupied by Ge clusters. The underlying (7×7) surface periodicity and the hexagonal superstructures coexist. Finally, the highly regular hexagonal superlattice forms at a Ge coverage of ~ 0.5 ML and covers the entire (7×7) surface (Fig. 17.19e).

Although the Ge cluster structures on the Si(111)-7×7 surface represent the majority, lots of individual Ge atoms can still be resolved with bright spots at positions of some Si adatoms. The identification of Ge atoms relative to Si adatoms can be confirmed by profile lines that clearly show a distinct height difference of 0.2

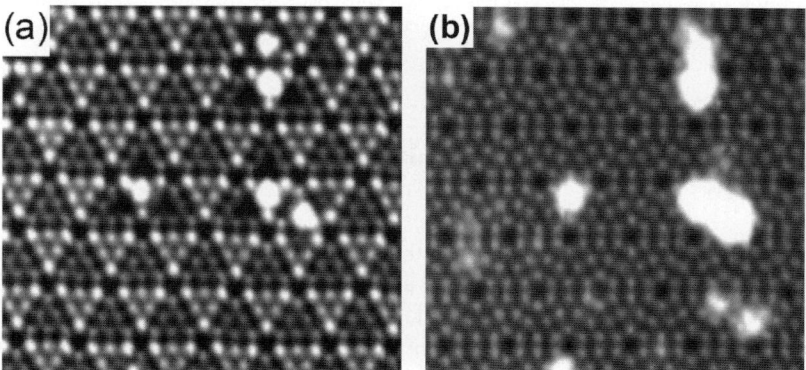

Fig. 17.18. STM images of ~ 0.12 ML Ge deposited on the Si(111)-7×7 surface with the substrate temperature of $\sim 150\,°$C during deposition. (**a**) Filled states. (**b**) Empty states

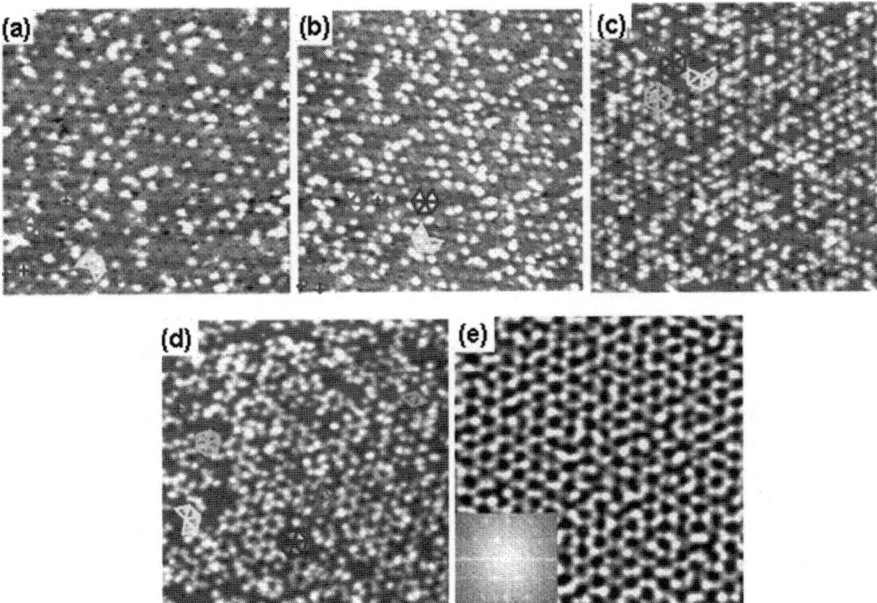

Fig. 17.19. (**a, b, c, d** and **e**) Series of STM images of a Ge-deposited Si(111)-7×7 surface showing the formation process of the hexagonal superlattice with increasing Ge coverages ranging from 0.15 ML, 0.2 ML, 0.3 ML, 0.4 ML to 0.5 ML, respectively. The substrate temperature is held at ∼ 150 °C. All image sizes are 50 × 50 nm. The *inset* in (**e**) shows a Fourier transform of the hexagonal arrays

Å. This means single Ge atoms and some Ge clusters coexist on the Si(111)-7×7 surface at proper substrate temperature and Ge coverage.

Compared to the regular hexagonal structures obtained at room temperature, one obvious feature is that Ge clusters formed at elevated temperatures have more uniform size and shape. These Ge clusters show a distinct electronic structure, and we believe this plays an important role in the evolution of the self-organized hexagonal superlattice of Ge clusters. We will discuss them in detail in the next section.

17.4.5
Electronic Structures of Ge Clusters and Evolution of the Hexagonal Superlattice

From the STM images, we find that the formed Ge clusters have strong bias-dependent features. Figure 17.20 shows a series of high-resolution STM images of a single Ge cluster at different bias voltages. The Ge cluster appears as dark sites at low positive or negative voltage in Fig. 17.20a and d, while the clusters become visible when the bias voltage is at a higher value. These results are in good agreement with the previous studies [61, 62], which indicate that the Ge cluster displays semiconductor behaviors and has a band gap near the Fermi level (E_F).

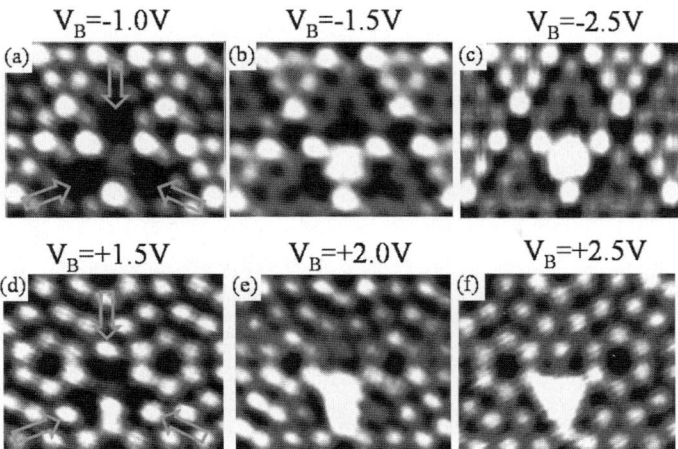

Fig. 17.20. (**a, b,** and **c**) Filled-state STM images of a single Ge cluster on a Si(111)-7×7 surface. Sample bias: −1.0 V, −1.5 V, −2.5 V, respectively; tunneling current: 0.02 nA. (**d, e,** and **f**) Empty-state STM images of the same Ge cluster. Sample bias: +1.5 V, +2.0 V, +2.5 V, respectively; tunneling current: 0.20 nA. Ge cluster appears as *dark* sites at low positive or negative bias voltage in (**a, d**). The neighboring center Si adatoms, as indicated by the *arrows*, are invisible in filled-state images (**a, b** and **c**)

Another obvious feature is that Ge clusters look more compact in the filled-state STM image (Fig. 17.20b and c) than in the empty-state image. The Ge cluster shows strong brightness in the center region of the FHUC, and also has a strong effect on its three neighboring UHUCs. The closest Si center adatoms in the nearest-neighbor UHUCs are invisible in the filled-state STM image. However, these Si adatoms do exist at their original places in the (7×7) reconstruction, as shown by the empty-state STM image. This fact suggests that the Si center adatoms in the nearest-neighbor UHUCs transfer charge to the Ge clusters. The charge transfer is visible both in the filled-state images as the darkened areas surrounding the clusters, indicating absence of electrons and giving the clusters a compact appearance with sharp outline, and in the empty-state images as extended brightness, making the clusters appear less compact and blurring their outlines. The possibility of absence of center adatoms in the nearest-neighbor UHUCs could be excluded. Thus, the STM measurements demonstrate lateral charge redistributions in the Ge–Si system.

The charge transferring from center adatoms to the Ge cluster is further revealed by first-principles density-functional calculations. For all the minimum-energy configurations Ge clusters contain 6–12 Ge atoms in the FHUC of a Si(111)-7×7 unit cell, and the dangling bond state of the center Si adatoms nearby the FHUC is almost empty, indicating charge transfer of neighboring Si adatoms. Figure 17.21a shows the projected electronic density of states (local DOS) onto the center Si adatom in an adjacent UHUC before and after the formation of a nine-Ge cluster in a FHUC, see the minimum-energy configuration in Fig. 17.21b. For a clean Si(111)-7×7 surface before Ge deposition, the dangling-bond state of the center Si adatom is partially occupied and crosses the Fermi level [96,97]. After the formation of Ge clusters, the occupation of the dangling-bond state is reduced significantly, confirming a charge

(a)   **(b)**

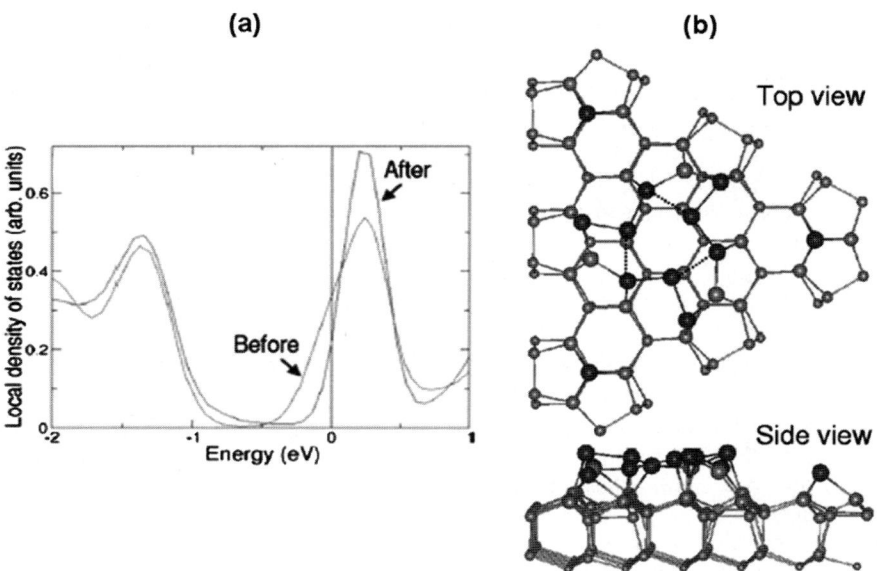

Fig. 17.21. (**a**) Local density of states projected onto a center Si adatom in a UHUC before and after Ge deposition. The Fermi level is at 0 eV. (**b**) The corresponding relaxed minimum energy configuration (only the FHUC is shown). The Si and Ge atoms are depicted by *gray* and *dark spheres*, respectively. *Spheres* of decreasing size represent the Si atoms with increasing distances from the surface. The *dotted lines* show weak bonds

transfer from the central Si adatom. Such a charge transfer occurs because it can lower the total energy of the system. The energy gain can be attributed to "local Madelung energy," which is used in determining the energy of a single ion in a crystal.

Previous results have reported the existence of hexagonal Ge nanostructures on Si(111)-7×7, however, the detail of the process and driving mechanism are still not clear. There are also many reports [98–100] on the self-assembled clusters of various metals formed on Si(111)-7 × 7, which have a similar network as Ge clusters. Several groups suggested that the interaction between the substrate and the metal clusters might play a role in the self-organization [26, 27], and other researchers emphasized the interaction between the clusters themselves [101]. Charge transfer and its role have not been reported before. Here by STM observations in association with density functional theory calculations, we would like to discuss the driving mechanism for the cluster evolution, ascribing to the charge transfer from Si center adatoms to Ge clusters.

From Fig. 17.19, a second Ge cluster may form preferentially at the center of an UHUC adjacent to a FHUC already containing a Ge cluster when local coverage is higher. Similar to the first one, the second Ge cluster darkens the center Si atoms in the two neighboring FHUCs in the filled-state image, which again indicates charge transfer, though the charge transfer is not as effective as in the UHUC. Charge transfer helps us understand the formation of a cluster in an UHUC nearby an existing particle in a FHUC, and further explains the evolution of Ge cluster patterns from an isolated one to closed hexagonal patterns [102]. The schematics in Fig. 17.22a–f simply depict the structures of these local cluster patterns, ranging from single Ge clusters, to pairs of clusters, to an open cluster ring, and finally a closed cluster ring with six clusters surrounding a hole in the Si(111)-7×7 surface. The histograms in Fig. 17.22g reveal the distribution features of six different local Ge nanostructures at varying coverages. The evident tendency is that the ratio of a simple cluster pattern reduces as complex ones increase.

Assuming that the amount of charge transferring from any of the center Si adatom is the same q_0, then a single cluster has a central charge of $-3\, q_0$ (Fig. 17.22a). The total energy is lowered by a local Madelung energy of the cluster, that is roughly $(-9/d_1 + 3/d_2)q_0^2$, where d_1 (~11 Å) is the distance between the Ge clusters and an adjacent Si adatom that has been depleted of charge, and d_2 (~19 Å) is the distance between two such Si adatoms. Because d_1 is smaller than d_2, the local Madelung energy is negative ($-0.66\, q_0^2$). When a second clusters forms in an adjacent UHUC, as in Fig. 17.22b, the charge on each cluster is reduced from $-3\, q_0$ to $-2\, q_0$ and the local Madelung energy is approximately $(-8/d_1 + 2/d_2 + 4/d_3)\, q_0^2 \approx -0.36\, q_0^2$, where d_3 (~15.5 Å) is the distance between the two Ge clusters. The substantial reduction in the Madelung energy is used to overcome the factors that inhibit the formation of isolated clusters in UHUCs, and the residual Madelung energy stabilizes the cluster pair. From the DOS curves of Fig. 17.21a, we estimate $q_0 \approx (0.3–0.5)\, e$, whereby the Madelung energy stabilizing a pair is ~0.5–1.3 eV.

The above results show that a new cluster forms adjacent to an existing FHUC cluster in a neighboring UHUC, and electron transfer occurs from the two new surrounding FHUCs. Similarly, with the emergence of one new Ge cluster, the net charge on each Ge cluster decreases gradually from $3\, q_0$ in the case of an isolated

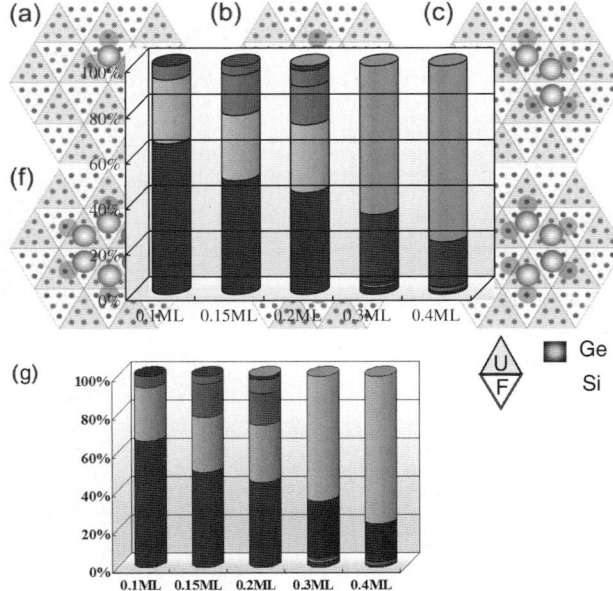

Fig. 17.22. (**a, b, c, d, e,** and **f**) Schematics illustrating the evolution of cluster structures from open to closed hexagonal rings. The Si *center* adatoms that transfer charge are *shaded in gray*. Labels (**g**) Histograms for the distributions of different local Ge nanostructures at varying coverages

cluster to 2 q_0, 5/3 q_0, 3/2 q_0, 7/5 q_0, and finally 1 q_0 if a complete isolated hexagon is formed as in the schematic of Fig. 17.22, and the effective Madelung energy per cluster also gradually reduces. So the reduction of the local Madelung energies contributes to the stabilization of the local cluster structures. Thus, the charge transfer can sustain the evolution of cluster patterns from isolated ones to ordered hexagonal arrays [102].

17.4.6
Formation of Ge Islands and Ge–Si Intermixing at High Temperature

Further increasing the substrate temperature causes the coarsening of clusters and intermixing between Ge and Si atoms simultaneously, which is believed to be due to the enhancing mobility of Ge atoms [103, 104]. When the substrate temperature is increased to about 300 °C or even higher temperatures, the epitaxial growth of Ge islands will begin. Deposition of Ge atoms on Si(111)-7×7 at room temperatures followed by annealing treatment also results in 2-D Ge nanostructures. As shown in Fig. 17.23, a typical surface morphology of Ge epitaxial islands appears with submonolayer Ge coverage. There are three distinct features in this image. First, the

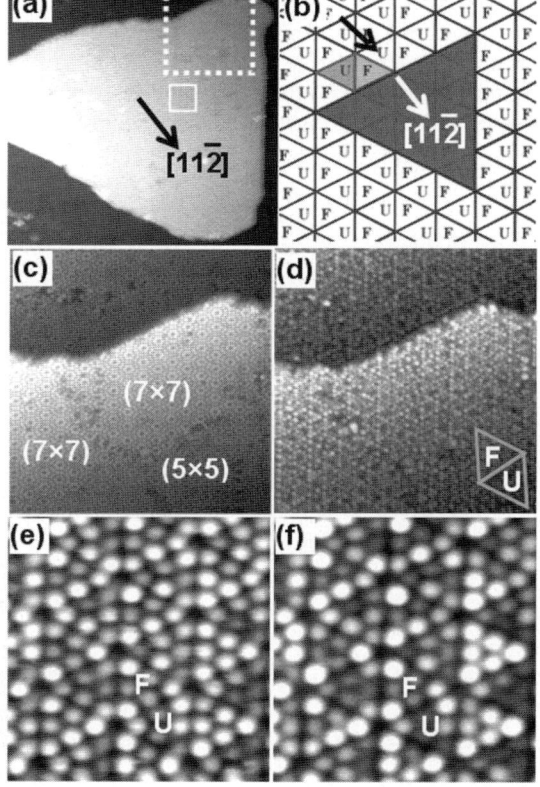

Fig. 17.23. STM images showing a typical Ge island on the Si(111)-7×7 surface with 1.2ML Ge coverage. The substrate temperature was kept at 300 °C. (**b**) The schematic drawing for the formation mechanism of the Ge island on the (7×7) reconstruction. (**c**) and (**d**) are amplified images of the area in (**a**) depicted with a *dot-line square*. (7×7) and (5×5) reconstructions coexist in the island. (**e**) and (**f**) are close-up images of the area in (**a**) depicted with a *solid-line square*. These images, with the irregular contrast difference of brighter atoms, illustrate the intermixing between Ge and Si atoms. Scanning parameters: $I_t = 0.15\,\text{nA}$, (**a**) $120 \times 100\,\text{nm}^2$, $U_b = 1.8\,\text{V}$; (**c**) $45 \times 45\,\text{nm}^2$, $1.2\,\text{V}$; (**d**) $45 \times 45\,\text{nm}^2$, $-1.2\,\text{V}$; (**e**) $9 \times 9\,\text{nm}^2$, $-1.0\,\text{V}$; (**f**) $9 \times 9\,\text{nm}^2$, $-1.5\,\text{V}$

reconstruction of the island is (7×7), the same as the configuration of the original substrate (Fig. 17.23c). Second, the dimer direction on Ge island is the same as the dimer orientation on the substrate, which reveals the supercell of the Ge island has the same orientation as the substrate. Third, the shape of Ge islands usually is close to a triangle, similar to the shape of a HUC triangle of the substrate. All these features reveal the modulation effect of the substrate to the epitaxy growth of Ge islands.

During the growth of an island, the substrate (7×7) reconstruction has to be removed and the substrate atoms will rearrange to the bulk (1×1) structure [105]. The energy barriers to remove the reconstruction of the unfaulted and faulted half cells are different [106]. To remove the reconstruction in the unfaulted triangle, only atoms in the topmost adatom layer rearrange, which is associated with a relatively low energy barrier. Removing the reconstruction of the faulted triangle requires the removal of the stacking fault (see schematic in Fig. 17.1) in the layer below the adatoms, this rearrangement of atoms in deeper layers is associated with a larger energy barrier. From STM images, we can find that the edge of the Ge islands is composed of unfaulted triangles and then surrounded by faulted triangles of the substrate. The schematic drawing of Ge island growth is shown in Fig. 17.23b. Faulted triangles on the substrate become high energy barriers to hinder the further lateral growth of the Ge island, so Ge will nucleate on the neighboring unfaulted triangles (the black arrow). The energy barrier of faulted triangles could thus be reduced by a gain of edge energy, and finally they can be conglutinated by epitaxial Ge atoms from the island edge and neighboring unfaulted triangles with Ge nucleation. As a result, the Ge island shows a lateral growth model along the edge of itself, and the shape of the Ge island usually is triangular, which is due to the modulation by the reconstruction of substrate.

Figure 17.23c shows the island involved in several domains with two different reconstructions, as indicated by (7×7) and (5×5) structures, and the domain boundaries (defect area) are very clear. The phenomena have two possible causes. One is that the defects on the substrate (like the absence of adatoms and vacancies) will deform the period of (7×7) reconstruction and give rise to strain between substrate and epitaxy island [107, 108]. In addition, the mismatch of the lattice constant of Ge and Si (Ge is 4% larger than Si) can also contribute to the formation of strain. The strain can be effectively released by formation of the domain boundaries and the different reconstruction structures. The (5×5) reconstruction can also be described by the well-known DAS model [67]. Each (5×5) unit cell includes one faulted and one unfaulted triangular half cell, and there are three adatoms and one rest atom distributing on the topmost layer in each half cell.

The close-up filled-state STM images in Fig. 17.23d and e show an irregular distribution of brighter adatoms in the (7×7) unit cell. Here, the brightness features lose order; the corner (or center) adatoms in the same half cell show different brightness and even the brightness at the center adatom sites is close to that at the corner adatoms sites, so it clearly illustrates the mixing condition of Ge and Si atoms [109]. According to the contrast feature of single Ge atoms on the Si surface at very low coverage, the brighter protrusions are Ge atoms, and the dimmer ones are Si atoms. These observations coincide with the results reported in earlier literature [110, 111], where Ge–Si exchange in Si(111)-5 $\times$ 5-Ge reconstructions was proposed. Most recently, results by Voigtländer et al. provided evidence for the exchange and intermixing of Ge–Si in a Si(111)-7×7 surface at high temperatures using their special techniques [103, 104]. They showed the chemical contrast images between Si and Ge in their STM observations (Ge is much brighter than Si) obtained on Bi-covered Ge/Si(111) surfaces. Thus, Ge–Si exchange and intermixing can exist at high temperature, and play an important role in epitaxial growth of Ge islands.

In our high-temperature deposition experiments, when Ge coverage was maintained within the range of about 0.2–0.5 ML, a novel local reconstruction with an ordered arrangement of Ge atoms over ~10 nm in size on the Si(111) surface was obtained. Figure 17.24 shows the STM images of such Ge-induced reconstruction, which coexists with the Si(111)-7×7 reconstruction. The reconstruction emerges not only on the Si(111) substrate (Fig. 17.24b) but also on the Ge island (Fig. 17.24a). Figure 17.24c illustrates the high-resolution image of the local atomic structure with a hexagonal arrangement. The atomic distance between the neighbor atoms is 0.65 ± 0.01 nm, that is, about $\sqrt{3}$ times the length 0.38 nm of the basis vector for the ideal bulk-terminated Si(111)-1 × 1 unit cell. In addition, we measured the angle between the main direction of the new local reconstruction and the boundary of the nearby (7×7) unit cells and found it to be 30°. Thus, the local reconstruction shows a $(\sqrt{3} \times \sqrt{3})R30°$ arrangement.

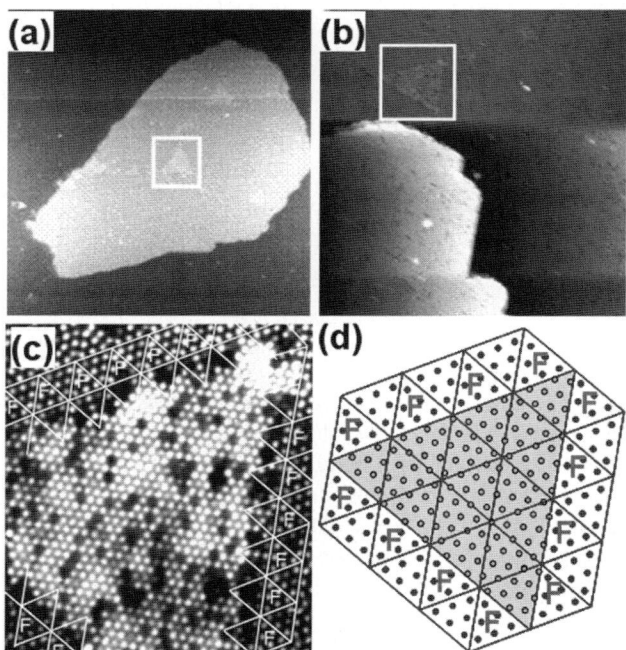

Fig. 17.24. STM images of 0.45ML Ge deposited on the Si(111)-7×7 surface with the substrate temperature held at 300 °C. Local $(\sqrt{3} \times \sqrt{3})R30°$ reconstruction emerges inside the Ge island (**a**) and Si(111)-7×7 substrate (**b**). (**c**) High-resolution image of triangle domain. (**d**) Schematic of the atomic arrangement of the $(\sqrt{3} \times \sqrt{3})R30°$ domain surrounded by the FHUC triangle. $U_b = 2.0$ V, $I_t = 0.10$ nA in (**a**), and 1.4 V, 0.20 nA in (**b**) and (**c**). Image sizes: (**a**) 236×236 nm^2, (**b**) 123×123 nm^2, and (**c**) 22×24 nm^2

The schematic in Fig. 17.24d shows the atomic arrangement of the local domain surrounding by unaffected (7×7) unit cells. Similar to the Ge island in Fig. 17.23b, the activation barrier for atom rearrangement in faulted triangles is higher than that in unfaulted triangles [106], thus the local ($\sqrt{3} \times \sqrt{3}$)R30° domain propagates energy-preferentially in unfaulted half cells. Note that while Ge-induced ($\sqrt{3} \times \sqrt{3}$)R30° reconstruction replaces some of the (7×7) unit cells, it does not cover the whole surface. The 2-D Ge nanostructures thus coexist with the (7×7) reconstruction. In addition, as shown in Fig. 17.24c, several darker features at some atom positions exist within the 2D reconstruction, suggesting that individual Si atoms are mixed with the Ge atoms.

We gain further insight into the bonding geometry of the local ($\sqrt{3} \times \sqrt{3}$)R30° arrangement with the support of first-principles calculations. When the top-layer atoms form a ($\sqrt{3} \times \sqrt{3}$)R30° reconstruction on Si(111), the underlying substrate changes its original (7×7) reconstruction to a (1 × 1) arrangement. On an ideal unreconstructed Si(111) surface, there are two types of threefold symmetric adsorption sites, known as T_4, a filled position directly above a second layer Si atom and H_3, a hollow site above a fourth-layer Si atom site [98], as shown in Fig. 17.25. The adsorbed atoms at either T_4 or H_3 sites are bonded to three first-layer Si atoms. When the dangling bonds of all the first-layer Si atoms are saturated in this way, the adsorbed atoms form a ($\sqrt{3} \times \sqrt{3}$)R30° reconstruction. Such a reconstruction could also be formed when the adsorbed atoms occupy the so-called S_5 site, in which an adsorbed atom substitutes a second-layer Si atom while the replaced Si atom is at the T_4 site directly above S_5 [112, 113]. The first-principles density-functional calculation was performed for a ($\sqrt{3} \times \sqrt{3}$)R30° reconstruction. In the case of the Ge-S_5

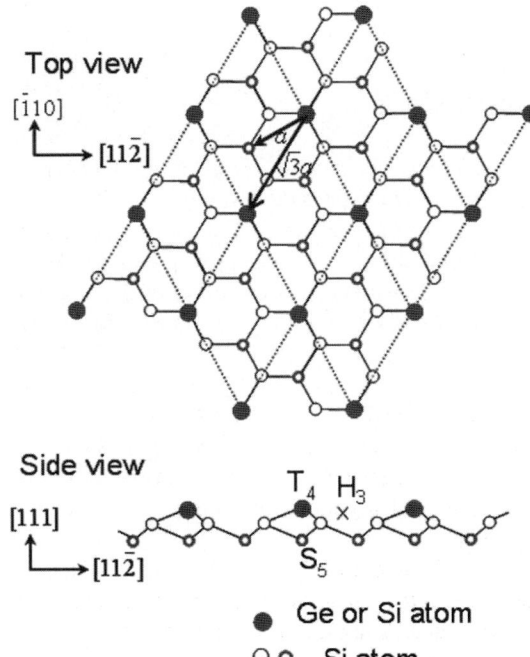

Fig. 17.25. Schematic *top* and *side views* of the atomic arrangement for the ($\sqrt{3} \times \sqrt{3}$)R30° reconstruction with the adatoms at the T_4 sites

configuration Ge or Si forms an adlayer with a Ge coverage of 1/3 monolayer for each of the three bonding configurations [114].

The calculations show that the T_4 configuration is the most stable structure. Its total energy is lower than both the H_3 and the S_5 configurations by 0.60 and 0.68 eV per unit cell, respectively. This is consistent with the general picture that the adatoms prefer to occupy the T_4 sites on almost all of the Si(111)–($\sqrt{3} \times \sqrt{3}$)$R30°$ surfaces induced by chemisorptions of groups III, IV, and V atoms. The occurrence of Ge atoms in the subsurface substitutional S_5 sites is usually adopted by small atoms such as boron and carbon [115–118], and is energetically unfavorable. Occupation of a Ge atom at the subsurface S_5 site would introduce significant strain energy due to its larger size than Si. In addition, for the Ge-S_5 configuration, fully filled Ge-associated bands do not warrant a charge transfer from the Si dangling bond to the subsurface to decrease the surface energy as observed in the boron-induced S_5 configuration [119]. Therefore, Ge atoms would prefer to stay on the surface.

The underlying substrate supporting the Ge-induced ($\sqrt{3} \times \sqrt{3}$)$R30°$ structure has an unreconstructed Si(111) configuration, and significant structural relaxation is also found. The Si atoms at the top layers show downward displacements upon the formation of the ($\sqrt{3} \times \sqrt{3}$)$R30°$ reconstruction with Ge at the T_4 sites, and the first layer Si atoms also move towards the threefold symmetry axis (0.17 A). The bonding energy of a Ge atom at the T_4 configuration is very large (4.8 eV), and the bonding of 1 Ge atom eliminates the dangling bonds of three first layer Si atoms. However, each Ge atom still has a dangling bond, resulting in a metallic feature of the Ge-induced ($\sqrt{3} \times \sqrt{3}$)$R30°$ reconstruction with a partially filled Ge dangling bond band at the Fermi level.

17.5
Conclusions

In this review, we have overviewed the STM studies of Si(111)-7×7 reconstruction structure and Ge nanostructure growth on Si(111) surfaces.

Since the invention of STM in the 1980s, a great amount of effort has been put into pursuit of the ultimate spatial resolution with STM. For the Si(111)-7×7 surface, classical STM images reveal only the 12 adatom spots to a large extent of the bias applied. We show, for the first time, clearly resolved STM images of the six rest atoms and the 12 adatoms on the Si(111) 7×7 unit cell simultaneously. These results are in good agreement with those obtained by first-principles calculations, revealing what should be the "ultimate" STM images for the Si(111)-7×7 surfaces. The strong dependence of imaging rest atoms on the bias voltage demonstrates that the STM can probe the real-space charge distribution far below the Fermi level at corresponding bias voltage if a geometric hindrance due to the finite size of the tip apex can be removed.

STM is also a unique tool for manipulating atoms and molecules on metal and semiconductor surfaces. These involve manipulating the single atom or molecule but also fabricating defined nanostructures on the surface. We performed a series of experiments in which groove structures with atomically straight edges and uniform lateral width of 7×7 unit cell size have been created on Si(111)-7×7 surfaces under the conditions of high tunneling current and low bias voltage. The grooves are formed by removing Si atoms one by one from the Si substrates. The critical current under various voltages for fabricating such grooves is measured.

The bonding nature and formation mechanisms of Ge atoms/nanostructures are essential issues for the controlled fabrication of desired nanostructures on the Si(111) surface. Our STM observations demonstrated individual Ge atoms can replace the Si adatoms rather than adsorb directly atop of Si adatoms at elevated temperature ($\sim 150\,^\circ$C), which is validated by the theoretical calculations that Ge–Si substitution configuration is more energetically favorable. However, at RT deposition, single Ge atoms are found to rapidly diffuse in the half-unit cell of the surface, and if the imaging temperature is reduced to 78 K, these Ge atoms are found to reside at the G sites near the corner Si adatoms or still hop among the G′ sites near the center Si adatoms. With increasing Ge coverage, distinct patterns emerged and evolved, until small clusters formed on the substrate. Ge nanoclusters are formed initially with a preference in the faulted halves, and ultimately self-organize into the formation of a well-ordered hexagonal superlattice on the (7×7) surface. A proper increase in the substrate temperature during the deposition will be helpful to enhance the average mobility of the additional atoms and form more uniform Ge clusters. The STM measurements reveal lateral charge redistributions in the Ge–Si system, and the Si center adatoms in the nearest-neighbor UHUCs transfer charge to the Ge clusters. Charge transfer from Si adatoms to Ge nanoclusters plays a key role in the self-organization, which is proved by experimental observations and theoretic calculations.

When the substrate temperature is increased to a higher temperature ($\sim 300\,^\circ$C), the epitaxial growth of two-dimensional triangular Ge islands will begin. The exchange of Ge atoms with the substrate Si atoms conduces an intermixing condition of Ge–Si components in the islands. Several local domains with different local reconstruction, like (5×5) and ($\sqrt{3} \times \sqrt{3}$)$R30^\circ$ arrangements are found on the substrates.

Acknowledgments. We are grateful to H.F. Ma, Z.H. Qin, H.W. Liu, D.X. Shi, Q.J. Gu, Z.L. Ma, J.N. Gao, N. Liu, H.Q. Yang, L. Yan, Y.P. Zhang, and S.J. Pang for experimental assistance, and I.G. Batyrev, W.E. McMahon, S.B. Zhang, A.S. Rao, S.W. Wang, and S.T. Pantelides for theory simulations and calculations. This research is supported by the Natural Science Foundation of China and the Chinese National "863" and "973" projects.

References

1. Takayanagi K, Tanishiro Y, Takahashi M, Takahashi S (1985) J Vac Sci & Technol 3:1502–1506
2. Binnig G, Rohrer H, Gerber C, Weibel E (1982) Phys Rev Lett 49:57
3. Binnig G, Rohrer H, Gerber C, Weibel E (1983) Phys Rev Lett 50:120
4. Avouris P, Wolkow R (1989) Phys Rev B 39:5091
5. Nishikawa O, Tomitori M, Iwawaki F, Hirano N (1990) J Vac Sci & Technol A 8:421
6. Lantz MA, Hug HJ, Schendel P, Hoffmann R, Martin S, Baratoff A, Abdurixit A, Güntherodt HJ, Gerber C (2000) Phys Rev Lett (2000) 84:2642
7. Giessibl FJ, Hembacher S, Bielefeldt H, Mannhart J (2000) Science 289:422

8. Giessibl FJ (1995) Science 267:68
9. Sutter P, Zahl P, Sutter E, Bernard LE (2003) Phys Rev Lett 90:166101
10. Hamers RJ, Tromp RM, Demuth JE (1986) Phys Rev Lett 56:1972–1975
11. Becker RS, Swartzentruber BS, Vickers JS, Klitsner T (1989) Phys Rev B 39:1633
12. Chen L, Pan BC, Xiang HJ, Wang B, Yang JL, Hou JG, Zhu QS (2007) Phys Rev B 75:085329
13. Wiesendanger R (1994) Scanning probe microscopy and spectroscopy: methods and applications, Cambridge University Press, Cambridge, England
14. Zhang SB, Cohen ML, Louie SG (1986) Phys Rev B 34:768
15. Kohn W, Sham LJ (1965) Phys Rev 140:A1133
16. Kresse G, Furthmüller J (1996) Comput Mater Sci 6:15
17. Vanderbilt D (1990) Phys Rev B 41:7892
18. Tersoff J, Hamann DR (1985) Phys Rev B 31:805
19. Wang YL, Gao HJ, Guo HM, Liu HW, Batyrev IG, McMahon WE, Zhang SB (2004) Phys Rev B 70:073312
20. Eigler DM, Schweizer EK (1990) Nature 344:524
21. Lyo IW, Avouris P (1991) Science 253:173
22. Uchida H, Huang DH, Grey F, Aono M (1993) Phys Rev Lett 70:2040
23. Hosoki S, Hosaka S, Hasegawa T (1992) Appl Surf Sci 60–61:643
24. Salling CT, Lagally MG (1994) Science 265:502
25. Kobayashi A, Grey F, Williams RS, Aono M (1992) Science 259:1724
26. Gu QJ, Liu N, Zhao WB, Ma ZL, Xue ZQ, Pang SJ (1995) Appl Phys Lett 66:1747
27. Ma ZL, Liu N, Zhao WB, Gu QJ, Ge X, Xue ZQ, Pang SJ (1995) J Vac Sci Technol B 13:1212
28. Behm RJ, Garcia N, Rohrer H (1990) Scanning tunneling microscopy and related methods, Kluwer Academic, Dordrecht, pp 113–141
29. Lang ND (1992) Phys. Rev. B 45:13599
30. Miskovsky NM, Wei CM, Tsong TT (1992) Phys. Rev. Lett. 69:2427
31. Yang HQ, Liu N, Gao JN, Jiang YS, Shia DX, Ma ZL, Xue ZQ, Pang SJ (1998) Appl Surf Sci 126:337
32. Brune H, Giovannini M, Bromann K, Kern K (1998) Nature 394:451
33. Hwang IS, Ho MS, Tsong TT (1999) Phys Rev Lett 83:120
34. Li JL, Jia JF, Liang XJ, Liu X, Wang JZ, Xue QK, Li ZQ, Tse JS, Zhang ZY, Zhang SB (2002) Phys Rev Lett 88:066101
35. Jia JF, Wang JZ, Liu X, Xue QK, Li ZQ, Kawazoe Y, Zhang SB (2002) Appl Phys Lett 80:3186
36. Jia JF, Liu X, Wang JZ, Li JL, Wang XS, Xue QK, Li ZQ, Zhang ZY, Zhang SB (2002) Phys Rev B 66:165412
37. Voigtländer B, Kastner M, Smilauer P (1998) Phys Rev Lett 81:858
38. Wu KH, Fujikawa1 Y, Nagao T, Hasegawa Y, Nakayama KS, Xue QK, Wang EG, Briere T, Kumar V, Kawazoe Y, Zhang SB, Sakurai T (2003) Phys Rev Lett 91:126101
39. Vitali L, Ramsey MG, Netzer FP (1999) Phys Rev Lett 83:316
40. Hibino H, Ogino T (1997) Phys Rev B 55:7018
41. Chizhov I, Lee G, Willis RF (2997) Phys Rev B 56:12316
42. Sonnet P, Stauffer L, Minot C (1998) Surf Sci 407:121
43. Gómez-Rodríguez JM, Sáenz JJ, Baró AM (1996) Phys Rev Lett 76:799
44. Lin XF, Chizhov I, Mai HA, Willis RF (1996) Surf Sci 366:51
45. Yoon M, Lin XF, Chizhov I, Mai HA, Willis RF (2001) Phys Rev B 64:085321
46. Zhang YP, Yang L, Lai YH, Xu GQ, Wang XS (2003) Surf Sci 531:L378
47. Hsu HF, Chen LJ, Hsiao HL, Pi TW (2003) Phys Rev B 68:165403
48. Voigtländer B, Weber T (1996) Phys Rev Lett 77:3861

49. Petro PM, Lorke A, Imamoglu A (2001) Physics Today 54:46
50. Voigtländer B (1999) Micron 30:33
51. Voigtländer B (2001) Surf Sci Rep 43:127
52. Voigtländer B, Kawamura M, Paul N, Cherepanov V (2004) J Phys: Conden Mat 16:S1535
53. Ratto F, Locatelli A, Fontana S, Kharrazi S, Ashtaputre S, Kulkarni SK, Heun S, Rosei F (2006) Phys Rev Lett 96:096103
54. Sekiguchi T, Yoshida S, Itoh KM, Mysliveèek J, Voigtländer B (2007) Appl Phys Lett 90:013108
55. Ansari ZA, Tomitori M, Arai T (2006) Appl Phys Lett 88:171902
56. Boscherini F, Capellini G, Gaspare L, Rosei F, Motta N, Mobilio S (2000) Appl Phys Lett 76: 682
57. Zhang YP, Yan L, Xie SS, Pang SJ, Gao HJ (2002) Surf Sci 497:L60
58. Beben J, Hwang IS, Chang TC, Tsong T (2000) Phys Rev B 63:033304
59. Zhang YP, Yan L, Xie SS, Pang SJ, Gao HJ (2001) Appl Phys Lett 79:3317
60. Lobo A, Gokhale S, Kulkarni SK (2001) Appl Surf Sci 173:270
61. Yan L, Yang HQ, Gao HJ, Xie SS, Pang SJ (2001) Surf Sci 498:83
62. Yan L, Zhang YP, Gao HJ, Xie SS, Pang SJ (2001) Surf Sci 506:L255
63. Ratto F, Rosei F, Locatelli A, Cherifi S, Fontana S, Heun S, Szkutnik P, Sgarlata A, Crescenzi M, Motta N (2004) Appl Phys Lett 84:4526
64. Motta N (2002) J Phys: Conden Mat 14:8353
65. Guo HM, Wang YL, Liu HW, Ma HF, Qin ZH, Gao HJ (2004) Surf Sci 561:227
66. Zhao A, Zhang X, Chen G, Loy M, Xiao XD (2006) Phys Rev B 74:125301
67. Köhler U, Jusko O, Pietsch G, Müller B, Henzler M (1991) Surf Sci 248:321
68. Patel JR, Golovchenko JA, Bean JC, Morris RJ (1985) Phys Rev B 31:6884
69. Dev BN, Materlik G, Grey F, Johnson RL, Clausnitzer M (1986) Phys Rev Lett 57:3058
70. Kajiyama K, Tanishiro Y, Takayanagi K (1989) Surf Sci 222:47
71. Carlisle JA, Miller T, Chiang TC (1994) Phys Rev B 49:13600
72. Castrucci P, Gunnella R, Crescenzi M, Sacchi M, Dufour G, Rochet F (1999) Phys Rev B 60:5759
73. Takaoka GH, Seki T, Tsumura K, Matsuo J (2002) Thin Solid Films 405:141
74. Grodzicki M, Wagner M (1989) Phys Rev B 40:1110
75. Stauffer L, Vana S, Bolmonta D, Koulmanna J, Minotb C (1994) Surf Sci 307:274
76. Stauffer L, Sonnet P, Minot C (1997) Surf Sci 371:63
77. Cho K, Kaxiras E (1998) Surf Sci 396:L261
78. Tong SY, Huang H, Wei CM, Packard WE, Men FK, Glander G, Webb MB (1988) J Vac Sci & Technol A 6:615
79. Wang SW, Radny MW, Smith PV (2001) J Chem Phys 114:436
80. Becker RS, Golovchenko JA, Swartzentruber BS (1985) Phys Rev B 32:8455
81. Fukuda T (1996) Surf Sci 351:103
82. Rosei F, Motta N, Sgarlata A, Capellini G, Boscherini F (2000) Thin Solid Films 369:29
83. Perdew JP, Wang Y (1992) Phys Rev B 45:13244
84. Cho K, Kaxiras E (1997) Europhys Lett 39:287
85. Lyo I-W, Avouris P (1990) J Chem Phys 93:4479
86. Wang YL, Gao HJ, Guo HM, Wang SW, Pantelides ST (2005) Phys Rev Lett 94:106101
87. Custance O, Brochard S, Brihuega I, Artacho E, Soler JM, Baró AM, Gómez-Rodríguez JM (2003) Phys Rev B 67:235410
88. Custance O, Brihuega I, Gómez-Rodríguez JM, Baró AM (2001) Sur Sci 482–485:1406
89. Zhang C, Chen G, Wang K, Yang H, Su T, Chan C, Loy M, Xiao XD (2005) Phys Rev Lett 94:176104
90. Wang K, Zhang C, Loy M, Xiao XD (2005) Phys Rev Lett 94:036103
91. Sato T, Kitamura S, Iwatsuki M (2000) J Vac Sci Technol A 18:960

92. Kresse G, Furthmüller J (1996) Phys Rev B 54:11169
93. Kresse G, Hafner J (1993) Phys Rev B 47:R558
94. Ansari ZA, Arai T, Tomitori M (2005) Surf Sci 574:L17
95. Rotta F, Rosei F (2003) Surf Sci 530:221
96. Northrup JE (1986) Phys Rev Lett 57:154
97. Brommer KD, Galvan M, Dalpino A, Joannopoulos JD (1994) Surf Sci 314:57
98. Lai MY, Wang YL (2001) Phys Rev B 64:241404
99. Chang H, Lai MY, Wei JH, Wei CM, Wang YL (2004) Phys Rev Lett 92:066103
100. Zilani M, Xu H, Liu T, Sun Y, Feng YP, Wang XS, Wee A (2006) Phys Rev B 73:195415
101. Vasco E, Polop C, Rodriguez-Canas E (2003) Phys Rev B 67:235412
102. Ma HF, Qin ZH, Xu MC, Shi DX, Gao HJ, Wang SW, Pantelides ST (2007) Phys Rev B 75:165403
103. Kawamura M, Paul N, Cherepanov V, Voigtländer B (2003) Phys Rev Lett 91:096102
104. Paul N, Filimonov S, Cherepanov V, Çakmak M, Voigtländer B (2007) Phys Rev Lett 98:166104
105. Suzuki M, Shigeta Y (2003) Surf Sci 539:113
106. Shimada W, Tochihara H (1994) Surf Sci 311:107
107. Suzuki M, Negishi R, Shigeta Y (2005) Phys Rev B 72:235325
108. Itoh M, Tanaka H, Watanabe Y, Udagawa M, Sumita I (1993) Phys Rev B 47:2216
109. Andersohn L, Berke T, köhler U, Voigtländer B (1996) J Vac Sci & Technol A 14:312
110. Motta N, Sgarlata A, Calarco R, Nguyen Q, Castro Cal J, Patella F, Balzarotti A, Crescenzi M (1998) Surf Sci 406:254
111. Boscherini F, Capellini G, Gaspare L, Rosei F, MottaN, Mobilio S (2000) Appl Phys Lett 76:682
112. Mönch W (2001) Semiconductor surfaces and interfaces, Springer-Verlag, New York
113. Wang S, Radny MW, Smith PV (1998) Surf Sci 396:40
114. Qin ZH, Shi DX, Ma HF, Gao HJ, Rao AS, Wang SW, Pantelides ST (2007) Phys Rev B 75:085313
115. Castrucci P, Sgarlata A, Scarselli M, Crescenzi M (2003) Surf Sci Lett 531:329
116. Peng XY, Ye L, Wang X (2004) Surf Sci 548:51
117. Pignedoli CA, Catellani A, Castrucci P, Sgarlata A, Scarselli M, Crescenzi M, Bertoni CM (2004) Phys Rev B 69:113313
118. Profeta G, Ottaviano L, Continenza A (2004) Phys Rev B 69:241307
119. Wang S, Radny MW, Smith PV (1999) Phys Rev B 59:1594

Subject Index